AF598053

Design for Manufacturability

Artur Balasinski

Design for Manufacturability

From 1D to 4D for 90–22 nm Technology Nodes

Artur Balasinski
Cypress Semiconductor
San Diego, CA, USA

ISBN 978-1-4614-1760-6 ISBN 978-1-4614-1761-3 (eBook)
DOI 10.1007/978-1-4614-1761-3
Springer New York Heidelberg Dordrecht London

Library of Congress Control Number: 2013947681

Printed on acid-free paper

Springer is part of Springer Science+Business Media (www.springer.com)

About the Author

Artur Balasinski works for Cypress Semiconductor as Principal Technology Development Engineer, responsible for all aspects of Design-for-Manufacturability in advanced technology nodes. He graduated from Warsaw University of Technology in Poland where he subsequently completed his Ph.D. on radiation effects in MOS devices. He then worked there as Assistant Professor before joining Yale University, New Haven, CT, as a Researcher. He moved on to ST Microelectronics to develop new TFT technologies and manage process transfers. In 1997, he joined Cypress Semiconductor in San Jose, CA, to work on design rules and run OPC/DfM team. This is his second book publication, in addition to about 100 papers and 15 patents.

Contents

Chapter 1
Preface

1.1 Motivation and Approach

What are the motivation and the approach for this book? Firstly, we intend to discuss the increasing number of key aspects of Integrated Circuit Design for Manufacturability in the early 2010's. Because the speed of information in this area is critical for making money in IC manufacturing, DfM is a popular topic for conferences and journals, and a directional summary is usually welcome by the experts in the field.

But, secondly and perhaps more importantly, the scope of DfM is changing, along with the IC industry, and single-topic publications would not always bring this out to the attention of the IC decision makers. One of the Academia leaders, Prof. J. Meindl, noted that developments in DfM coincide with the ongoing transition from micro- to nanoelectronics. Throughout the years of 1960–2012, the progress in IC DfM enabled the most significant economic development: The Information Revolution, by becoming its most potent technological driver, due to the Silicon Microchip productivity [1].

While the multibillion dollar industry would see to it that the integrated circuit (IC) progress is far from being stalled, as evidenced by the list of metrics to keep monitoring it, such as the count of transistors per chip:

$$N = F^{-2} \cdot D^2 \cdot PE \quad (1.1)$$

and performance measured by instructions per second:

$$IPS = IPC \cdot f_c \quad (1.2)$$

A. Balasinski, *Design for Manufacturability: From 1D to 4D for 90–22 nm Technology Nodes*,
DOI 10.1007/978-1-4614-1761-3_1, © Springer Science+Business Media New York 2014

Table 1.1 Basic parameters of IC starting material as one of the important factors for DfM considerations. Many more material parameters are required for performance and reliability modeling, esp. of the impact of device proximity and packaging stress after [1]

Wafers	300 nm diameter
Ingot body length	>1 m
Silicon	5×10^{22} atoms/cc
	0.236 nm atomic spacing
Circuit assembly	Entirely self-assembled

and it increased due to incremental improvements of the following factors:

N	Transistors per chip
F	Minimum feature size (scaling increases N)
D^2	Chip size (x times y)
PE	Transistor packing efficiency
IPS	Instructions per second
IPC	Instructions per cycle (depending on number of Cores, Architecture & Software)
f_C	Cycles per second (Clock frequency)

it is critical that these incremental improvements are not met by a brick wall of limitations of semiconductor physics. DfM is to prevent such scenario.

So far, productivity and performance, expressed by N and f_C, have always been most efficiently increased by scaling, the potent "fuel" energizing the microchip engine. In the same 1960–2012 timeframe, they increased by six orders of magnitude each, as manufacturing with optical micro/nanolithography and DfM enabled scaling of printed gate length from 25 μm in 1960 to 25 nm in 2012, so by 1,000× per dimension, in x and y. The scaling in z – dimension has been mostly consequential as it did not bring in extra devices so far. This is changing now. New DfM approaches are needed to expand in the third dimension.

Silicon chips have entered the nanotechnology domain (0.1–100 nm) by exploiting a "top-down" approach i.e., following the ITRS roadmap [2]. Since 2011, they started to exploit a fusion of top-down and bottom-up nanotechnology. This meant, understanding the intrinisic limitations of technology nodes and improving on them. How? One way was to expand the material portfolio used by IC Manufacturing in 2012 (Table 1.1). It has been evolving, and with this, the criteria for design rules. Another way was by getting prepared to what the design may want to use for mutual advantage of layout and manufacturing: the correct-by-construction template cells.

1.2 Design Rule Criteria for DfM

Many designers and fab engineers think of a design rule as of a number. For example, minimum spacing of feature on layer A to a feature on layer B must be X nm. Such paradigm is not valid for DfM.

This simplistic approach can lead to confusion and delays as the technology shrinks or gets ready to enter unfamiliar territories. For example, 90 % of die features would easily meet that rule, but it would be beneficial to change it to a higher value for 10 % of devices. Without complete understanding of the physical reasons underlying the design rule, it is not possible to rationalize its value to build a manufacturable device.

In the recommended rule (RR) system, especially based on a grid, it would be particularly problematic to modify a number associated with a rule without impacting multiple layout features. For that reason, one should present, sometimes in painful detail, the physical mechanisms behind any DfM recommendation. The prevailing criterion for any design rule, be it standard design rule checks DRC, recommended rules RR, or DfM, is to ensure that the feature designed on the IC would follow the design intent. It is therefore important that the design intent is transparent throughout design rule manuals and it is for the manufacturing to decide how to make it happen.

Simultaneously, there are new material parameters adopted by the IC makers to enable the "bottom-up" extension of nanotechnology. But further scaling, from 25 nm in 2011 to the stipulated 6 nm in 2026, may be impeded by the many barriers related to the desired increases of IC electrical performance (Table 1.2). This target performance has been communicated to the manufacturing, which has now to consider all kinds of opportunities to extend scaling or use new materials or device architectures (Table 1.3) to overcome these barriers. Firstly, fab engineering responded with 193 nm lithography innovations: immersion, optical proximity correction, phase shift masks, unidirectional gate layout, double patterning, computational lithography, and perhaps at some time in the future, EUV @ 13.4 nm.

Table 1.2 Key static IC parameters and their field of impact, the reduction of which could serve as DfM quality metric

Parameter	Impact
Gate tunneling current	Standby power for handheld devices
Subthreshold channel current	Tradeoff: leakage – speed
Device parameter variability	Analog circuit performance
Source/drain resistance	Low speed
Interconnect resistivity	Low speed and thermal effects

Table 1.3 Opportunities for extended scaling. Scaling is not DfM [3] but existing DfM solutions will have to adjust to support novel architectures of devices, processes, and design

Technology or design concept	Improvement
Channel strain	Higher carrier mobility (speed)
III-V (e.g. InSb) channel FETs	
High permittivity gate dielectrics (e.g. HiSiON) Metal gates (e.g. TaSiN)	Better tradeoff: leakage – speed
New FET structures (e.g. FDSOI, FinFET and Trigate FET)	More efficient process architecture (2D → 3D)
Low permittivity interconnect dielectrics	Reduced capacitive coupling
Power and clock gating	More flexible design architecture

They all require new DfM approaches, as we keep changing design methodology along with reducing the CD's.

Another option of IC manufacturing is exploiting the 3D structure. For example, switching from traditional planar transistor to 22 nm 3D-Gate Transistor (Fig. 1.1) would require not only a new set of design rules, but also architectural criteria based on the new manufacturing capabilities. As a reward, a 22 nm 3D-Gate transistor provides improved performance at high voltage and an unprecedented performance gain (Fig. 1.2).

Developing nano-DfM to convert from 2D to 3D includes chip stacking to enable:

- High density electrical I/O interconnects
- Wafer/chip back-side microchannels for liquid cooling (Fig. 1.3)
- Through silicon electrical and fluidic vias (TSVs)
- Optical I/O Interconnects
- Dense wavelength division multiplexing (DWDM)
- Direct cooling of chip back side to reduce thermal interface resistance
- Cooling on each stratum to extract >100 W/cm^2 from the die
- Microscale fluidic interconnection between strata.

This, of course, requires a whole new set of detailed design rule criteria i.e. statement of design intent, supported by simulation tools, followed by manufacturing response in design rules, etc.

If, on the other hand, CMOS fails to innovate, there are other technologies, e.g. the ones capable of manufacturing graphene FETs, ready to take over the IC development (Fig. 1.4, Table 1.4).

One can expect that nano-DfM in 3D should expand the IC lifetime for another 50 years, offering the extension to nanoelectronics, the most promising opportunity for sustaining the exponential rate of advance of the ongoing Information revolution for another half century. The implications of continuing this exponential rate of advance until the middle of the twenty-first century and beyond are difficult to imagine, but are going to be profound, to the global information infrastructure, education system, and economy [1],

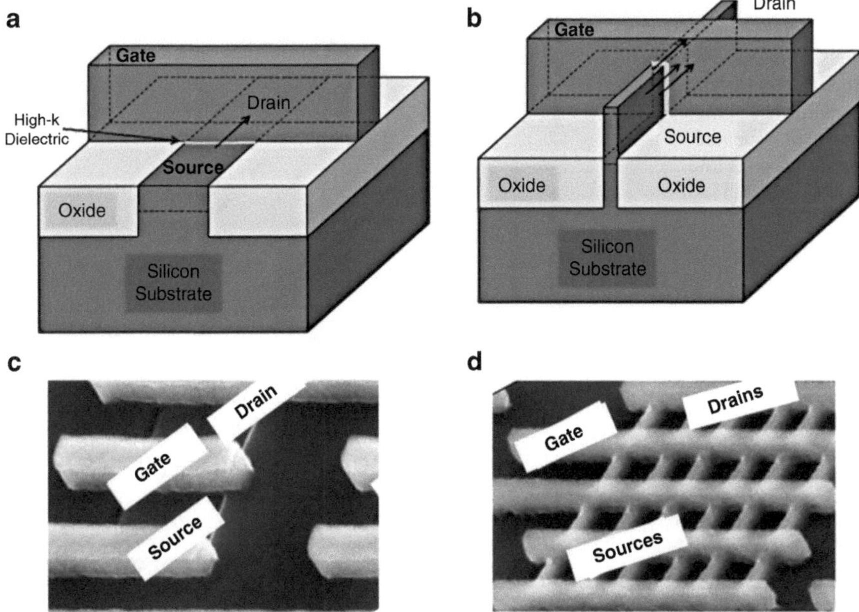

Fig. 1.1 MOS Transistors: (**a**) planar, with flat conducting channel, (**b**) 3D Tri-gate, with channel on three sides of a fin structure, for 22 nm technologies and beyond. MOSFET pattern on silicon: (**c**) planar gates for 32 nm design, (**d**) gates and fins for 22 nm design (after [1])

DfM, or Integrated Circuit Design for Manufacturability (IC DfM), is needed to take us there.

1.3 A Historical Perspective

DfM started not long after 1960, when layout enhancements were identified to be a good way to help design scaling regardless of the functionality of the circuits being made. Thus, designers helped manufacturing achieve their intent on silicon. Since then, the scope of IC DfM has evolved and expanded from assisting CD reductions to a tool correlating all aspects of design methodology with product reliability and yield. Because all design information is encrypted in the layout, the role of DfM is changing as the device shrinkpath supported by Moore's law, which was the engine of IC growth, would gradually run out of steam. From the initial auxiliary concept, IC DfM becomes one without which progress in device integration, especially 3D capabilities, at new levels of cooperation among the IC design, process, and packaging engineers, would not be possible. In this capacity, DfM enables now the transition from electrical to mechanical design rules and adds the aspects covered by the generic engineering DfM principles. It is becoming a profit engine, too.

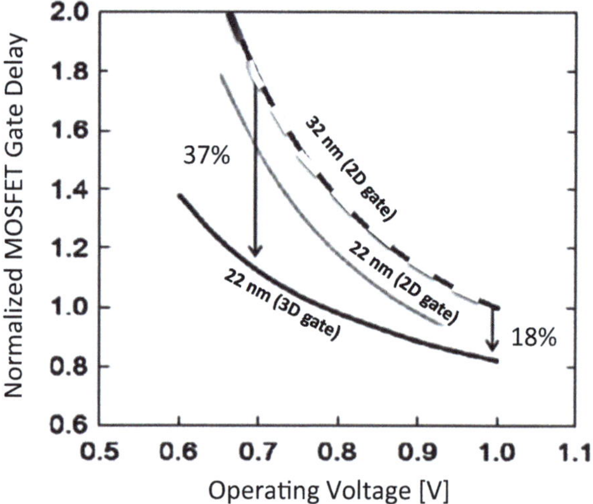

Fig. 1.2 Transistor gate delay (normalized) for 32 and 22 nm devices, the latter showing improved performance at high voltage and unprecedented performance at low voltage (after [1])

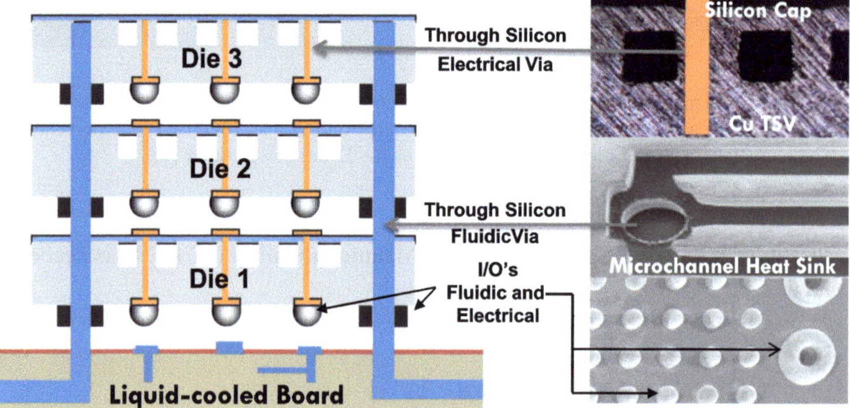

Fig. 1.3 Inter-layer Microfluidic Cooling Design for 3D ICs (after [1])

As a result, the current scope of IC DfM expands into design methodology development from 1D to 4D (3D plus reliability = time dependence), involving IC design flow setup, best practices, links to manufacturing and product definition, for process technologies down to 20 nm node (at this time) and product families including memories, logic, SoC, SiP, for ever more increasing range of packages and applications. For this reason, a new summary work in the area of DfM is needed every few years, in order to:

1. Provide guidelines about layout techniques for most advanced technology nodes
2. Help understand DfM at product level (including in – package reliability)
3. Guide the IC manufacturers through the large volume of publications.

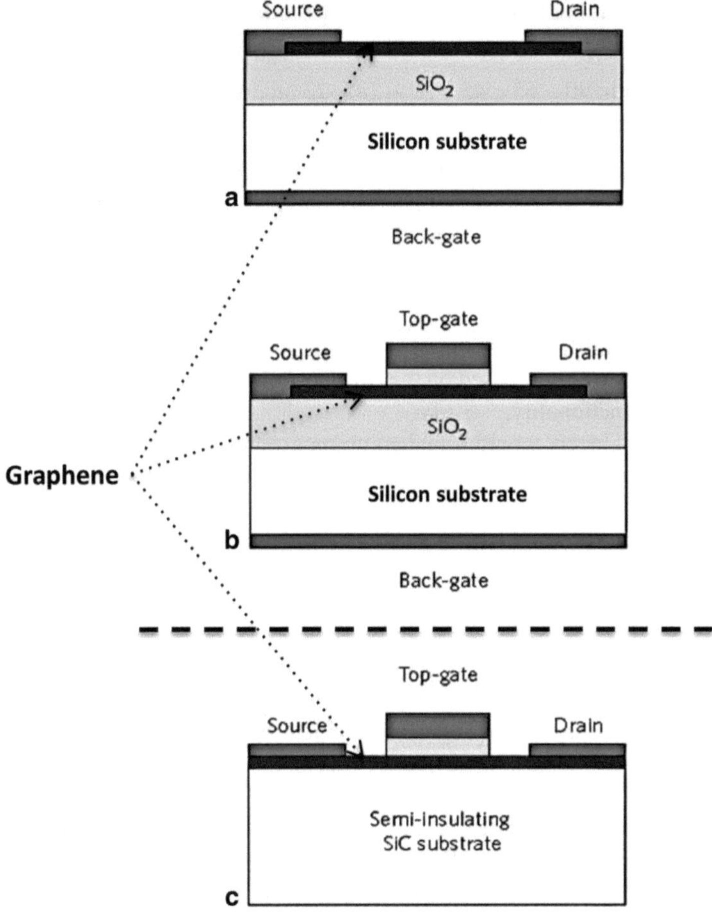

Fig. 1.4 Architectures of graphene FETs (after [1])

Table 1.4 If silicon DfM fails to deliver sustainable RoI, Graphene based devices may perform better than existing CMOS Technology

Property	Graphene	Existing Si	Application
Mobility $cm^2/V\text{-}s$	40,000	500	Switches
Resistivity μohm·cm	1.05	5.0 (Copper, 20 nm)	Interconnects
Thermal conductivity W/m·K	5/100	400 (Copper)	Heat spreaders
Young's modulus GPa	500	150 (Silicon)	MEMS'S
Current-carrying capacity A/cm^2	10^9	10^6 (Copper)	Interconnects

In particular, this book should be interesting for design, technology, and packaging engineers who follow two directions of IC development:

- The shrinkpath: 90 – 65 – 45 – 32 – 28 – 22 nm, or
- Expansion into new applications.

Each of these directions calls for its own set of DfM approaches for core cells, layout architectures, new devices, materials, and processes. This book should make it easier to compare them from the standpoint of both technical issues and return on investment (RoI).

What are the benefits of this book for the reader?

- Layout designers would learn what are the factors to optimize layout into the next technology nodes, based on the overview of the recent issues related to scalability and functionality,
- Packaging engineers would found common ground with process developers and layout designers based on the understanding of device architecture and product development issues,
- Quality engineers would understand how to integrate reliability methodology into design flow.

This book consists of the following chapters:

- A review of "classic" 2D and 3D DfM developments, aimed at cost reduction. Progress in IC Manufacturing requires new tools of immense complexity and cost. What if we try to use engineering ingenuity to mitigate it, based on recent disclosures, including new device architectures, packaging options, and layout enhancement methodologies.
- An introduction to 2D DfM beyond 28 nm based on design streamlining, supported by Restricted Design Rules. Layout freedom is over, how do we control designers' hand? If it is still 2D, it better be simple.
- A look into the mechanical stress aspects of IC manufacturing and reliability (3D – to – 4D). The importance of IC mechanical aspect is growing but it is very non-intuitive. It requires profound material knowledge and new simulation engines.

How does the information contained in this book compare to the one on the publication market:

1. DfM books till date deal with flat layout and post-layout processing. No impact of design flow setup and 3D analysis are provided
2. The few large-scope publications about DfM do not discuss for process nodes more advanced than 65 nm
3. No guidelines of DfM for packaging address critical aspects of IC manufacturability and applications.

Going back to the opening line, what is the approach for a review publication like this book? Relying on the landmark Ph. D. dissertations [4, 5] as well as on methodology monographies (e.g. [6–10] was a good starting point, with the intent to demonstrate how the information contained in these publications creates a system to support future DfM needs. The significant novelty here was the grouping and correlating the DfM trends to visualize the new challenges.

Acknowledgments The author extends special thanks to Mrs. Agnieszka Baranowska. She not only helped with the editing process but was indispensable in keeping this work together, in terms of motivation, chapter integrity, and clarity of message. Thank you, Agnieszka!

References

1. Meindl, J.D.: Nanoelectronics in retrospect, prospect, and principle, IEEE ISSCC 2010, February (2010)
2. International Technology Roadmap for Semiconductors, http://public.itrs.net/
3. Balasinski, A.: Semiconductors: Integrated Circuit Design for Manufacturability. CRC Press/ Taylor and Francis, Boca Raton (2011)
4. Yang, J.: Manufacturability aware design. Ph.D. thesis, University of Michigan (2007)
5. Zhang, X.: Chip package interaction (CPI) and its impact on the reliability of flip-chip packages, Ph.D. thesis, UT Austin (2009)
6. Abercrombie, D.,Elakkumanan, P., Liebmann, L.: Restrictive design rules and their impact on 22 nm design and physical verification, E.DPS (2009)
7. Wong, N.P., Mittal, A., Starr, G.W., Zach, F., Moroz, V., Kahng, A. Nano-CMOS design for manufacturability: robust circuit and physical design for sub-65 nm technology nodes. Wiley-Interscience (2008)
8. Chiang, C., Kawa, J.: Design for Manufacturability and Yield for Nano-scale CMOS/Integrated Circuits and Systems. Springer, Dordrecht (2007)
9. Bralla, J.G.: Design for Manufacturability Handbook. McGraw-Hill Handbooks, New York (1998)
10. Orshansky, M., Nassif, S.R., Boning, D.: Design for Manufacturability and Statistical Design: A Constructive Approach. Series: Integrated Circuits and Systems XIV, vol. 316. Springer, London (2008)

Chapter 2
Classic DfM: From 2D to 3D

"Manufacturability" is the ability to make large numbers of identical products (IC devices), with substantially reproducible parameters in time and in space. Of course, these devices must perform a useful function, but that is ensured by their prototyping.

Variability is the anti-thesis to manufacturability. Therefore, DfM should reduce variability in all accessible dimensions, from 1D (fixed pitch), to 2D (wafer patterning and yield), 3D (stress and package issues), and to 4D (reliability failures in time or temperature). Out of these, 2D and 3D variabilities are the most consequential ones at the time of manufacturing, as they cover the critical aspects of variability in planar technologies and assembly houses. They are also considered as core concerns of the "classic" DfM, i.e., the methodology working for several generations but not necessarily towards future solutions. At the same time, 4D variability is the most important one, according to the Rule of 10 [1].

Manufacturing variability impacts all hierarchical aspects of DfM approach:

- Reliability
- Performance
- Parametric deliverables.

Variability can be dealt with during at least three stages of design-to-wafer process: electrical design, physical design (layout) and mask data preparation (MDP). The earlier it is done in the design cycle, the better are the process and its cost controlled.

2.1 Variability Reduction in Design Phase

Historically, the "mother" of all variability issues was considered to be pattern transfer from design to wafer. The factors driving parametric and yield variability include all aspects of design rules translating design intent into process capabilities of lithography and mask making. The goal of DfM is to mitigate pattern transfer issues in design phase, i.e., before the problem enters the fab.

A. Balasinski, *Design for Manufacturability: From 1D to 4D for 90–22 nm Technology Nodes*,
DOI 10.1007/978-1-4614-1761-3_2,

2.1.1 Litho Process and Its Challenges

To start IC DfM analysis from the basic challenge of device manufacturability, i.e. pattern printability on silicon wafer, one has to acknowledge a remarkable progress in patterning techniques. Due to the countless millions of dollars spent on lithography tools, the circuits contain now a billion components on the area no larger than a fingernail. While the invention and innovation of lithography process steps has been the back bone of Moore's law's, the classical photolithography for the 20 mm technology node has been pushed the furthest in recent technology generations. The optical lithography issues are best summarized in the Rayleigh equation [2]:

$$R = k_1 \cdot \frac{\lambda}{NA} = k_1 \cdot \frac{\lambda}{n \sin \alpha} \tag{2.1}$$

R	Minimum feature width the process is capable of resolving to satisfy product specifications
k_1	Process dependent adjustment factor $k_1 = 0.25$ is the practical lower limit for single exposure optical lithography allowing for pattern reproduction
λ	Wavelength of light used to pattern an integrated circuit layout onto silicon
NA	Numerical aperture i.e., the opening angle of a lens that is used to project a mask (reticle)
n	Refractive index of the medium surrounding the lens and wafer environment
α	Acceptance angle of the lens

Over the years, the λ/R, i.e. the ratio of target linewidth resolution to optical resolution has been continuously shrinking. The wavelength of light has changed several times, supported first by Hg-arc lamps, then by excimer laser-based systems, and now operating in DUV (Deep-UV) at 193 nm. For many years already, starting at the 350 nm generation, litho engineers faced the challenges of dealing with features printed at critical dimensions (CD's) less than λ, with k_1 factors of 0.3, "close enough" to design intent. The projecting light passing through a diffraction limited system results in severe distortion of patterns printed on silicon compared to geometries drawn in the design process. The poor pattern fidelity can be expressed as edge placement errors (EPE) on silicon, and can even eliminate patterns entirely, due to low contrast. The locally improved pattern fidelity commonly uses resolution enhancement technologies (RETs) such as OPC (optical proximity correction) [3].

Successive technology nodes are expected to see increased process variation resulting in decreased predictability of nanometer scale circuit performance. Despite the relaxation of some 3σ tolerances, there are no known solutions for multiple variability control requirements, as admitted by the ITRS (International Technology Roadmap for Semiconductors) [4]. Even though such variability is charged against the value of semiconductor products, the IC enterprises prefer to trade the different risks based on ROI (Return on Investment), e.g., improving the control of gate oxide thickness (T_{ox}) or effective channel length (L_{eff}) CD for new design for value technologies (Table 2.1) versus revising performance targets for existing products. The investment part of this ROI equation is driven by correction techniques and tools,

Table 2.1 The ITRS requirement of gate dimension variation control is becoming more stringent as the technology scales (after [4]), (MPU – microprocessor unit)

Year	2005	2007	2010	2013
Technology node	90 nm	65 nm	45 nm	32 nm
MPU gate length	32 nm	25 nm	18 nm	13 nm
MPU gate CD 3σ	3.3 nm	2.6 nm	1.9 nm	1.3 nm

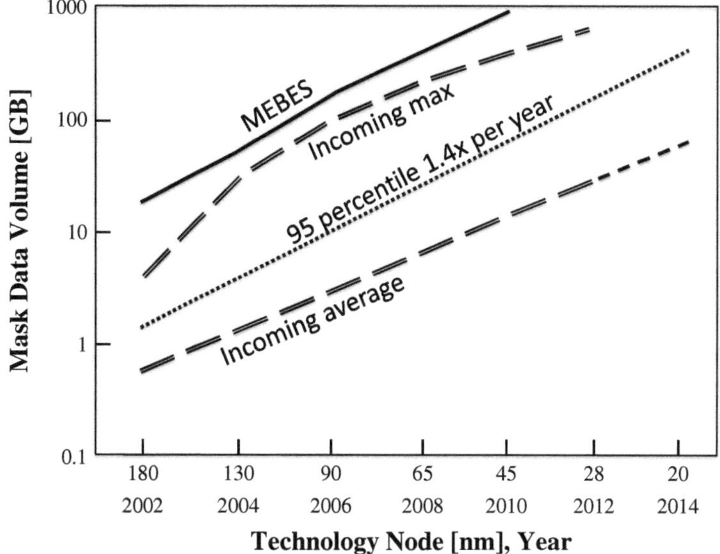

Fig. 2.1 Mask data volume as a function of technology node (after [7])

which enhance resolution and avoid critical path performance variation [5]. Resolution enhancement techniques (RET's) that address three degrees of freedom in lithography: aperture, phase, and pattern uniformity, have been adopted in nanometer-scale design (i.e. 90 nm processes and beyond) on an increasingly higher number of mask levels. A variety of techniques of multiple patterning and direct write also support this effort.

Due to the challenges of controllably printing very small features, the non-recurring engineering expenses (NRE) and turn-around time (TAT) costs of RET-based correction (OPC, phase-shifting, dummy features) are very high in terms of design time and mask yield/verification. Many cost-related parameters (yield, mask writing time, data volume, etc.) are directly proportional to the complexity of the shapes required on the masks for effective design pattern transfer (Fig. 2.1). Average mask data volume for a 45 nm design is 33× larger than for a 180 nm design. Mask writing time has increased from days to sometimes over a month due to RET complexity [6]. The relationship between cost of design and lithography reduces the ROI of low-volume products (such as ASIC) where mask price may dominate over the expenses for design or wafers [7].

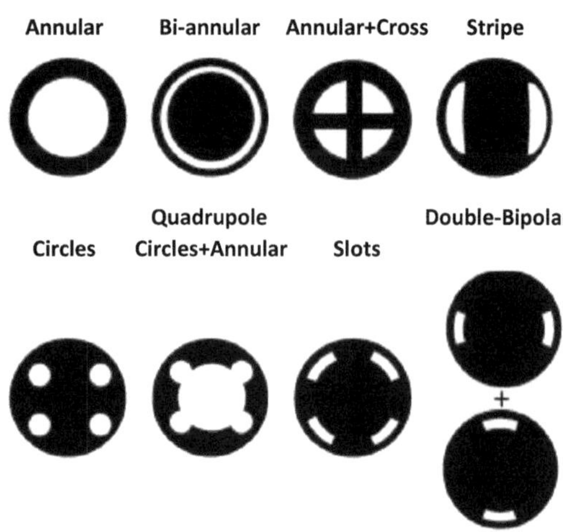

Fig. 2.2 Shapes of illumination sources for pattern – dependent best resolution (after [2])

2.1.1.1 The Physics of RETs

Four basic features of the patterning wave: the wavelength, the amplitude, the phase, and the direction, are engineered in the RET (resolution enhancement technology) to modify the wavefront and to enhance lithography resolution, R (Eq. 2.1) [8]:

- Wavelength. It is determined by the light source, reduced from 365 to 193 nm throughout the years of IC manufacturing. However, as λ decreases to push the limits of resolution, the light is increasingly more absorbed by materials of the optical system so that it is impossible to build refractive optics for small wavelength. While with immersion tools operating at 193 nm, a resolution of 50 nm and below is being achieved, the EUVL using λ of 13.5 nm cannot take off the ground in volume production due to the absorption problems.
- Direction. Different shapes of illumination sources (Fig. 2.2) change the direction of the wavefront to achieve the practical ultimate resolution for acceptable CD variation of $k_1=0.25$. Referred to as OAI (off-axis illumination), it increases the diffracted angle resulting from the first-order-light. As the resolution on the mask becomes small (beyond the acceptance of the exposure lens, making the image contrast nearly zero), OAI improves the contrast of the image by transmitting more diffracted orders through the lens (Fig. 2.3). The angle of OAI is a function of feature pitch: lithographic benefit erodes as feature pitches deviate from the one that the illumination angle has been optimized for. To prevent this loss of process window, dummy features (or subresolution assist features, SRAFs) are added to the layout, to lithographically emulate the primary pitch.
- Amplitude. Amplitude is controlled by the shapes of the geometry openings, corrected by rule-based and model-based OPC. Rule-based (RB) OPC applies corrections to wavefront through the mask based on a predetermined set of rules. Simple RB OPC i.e. iso-dense biasing, line end extensions or serif additions, and SRAFs insertion (Fig. 2.4) worked well for light refraction controlled by mask

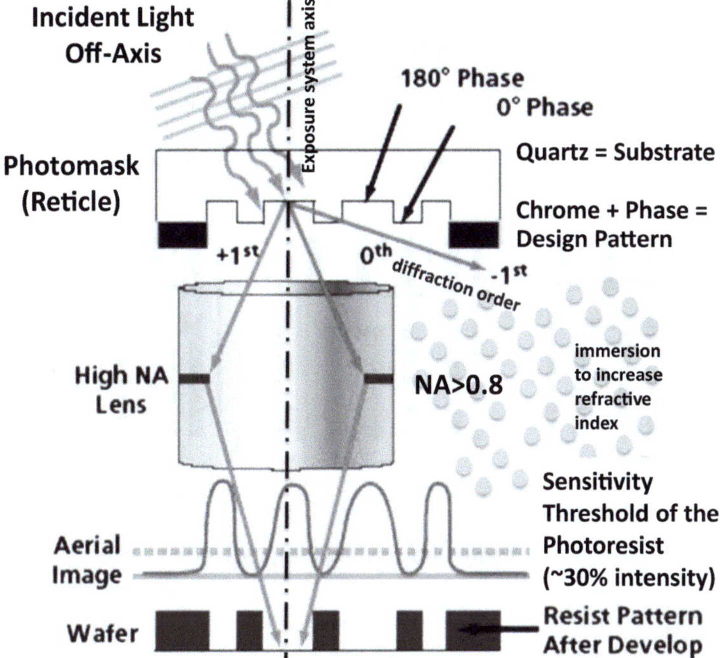

Fig. 2.3 Off-axis illumination (OAI) (after [2])

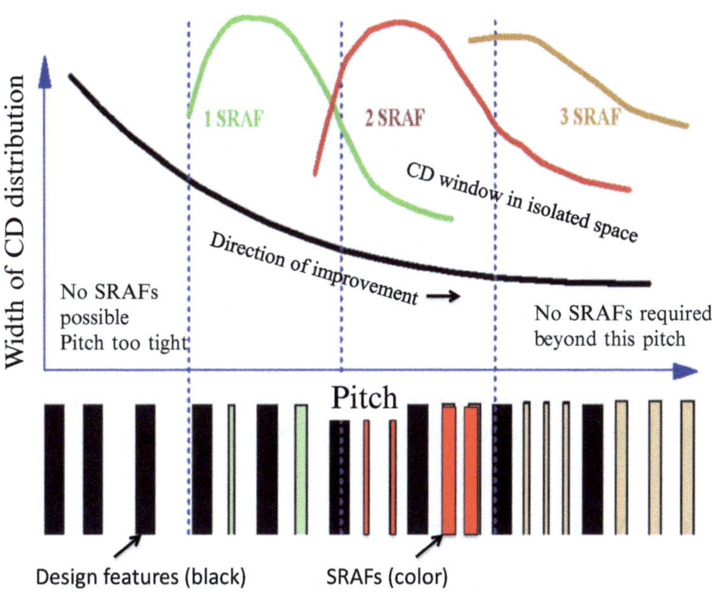

Fig. 2.4 Sub-resolution assist features (SRAF) improve across-pitch process window (after [2])

Table 2.2 Rule-based (RB) versus model based (MB) approach

Approach	RB	MB
Source information for design	Rules	Design kit
Implementation	Layout constraints	Process + material constraints
Handshake with Fab	Integration rules	Qual data
Verification	DRC	Simulation

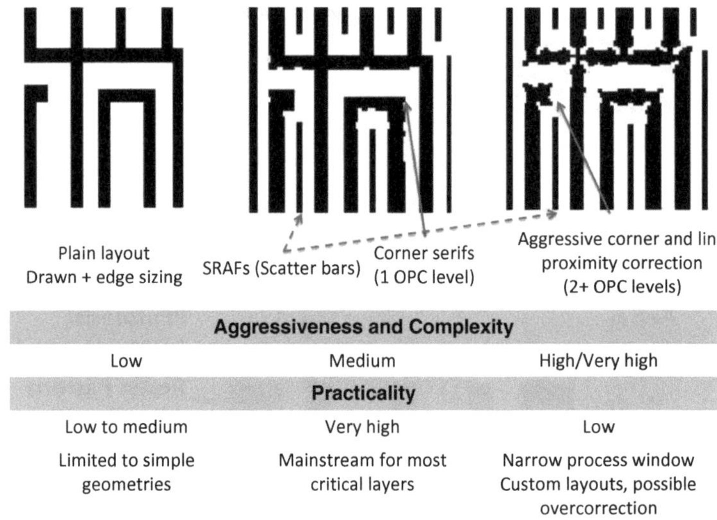

Fig. 2.5 OPC and SRAF implementation on random layout: rule-based edge bias combined with SRAF insertion (after [3])

apertures close to the wavelength. It has given way to model-based (MB) OPC for nanometer – scale designs which uses process simulation to determine far more complex corrections to the masks of arbitrary shapes. These corrections provide more accuracy and higher yield at the expense of higher cost and longer run time (Table 2.2). A hybrid, even more complex flow of RB OPC followed by MB OPC may be adopted for custom performance (Fig. 2.5).

- Phase. Phase shifting masks (PSMs) create interference fringes on the wafer to boost contrast and enable deep subwavelength features, based on the many techniques for phase shifting and assignment algorithms (binary or attenuated PSM) [9]. Phase shifting may result in conflicting phase assignment on layout regions, to be resolved by restricted design rules (RDRs).

2.1.1.2 OPC Versus Variability

The various RETs that, on the one hand, enhance feature resolution and improve chip yield, may degrade chip performance on the other. Their variations, due to the four-parameter wavefront engineering, add to process variations of gate CD, oxide thickness, metal width and thickness, temperature, voltage, etc., impacting circuit tolerance at the design stage. They also require modeling [10–12]. The most common speed/frequency model is often based on "worst case scenarios" (corner cases) assuming all parameters are independent and can be all in the "bad" corner at the same time. This provides overly pessimistic simulations, making the design unnecessarily complex [13, 14]. More "intelligent" approaches may provide more accurate predictions of process variations using a probabilistic framework to adequately reflect their correlations [14, 15]. Over 50 % of L_{gate} variation is due to systematic sources, which can be identified on the physical layout [16]. Assumptions about the L_{gate} distribution in Monte Carlo simulations and statistical timing analysis could be rigorous, by considering realistic contributions (mostly due to proximity effects) to the overall process variations [17, 18]. A systematic intra-chip variability leading to large circuit path delay variation and the location-dependent variability of L_{gate} divides the layout patterns into several categories [19]. Aerial image lithography simulations may reflect systematic L_{gate} variations but are limited to fixed layout and cannot be expanded to full-chip timing analysis. A systematic variation-aware static timing methodology using library-based OPC may not work at the full-chip level requiring full knowledge of neighboring geometries [20]. Post IP additions to design (especially OPC) may be evaluated at both IP and chip level to optimize their point of insertion depending on cost and accuracy of reproduction of design intent.

So far, adding RETs has been a post-layout procedure, but it now needs to become part of design flow in which libraries and layouts are optimized based on conflicts discovered by the RET tool [15]. This "trickle up" effect of RETs within the design process can be mitigated by more conservative (restrictive) design rules.

To print very tight pitches at a wide CD range compromises the resolution for lithographic systems. One can limit the range of allowed pitches by restricted design rules (RDR) to enforce highly manufacturable layout. Any design rule is a tradeoff between manufacturability and intended design performance (measured as layout density, delay, power, etc.). RDRs push the tradeoff in favor of the manufacturing side to the direction that may look like sacrificing layout quality (pattern density) and when over-constraining, may compromise the benefits of technology scaling. Making design rules flexible outside the critical regions relaxes the RDR-compliant layout to recover die area in the printability sweet spot. Tightening the rules in the hotspot regions should help when RDRs are not sufficient to guarantee high yield or waste too much area.

The impact of OPC on variability must be understood in the context of overall impact of process variations on circuit performance and product ROI. Accurate device models and circuit simulation (e.g., Monte-Carlo analyses) help estimate yields and correlations among process parameters. The impact of process variation range and control on parametric yield at selling point can be mitigated by investing into design guardbanding.

Gate CD variation is considered the most important parameter, to which product performance shows most sensitivity. To meet the ITRS requirement of gate 3σ CD control while obtaining the lowest cost of ownership (CoO), a minimum cost of correction (MinCorr) methodology was proposed [2]. The Min Corr determines the level of correction for each layout feature such that the prescribed parametric yield is attained with minimum RET cost. This approach implemented with model-based OPC driven by timing constraints, uses a programming-based slack budgeting algorithm to determine RET level for all gate geometries to reduce MEBES data volume and OPC runtime. The reduction may range from 20 % to 30 % without compromising product performance [2].

Process control techniques help evaluate the impact of residual correction errors on a microprocessor's speedpath skew. For diagnosing and improving OPC quality for critical gates or matching transistors, timing analysis highlights the necessity of an embedded post-OPC design flow. Manufacturability, mask cost and performance are impacted by globally applying restrictive design rules (RDRs) to reduce pitch induced process variations or by a hybrid method for adapting standard and restricted design rules based on pattern matching.

2.1.2 *Design for Low Variability*

In addition to lithography techniques, aggressive technology scaling is introducing new variability sources with every technology generation. This makes process control more difficult especially with respect to the CD variation of printed geometries. Despite the relaxation of some 3σ tolerances, ITRS agrees that there are no good solutions for a number of variability control requirements [4]. Markets that drive the IC industry are aware of the impact of variability on the value of semiconductor products. Therefore, IC makers must be able to trade off the different risks and ROI opportunities, e.g., large ranges of T_{ox}, L_{eff}, CD control, versus new "design for value" technologies, revised performance targets etc. Through a combination of circuit simulators, analytical performance models, and Monte Carlo analyses, one can estimate parametric yields, based on:

- Statistical modeling of process variation die-to-die (D2D) and within-die (WID) by the sources of origin
- Correlation models of variations
- Projection of the impact of variability on critical-path delays to future process nodes
- Analysis of the sensitivity of performance variation to control parameters, measured by a change in the number of "sellable" chips and the extent of guard-banding required to meet parametric yield target.

These methods should address not only circuit variability as such, but more importantly, its testing and performance implications.

2.1.2.1 Circuit Variability

Circuit variability is due to deviations of process within specified limits, translated into circuit parameters (e.g., L_{eff}, V_{dd}, thicknesses of conducting layers etc.) outside of model space. It is introduced either during chip fabrication or due to circuit operation. Depending on the origin, it may be characterized as die-to-die (D2D) variation or within-die (WID) variation. The taxonomy of variability is as follows:

- D2D variation may affect each element on a chip and adds a random component across the wafer. It determines the nominal values of die parameters differing among chips on the same and on different wafers. D2D variation may be responsible for up to 50 % of the total CD variation and it is considered design-independent, but related to equipment properties, wafer placement, processing temperatures, etc. [2]. We only model D2D variations due to random effects, ignoring predictable systematic issues that can be designed around.
- WID variations:
  - Systematic WID variation is layout-dependent and may cause chip malfunction. It is predictable in the sense that any given pattern should yield placement-independant characteristics. Because pattern variation is unavoidable within a large layout context, systematic variation components have to be compensated at the design and reticle stages. Optical proximity corrections applied to identical gates in a standard cell may not result in identical physical dimensions due to the impact of different local feature densities.
  - Random WID variation, due to the inherent unpredictability of the fabrication process, such as fluctuations in channel doping, gate oxide thickness, the ILD permittivity, etc. This type is likely to have spatial correlation (nearby devices more similar than ones across the die) and may eventually pose significant challenge to design because random phenomena cannot be compensated for.

2.1.2.2 Experimental Setup and Methodology

Experimental setup for circuit variability characterization consists of:

1. A parameterized, scalable, single critical path circuit model and a multi-critical path model composed of a user-selectable number of independent single critical paths
2. Correlation data for parameter variations
3. Variation mitigation methodology by tuning up underlying sources
4. Modeling and Monte-Carlo circuit simulation.

ITRS Roadmap details the agreeable variation levels for each parameter, depending on the type of the circuit.

Single parameterized critical path (Fig. 2.6) can be composed of "l" identical local stages and one long top-level buffered global interconnect. The parameter "l" can be e.g. set to 10 at the 130 nm technology and reduced by one in each subsequent technology generation, to reflect pipelining and other micro-architectural advancements. In each local stage, a two-input NAND gate drives a short local line and is sized to optimize the speed-power tradeoff (fanout = 2), i.e. the knee of the delay versus sizing curve [2]. The global line length remains constant at 10 mm, consistent with the ITRS projections of fixed die size for future microprocessors [4]. Optimal inverting repeaters are inserted at even intervals into the global line to minimize delay. For each

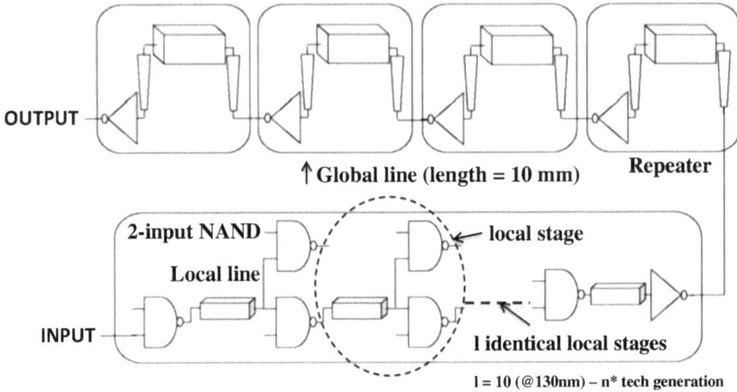

Fig. 2.6 Example of critical path for DfM performance evaluation (after [2])

local stage and the global line, two quiet parallel neighboring lines may be added to provide a more realistic capacitance environment. Each line is modeled as a sufficiently long chain of l segments to determine the distributed RLC characteristics. We then combine n identical single critical paths to analyze the impact on chips with multiple critical paths. Input transition times to initial stages can be set e.g., at 20 % of the clock period and the nominal dimensions taken from the models [22]. The critical path should include sources of delay such as local and global communication, important in generic paths.

It is often assumed that variation sources are either independent of each other or perfectly correlated [23]. Instead, several less-than-perfect, but significant correlations, should be recognized:

- V_{th} as a function of T_{ox}, N_{ch}, L_{eff} and X_t, calculated from a delta-doping approximation and BSIM3v3 models (X_t is the retrograde channel depth and N_{ch} is the effective channel doping [24–26]).
- Corresponding NMOS and PMOS parameters should have a correlation coefficient of one, i.e. NMOS and PMOS using the same gate should exhibit the same deviations from respective sources.
- For a fixed wire pitch, wire spacing variation should be the negative of wire width variation. Metal thickness (T) and underlying interlevel dielectric (ILD) thickness (H) are negatively correlated (with a correlation coefficient of −1) from the relationship of trench etch depth in damascene processes and chemical–mechanical polishing.
- Spatially proximate devices and interconnections (e.g., in local stages) have similar variations.
- Spatial correlation among repeaters inserted along the global line is modeled by incorporating a distance-dependent correlation factor. This correlation decays with distance to complete independence over a parameterizable length of approx 1 cm [27].
- Modeling of interconnect spatial correlation is done by dividing global line into segments (e.g., 100 μm long). Interconnect parameters within each segment are

Table 2.3 Comparison of RLC model with perfect correlations, spatial correlations and no correlations for 100 nm technology node (after [17])

Delay (ps)	Mean	3σ	Normalized $\frac{3\sigma}{mean}$
0 correlation	1453.9	133.9	0.9808
Spatial correlation	1452.5	136.4	1
Perfect correlation	1454.3	139.5	1.0216

perfectly correlated, but correlation between segments decays linearly with separation. For interconnect width and space, the zero-correlation separation distance may depend on the parameter. It may be set at 4 mm for all the technology nodes [28] or at 2 mm for metal and ILD thickness [29]. The line width is set by the dielectric etching while ILD and metal thickness are set by the CMP damascene processes. CMP planarization length relates to the distances over which features can be correlated due to pad deformation and other physical phenomena, typically also on the order of 2 mm, to avoid the overestimation of interconnect variation.

- WID and D2D variation are considered to be similar [2]. Systematic WID variation equally affects all critical paths, random WID variation adds random devices and interconnects inside the circuit.

Parameter variations are assumed to be normally distributed [24] (the variability in physical gate length is allowed to be 10 %). The physical gate length is ~50 % of the technology pitch (e.g. the DRAM half-pitch). Translating this to effective channel length (a fraction of physical gate length due to source-drain underdiffusion), we expect a 3σ for $L_{eff} > 10$ % which can be approximated as $L_{eff} = 0.6 \times L_{physical}$, corresponding to a 3σ process tolerance of 16.7 %.

One can model variabilities in a Monte Carlo analysis with prespecified number (n) of nominally identical, independent critical paths, with respect to WID variation and identical D2D components. Systematic layout dependent WID variation should be applied across different critical paths, with different dice having the same distribution, which shifts the nominal value of the parameters for different critical paths. Systematic WID variation should be modeled by generating n samples from a Gaussian distribution before running Monte-Carlo simulations.

Circuit simulation should be performed for a single critical path with a distributed-lumped RLC interconnect model and all correlations. Comparison of delay distributions obtained using Monte Carlo simulation for RLC interconnect model with:

1. Perfect correlation (correlation = 1),
2. No correlation (correlation = 0),
3. Spatial correlation,

in Table 2.3, shows that the upper/lower bounds for total delay variation due to interconnect fluctuations are set by (1) and (2).

In contrast with the linear regression analysis for Monte Carlo approach with 1,000 trials, where the variation sources all vary simultaneously [5], each model of process variability gives rise to 1,000 sets of random parameter values within the single critical path model, which one can simulate using HSPICE. The maximum delay is obtained for each die, and performance and yield analyses are performed on samples on the whole wafer.

2.1.2.3 Impact on Circuit Performance

The effectiveness of DfM can be measured by the response of the critical circuit parameters, i.e., the ones that impact its RoI. These parameters can be further converted to the selling point yield and the parametric tolerance around it.

(A) Selling Point Parametric Yield
Assume target parametric yield to be 99.7 %, which corresponds to the mean $+3\sigma$ point on the delay distribution and is the selling point of the chip. The selling point is calculated from the baseline results for all technologies. The change in parametric yield at the selling point is then taken as a measure of impact of process variation.

(B) Guardbanding Analysis
Guardbanding is the typical approach to account for variability. The expected ("designed for") value of performance is given by the mean of the delay distribution. The difference between the selling point and the mean gives the amount of guardbanding required: $\frac{3\sigma}{mean}$ expressed in percentage.

In turn, these two parametric distributions of 3σ and mean should be considered for two key types of circuits: with single – and with multiple critical paths [2].

2.1.2.4 Single Critical Path

Cumulative Effect of All Parameter Variations

One can simulate a single critical path and measure delay with all the parameters varying with 3σ and mean values of "baseline" $\frac{3\sigma}{mean}$ value, which may drop with technology node by e.g., 5 %. Assuming a critical path formed by N identical stages, with either perfect correlation (n = 1) or no correlation (n = 0), the $\frac{3\sigma}{mean}$ value of the total path delay is given by [2]:

$$\left(\frac{3\sigma}{mean}\right)_{path} = \begin{cases} \left(\frac{3\sigma}{mean}\right)_{stage} & ,correlation = 1 \\ \frac{1}{\sqrt{N}}\left(\frac{3\sigma}{mean}\right)_{stage} & ,correlation = 0 \end{cases} \tag{2.2}$$

Table 2.4 Trends of delay variation for a single local stage (after [17])

Delay (ps)	130 nm	100 nm	70 nm
Mean	92.15	82.53	77.04
3σ/Mean (%)	30.79	29.68	31.50

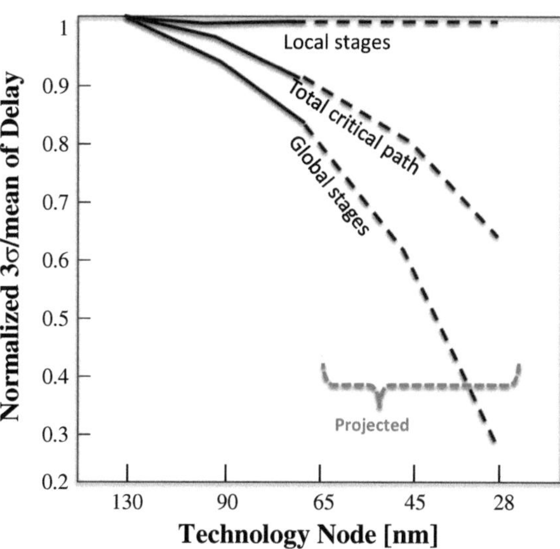

Fig. 2.7 Example trends of normalized required guardbanding for global and local stages, and total critical path (after [2])

Parameter N increases for global stages due to more aggressive buffering, but it decreases for local stages. The zero correlation leads to a $\left(\frac{3\sigma}{mean}\right)_{path}$ determined by both $\sqrt{N}$ and $\left(\frac{3\sigma}{mean}\right)_{stage}$; for perfect correlation, it is a function of $\left(\frac{3\sigma}{mean}\right)_{stage}$ only. In order to investigate the trend of $\left(\frac{3\sigma}{mean}\right)_{path}$, it is essential to know what trend would $\left(\frac{3\sigma}{mean}\right)_{stage}$ follow. The delay variation of a single local stage (Table 2.4) could remain fairly constant though technology nodes. Although a smaller average stage delay is expected as the devices scale down, a constant trend for $\left(\frac{3\sigma}{mean}\right)_{stage}$ versus technology is possible. In line with the ITRS expectations, a constant level of process variation should be achieved through advanced lithography tools. The delay variations for local stages show that spatial correlation and perfect correlation assumptions produce results within 4 % of each other. Therefore, a decreasing N does not impact path delay variation for local stages but shows a similar trend as obtained for a single stage: fairly constant for the next two technology generations. The delay variation for global stages is shown to decrease at a reasonably fast rate due to the increased number of repeaters (i.e. increased N in Eq. 2.3). This effect dominates for future technologies and causes the total delay variation to reduce (Fig. 2.7).

Table 2.5 A comparison of changes in delay variation when the nominal σ for an individual parameter changes from 0.5× to 2× (after [2])

Parameter	L_{eff}	T_{ox}	W
Increase in delay 3σ/mean	82.08 %	3.96 %	1.89 %

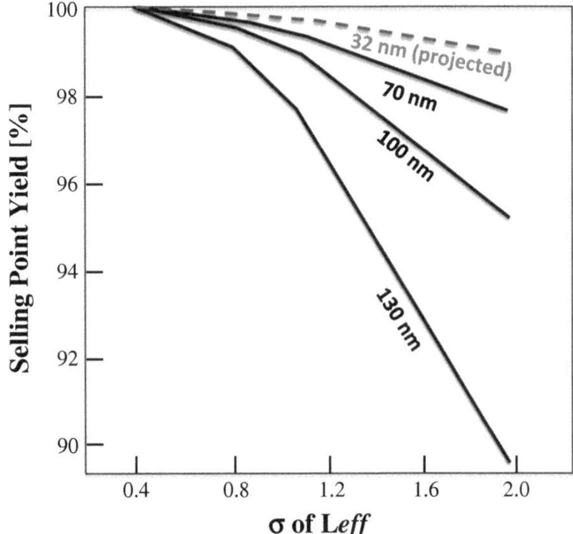

Fig. 2.8 Degradation of selling-point parametric yield with effective channel length tolerance (after [2])

Sensitivity to Process Tolerance

To determine the sensitivity to individual parameter tolerances, one can change the σ from 0.5 to 2 times their original values for each parameter individually while maintaining the σ unchanged for other parameters [2]. Delay variation is most sensitive to L_{eff} (Table 2.5). Loose L_{eff} control (on the order of 2× the current levels) can cause yield loss in the range from over 10 % to less than 2 % as the channel L decreases (Fig. 2.8). The roll-off of selling point yield versus process control around the nominal variation level implies that current levels of process control are near-optimal. However, slopes in the three curves differ, which indicates a smaller ROI for enhanced L_{eff} process control in future technologies. This is because the 28 nm devices are more velocity-saturated and their saturation drain currents are less dependent on channel length velocity increases from 1.2/65 to 0.9/28 nm, i.e., 74 % in average channel electric field. Switching speeds become less sensitive to channel lengths also through the use of strained silicon channels which improve carrier mobility.

Both D2D and WID variations play a key role in the overall circuit variability. Figure 2.9 shows the sensitivity of the required guardbanding to various controls on uncertainty sources, e.g., WID, D2D, and the cumulative effect of both on L_{eff},

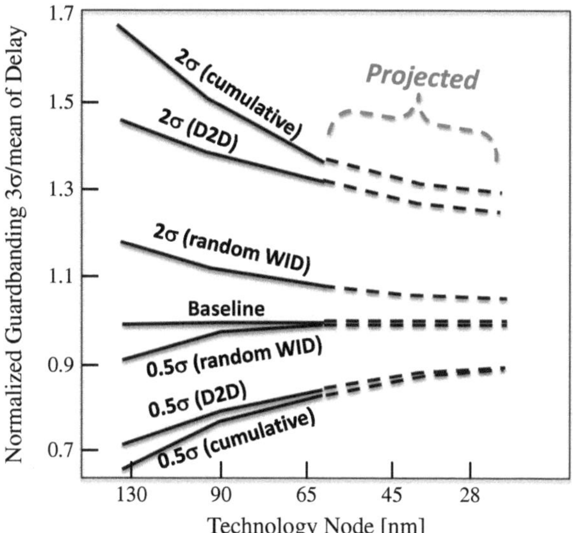

Fig. 2.9 Effect of variability control on normalized guardbanding required for 99.7 % parametric yield (after [2])

where for one critical path, delay variation is most sensitive to random D2D variation. One should note that while the sensitivity of the delay variation to CD fluctuations arising from all these sources decreases in future technologies, greater benefit in terms of reducing the required design guardbanding can be achieved by reducing the D2D variation rather than reducing the WID variation. In other words, reducing the scope of variations would be more important than reducing their size.

Tolerance of Process Variation Derived from Desired Guardbanding

Because process control involves high cost of data preparation (e.g., OPC), it is critical to know what level of CD control is necessary for adequate circuit variability performance, to maximize the ROI.

Assuming linear relationship between the magnitude of σ and the variation of total delay for an individual process parameter. Based on the simulation results for sensitivity analysis for the same design guardbanding, the requirement for L_{eff} accuracy relaxes as technology shrinks, meaning less effort of CD variation control. In other words, to maintain 30 % design guardbanding, levels of control dictated by the ITRS may be overly stringent, driven primarily by the predicted slow voltage scaling which results in device delays less sensitive to channel length. ITRS indicates a lack of known solutions to achieve the physical gate length control requirement in future technologies, but if these requirements would be relaxed, then existing, well-understood and cost-effective approaches should maintain manufacturability for the subsequent process generations. One should only notice that the historically slower scaling of voltage, than of CD control compromised device reliability in the end.

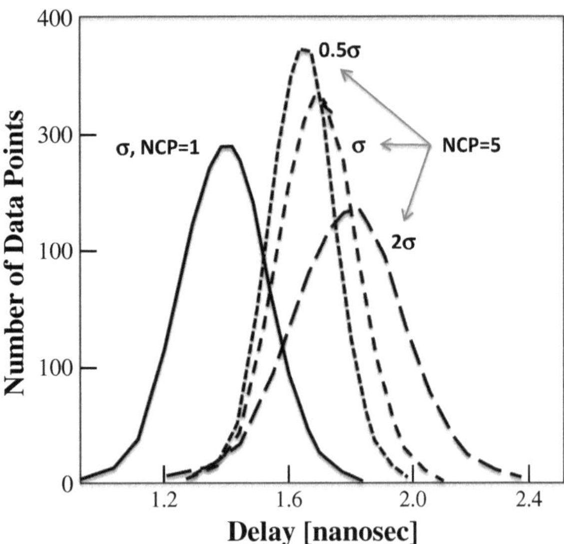

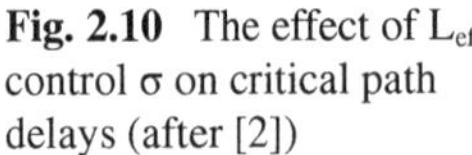
Fig. 2.10 The effect of L_{eff} control σ on critical path delays (after [2])

2.1.2.5 Multiple Critical Paths

Impact on Delay Distribution

Since systematic WID variation has the same impact across all dice, it does not affect yield when there is one dominant critical path on a die. However, when the number of critical paths (NCP) increases, systematic WID variation impacts chip performance causing delay distributions for different critical paths with identical designed-for delays (Fig. 2.10, [30]). The mean value of total delay changes rapidly when the number of critical paths increases from 1 to 10, beyond which it saturates (Fig. 2.11). Because for doubling of L_{eff}, its variation shows more impact on delay variation than for a similar reduction in its variation, for more critical paths, the delay becomes more sensitive to CD variation (note the vertical spread of the three data points at NCP=50 versus NCP=5). Therefore, for a large number of critical paths, tighter control of L_{eff} is desired. Design techniques to reduce power (such as dual-V_{th}, dual-V_{dd}), create multiple additional critical and near-critical paths.

Sensitivity to Process Tolerance

To achieve 99.7 % parametric yield, the guardbanding required under the same CD variation as a function of σ L_{eff} (Fig. 2.9) becomes smaller as NCP increases, and its sensitivity to NCP falls off for more than ten paths. For NCP > 100 in low-power designs, the guardbanding will be lower, but predictable from the results for fewer critical paths. In contrast to the mean delay, the guardbanding becomes less sensitive to L_{eff} variation as NCP increases (Fig. 2.12).

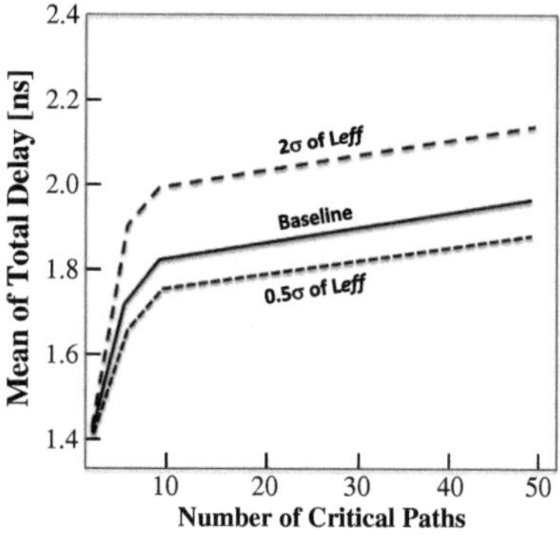

Fig. 2.11 Mean delay as a function of the number of critical paths (after [2])

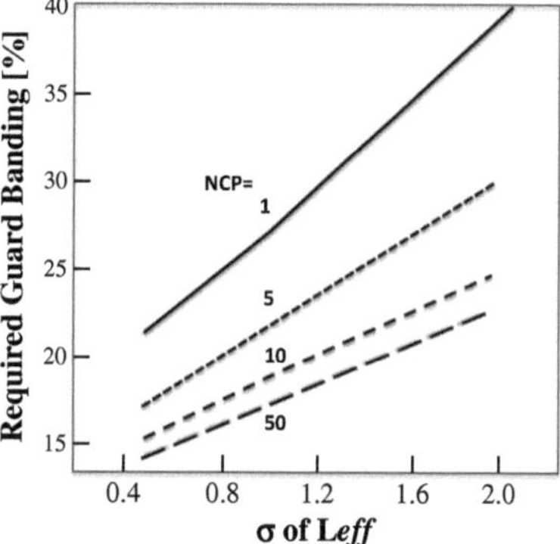

Fig. 2.12 Impact of *Leff* tolerance on guardbanding required to achieve 99.7 % parametric yield (after [2])

With more critical paths, the $\left(\frac{3\sigma}{mean}\right)_{stage}$ delay ratio becomes smaller. However, due to the shifting in the average, the selling point delay (i.e. mean +3σ) becomes worse. For the delay that gives 99.7 % yield for one critical path with L_{eff} varying at normal σ, i.e. the selling point delay, one can plot the parametric yield as a function of the number of critical paths under different process controls (Fig. 2.13). To reduce the yield loss caused by an increasing NCP, one can improve process control (increase manufacturing costs) or reduce the number of critical paths in circuit design (by adding design effort and power consumption).

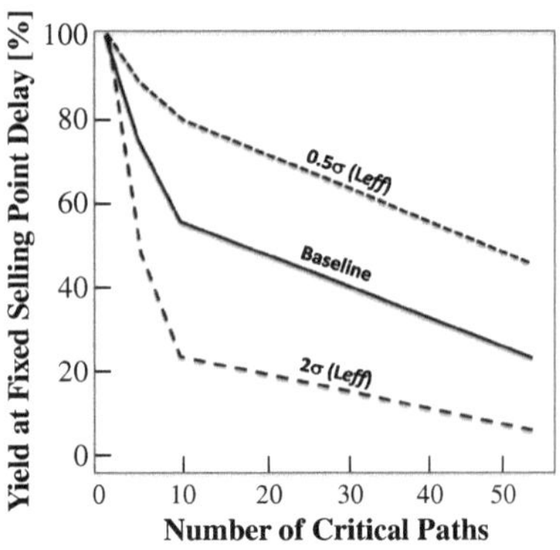

Fig. 2.13 Parametric yield as a function of the number of critical paths and process variation control (after [2])

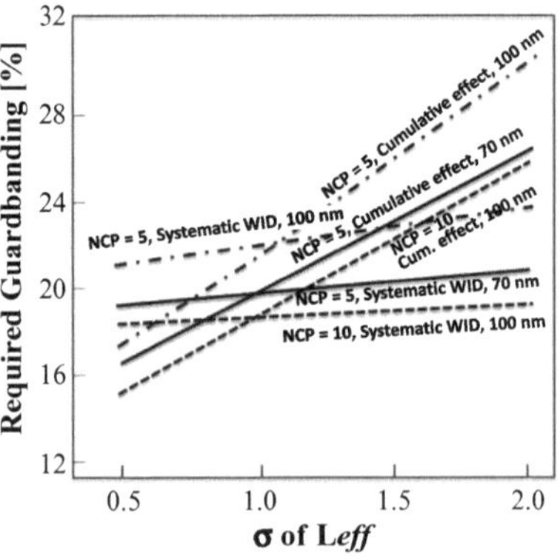

Fig. 2.14 Sensitivity of design guardbanding required for 99.7 % yield to variation with respect to its sources (after [2])

Simulations with NCP > 10, where design guardbanding becomes less sensitive to process variation control (Fig. 2.14), compare the sensitivity of the required guardbanding for 99.7 % yield to different controls on L_{eff} variation. The change in L_{eff} variation is manifested in systematic WID variation or in the cumulative effect of WID and D2D variations. In both cases, the required design guardbanding becomes less sensitive to CD control as NCP increases and the technology scales down.

In summary, the following correlations among NCP, guardbanding, and sensitivity are observed:

1. For multiple critical paths, both the required design guardbanding and its sensitivity to CD control become smaller with technology scaling. This important correlation allows for aggressive technology shrinks,
2. As NCP increases, a larger mean delay and a significantly increased selling point delay (mean + 3σ) will cause yield losses in ASIC designs unless the effect of CD variability is carefully premodeled,
3. Delay mean value is more sensitive to CD variation control for larger number of critical paths, while delay variation shows more sensitivity for smaller count of critical paths,
4. Both mean delay and delay variation are more sensitive to NCP for larger process variation, i.e. for the same level of CD control, these parameters vary widely for $1 < \text{NCP} < 10$, beyond which the sensitivity is reduced.

To improve yield at a fixed selling point delay, keeping NCP under ten is the most effective way to achieve a smaller delay mean value and to create a situation where process control is most valuable.

2.1.2.6 Variation-Centric Physical Design

According to the DfM rule of ten, the second most important DfM goal is performance optimization, or design for performance (DfP) [1]. As process variation leads to performance distributions, design for value (DfV) methodologies should maximize the yield by the desired structure of performance buckets. Performance is measured by critical path delay T, a function of design variables x_i and process parameters y_i, i.e. [2]:

$$T = f\left(x_1,\ldots,x_m,y_1,\ldots y_n\right) \tag{2.3}$$

Design for performance seeks values of x_i to minimize T, given the nominal values of y_i, and ignoring process variations, i.e.:

$$y_i = y_{i\_nom} \mid Minimize\, T \tag{2.4}$$

Worst-case values of y_i may represent a corner-based approach where all parameters are very pessimistically taken at their 3σ points, within a deterministic framework. We define value to be the total dollars earned from the chip sold on the market. A value function *v(f)* provides the market value of the chip for some performance measure *f* (e.g., speed, power). Thus, the total value of a given process is obtained as:

$$Value = \sum v\left(f\right) x\, yield\left(f\right) \tag{2.5}$$

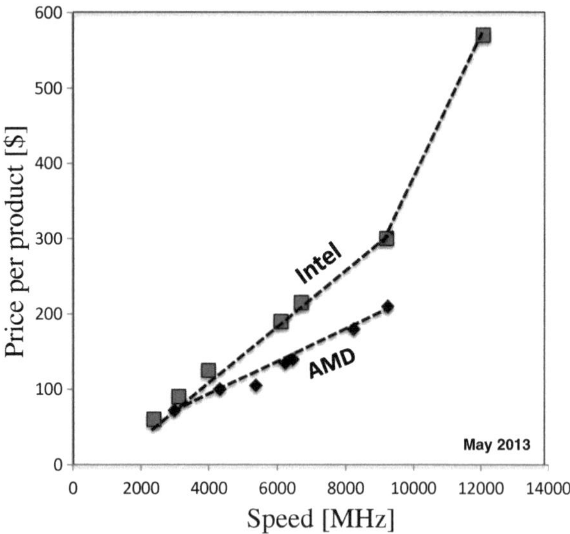

Fig. 2.15 Market prices of example microprocessor products (various sources)

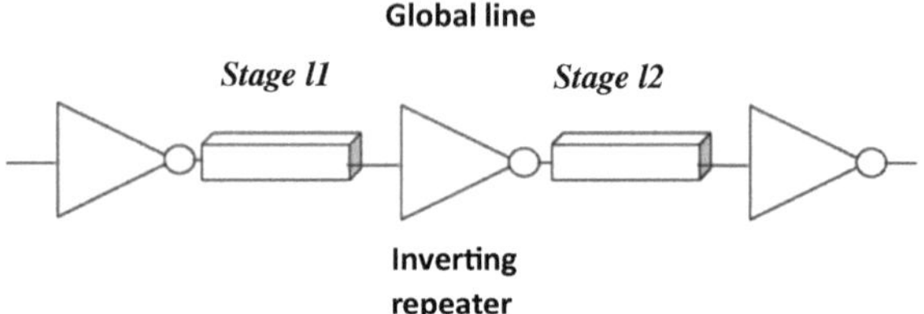

Fig. 2.16 Simulation structure for DFP versus DFV study (after [2])

For examples of microprocessor unit (MPU) for value functions (Fig. 2.15), design seeks to find values of x_i to maximize yield of $T<T_m$, for the variability distributions of parameters y_i, where T_m is the target delay, i.e.

$$y_i = N\left(\mu_i, \sigma_i\right), for 1 << n\, Maximize\, P_T\left(T_m\right) \tag{2.6}$$

To understand the difference between design for performance and design for value, one can conduct a MC simulation with an analytical delay model and the interconnect capacitance model from [31] for a simple example of a global line with a source, sink, and repeaters (Fig. 2.16). Device resistance is $0.8V_{dd}/I_{dsat0}$, source/drain series resistance $R_s=0.1\ V_{dd}/I_{dsat0}$ [32]. The RLC delay is used to calculate the inverter and interconnect delay. Normally distributed variation in L_{eff} is assumed, as the CD typically has the largest variation and impact on circuit performance. Correlation is assumed to decay linearly with distance and the correlation for distance greater than 10 mm is considered to be zero. Repeater location (distance from source) is varied from 0 mm to 4 mm in 0.1 mm steps. The DfV optimum is calculated for each selling point or threshold delay by running 1,000 Monte-Carlo simulations. The difference between yields of DfP and DfV optimizations can reach a few percentage points (Fig. 2.17). The parametric yield difference may range from 1.1 % to 5.6 %

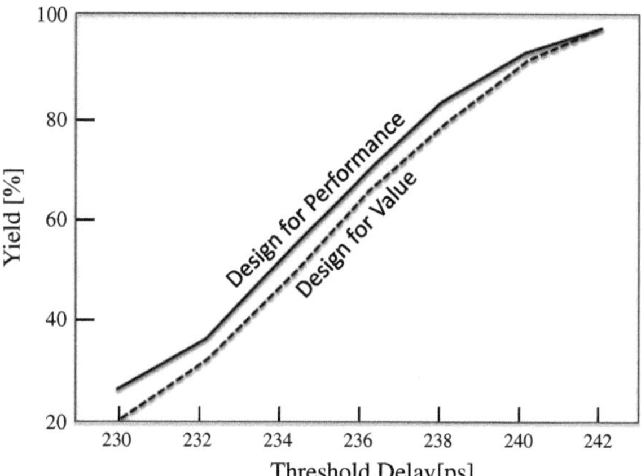

Fig. 2.17 Yield difference between DFP and DFV, based on MC simulations (after [2])

depending on the selling point delay and may be more pronounced for more complex value functions than the ones shown in Fig. 2.15 at DfV optimization.

2.1.3 Variability Control by Design: Conclusions

For sub – 100 nm designs, DfM methodologies and metrics need to include guardbanding, parametric yield at selling point, and inferred variation tolerance to allow for a comprehensive taxonomy of variations [2]. DfM approaches for the upcoming technologies, are being based on the cornerstone assumptions inherited and extrapolated from the analysis for previous technology generations, such as:

- Technology scaling and ITRS mandated levels of process variability to reduce delay variation
- For chips containing one dominant critical path, systematic WID variation does not affect yield
- Performance is very sensitive to L_{eff} variation but the sensitivity reduces with technology scaling due to enhanced velocity saturation and a growing number of critical paths
- Larger NCP results in a smaller delay variation but larger delay mean
- For the same NCP, looser control of CD variability leads to a larger required design guardbanding accompanied by a larger delay mean value, both of which show more sensitivity to relaxed than to tightened process specifications
- The delay distribution shifts to higher mean values but tighter sigma values as the number of critical paths increases and saturates beyond approximately ten critical paths
- For ASIC designs, reducing NCP is the most effective way to achieve a smaller average delay
- Variability impact can be restricted by innovative design, which is preferable to process improvement due to high cost.

Superior to taking variability into account during design optimizations, selling point optimization may be preferred over performance optimization. There may be multiple selling points with some prespecified values. The total design value is then given by $\sum v(f) * yield(f)$, for a given value function v of performance measure f and given parametric yield distribution $yield(f)$. DfV then seeks values of design parameters to maximize value function assuming normally distributed process parameters. Probabilistic optimizations should quantify the value and costs associated with both manufacturing and design solutions for the process variability.

2.2 Mask Data Corrections After Layout Closure

Next aspect of variability relates to layout and mask data optimization to ensure design performance. Even circuits with design pre-optimized for variability and yield, are subject to mask level enhancements without design control. Corrections to optical proximity and pattern density are the ones with most significant cost and performance impact.

2.2.1 Performance-Driven OPC for Mask Cost Reduction

ITRS Roadmap consistently considers microprocessor (MPU) gate lengths and highly controllable gate CD to be two critical issues for the continuation of Moore's Law cost and integration trajectories [4]. To meet ITRS requirements (Table 2.1), resolution enhancement techniques (RETs) such as optical proximity correction (OPC) and phase shift masks (PSM) are applied to an increasing number of design layers with increasing aggressiveness. Unfortunately, OPC also adds variability [33]. Not only it mixes in its own process-related performance range which varies by the location on wafer, but it narrows the window of its customer process i.e., the lithography. It also creates hardship for the mask making. All these factors contribute to the cost of OPC which should be minimized. The resulting steep increase in mask expenses and lithographic complexity has harmful impact on design starts and project risk. Cost of ownership (COO) is now a key consideration in adoption of various lithography technologies and has to be understood to ensure the cost-effective DfM approach as well as based on operating and non-recurring expenses (NRE).

2.2.1.1 OPC and Mask Cost

The total cost to build advanced but low-volume parts is dominated by mask expenses [7]. Significant portion of all masks are used on fewer than about 600 wafers, which translates roughly to production volumes of $\leq$ 100,000 units. The high costs of RETs cannot be fully amortized and the corresponding cost per die becomes very large, mostly due to the NRE component.

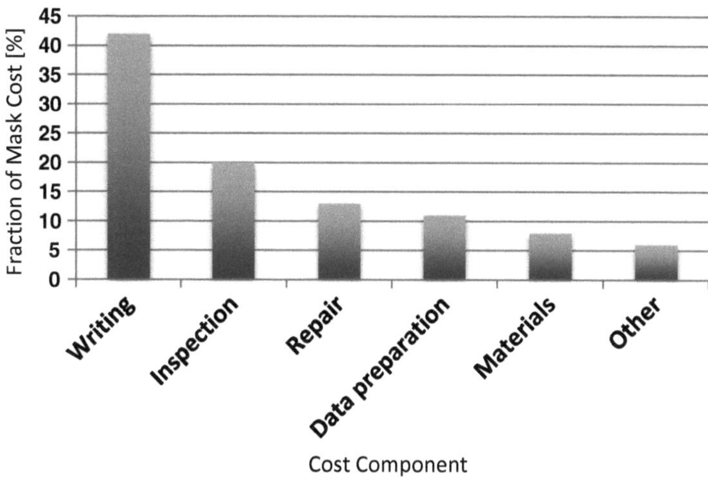

Fig. 2.18 Relative contributions of different components of mask cost (after [9])

OPC is a major contributor to both mask cost and design turnaround time (TAT). More than a 5× increase in data volume and several days of CPU runtime can be due to OPC insertion [34]. OPC affects mask data preparation (MDP), defect inspection and repair, and the mask-writing process (Fig. 2.18). Variable-shaped electron beam mask writers, in combination with vector scanning (with run time proportional to feature complexity), dominate the high-speed mask writing. In the standard mask preparation flow, the GDSII layout data is converted into the mask writer format by fracturing into rectangles or trapezoids of different dimensions. With OPC, the number of line edges increases by 4×–8× driving up the resulting GDSII file size as well as fractured (e.g., MEBES format) data volume leading to super-linear increases in mask writing and inspection time. One of the DfM goals is to reduce cost implications of OPC on product COO while improving the printability.

2.2.1.2 Role of Design in OPC Optimization

OPC correction and mask writer accuracy depends on the shapes being patterned. Metal lines in critical path should not be given the same error margin as less critical geometries e.g., the ones forming a company logo, potentially causing mask inspection tools to reject a mask. On the other hand, dividing layout features into different criticality groups may create confusion and miscategorization.

A performance-driven DfM – for – OPC approach would:

- Quantify CD error tolerance with a budgeting algorithm that outputs layout edge placement error tolerances,
- Integrate within a commercial MDP flow, with minimum cost of correction,
- Reduce OPC maintenance, in terms of the number of MEBES features and the runtime of OPC insertion.

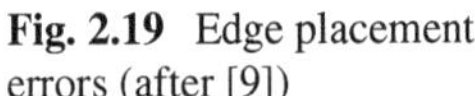

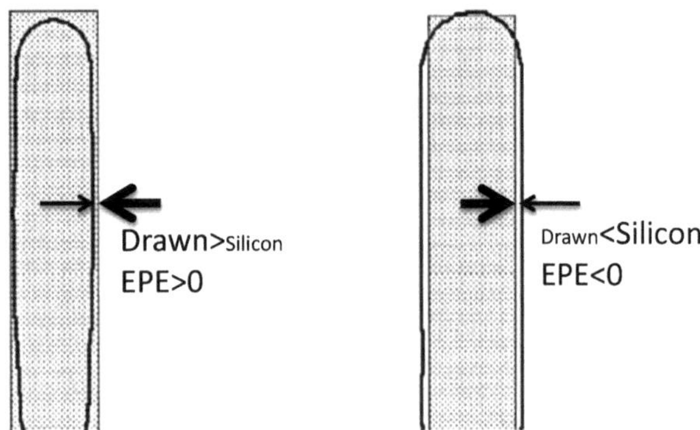

Fig. 2.19 Edge placement errors (after [9])

Because the mask writing time, a key component of mask cost, is a linear function of figure count, driven by layout style and extracted from post-OPC mask pattern, the OPC – dependent cost of correction flow would also be minimized [35].

Yield closure in the yield flow would be considered similar to flows for timing closure. In both flows, there are discrete allowed "sizes" that correspond to OPC aggressiveness (Fig. 2.5). For each instance in the design, there is a cost and delay penalty associated with levels of correction. The flow involves construction of cost/yield aware IP libraries and a selling point yield algorithm, which applies timing driven cost optimization. Accordingly:

- Different levels of OPC can be independently applied to different types of gates in the design to modulate effective channel length L_{eff} variation at the desired cost level,
- Field-poly, which does not impact performance, should be treated separately from gate-poly with different quality metrics (e.g., contact overlap),
- OPC corrects the layout for pattern-dependent through-pitch CD variation.

DfM optimization should find the adequate level of OPC correction for each feature such that prescribed circuit performance is attained with minimum total cost. Because OPC tools are driven by edge placement errors (EPEs) not by critical dimensions (CDs), the flow should pass design constraints on to the OPC tool, which breaks up edges into fragments that are then iteratively shifted outward or inward (with respect to the drawn feature boundary). The simulation shows if the estimated wafer image of each edge-fragment falls within the specified EPE tolerance (Fig. 2.19). That tolerance would translate into mask data volume (Fig. 2.20), e.g., within 20 % range, depending on the technology node. Restrictive layout methodology should help standardizing OPC in similar flavors of building block cells. However, even in the most restrictive design, there would be a share of cells with random OPC, which needs to be optimized on a case-by-case basis.

Since model-based OPC corrects for pattern-dependent, systematic and predictable CD variation, OPC should actually determine nominal timing. Therefore, OPC insertion can be correlated with corner-case instead of statistical timing analysis. A slack budgeting approach, as opposed to the sizing approach, would determine EPE tolerance, which for the proposed OPC flow would be only applied to gate-poly features.

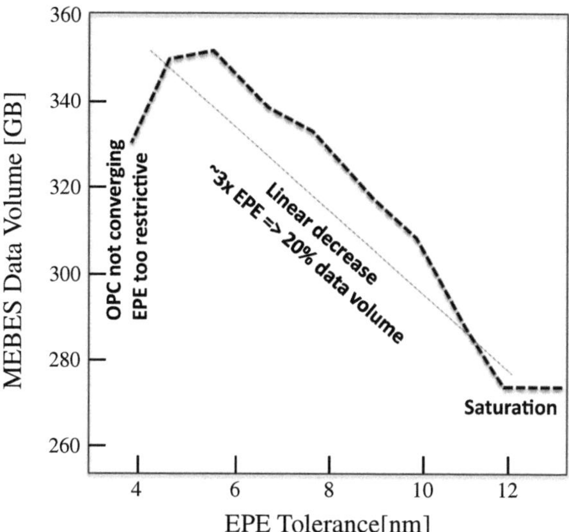

Fig. 2.20 Example of increased mask data volume with reduced EPE tolerance (after [3])

The metrics of results for the EPE correction flow (Fig. 2.21) are MEBES data volume and OPC insertion tool runtime.

One may notice that for advanced technology nodes, the cost of poly OPC may be comparable, if not smaller than of metal OPC. This is because poly is often drawn with restricted cells and its variability is lower. By contrast, metal is used for random routing with layout features largely hard to predict. Intelligent PNR tools are critical to avoid high cost of metal OPC.

2.2.1.3 Slack Budgeting

The slack budgeting would distribute slack at the primary input of combinational logic (i.e. sequential cell outputs) to various nodes in the design. The zero-slack algorithm (ZSA) [36] iteratively finds the minimum-slack timing path and distributes its slack equally among the nodes in the path. The MISA (Maximal Independent Set based Algorithm) [37] distributes the slack iteratively to an independent set of nodes. One should note that:

- MISA is not optimal. Budgeting problem can be due to convex programming. Full-chip MISA is too CPU-intensive,
- ZSA is faster than MISA and its weighted version can be formulated.

Full-chip programming can iteratively solve a sequence of linear programs (LPs), with slack budgeted among the top k available paths, repeated until all nodes have been assigned a slack budget or path slack is sufficiently large.

When budgeting is adopted in place of sizing, the accounting for changes in next-stage input pin capacitance becomes an open question. To be conservative, timing reports with pin input capacitances may correspond to the loosest tolerance

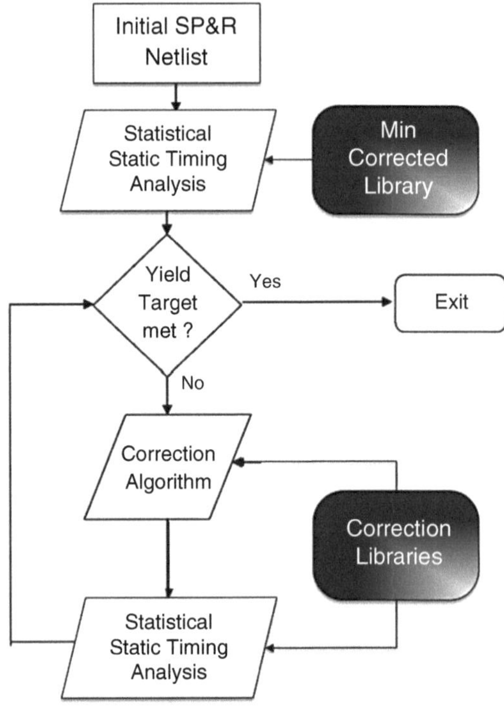

Fig. 2.21 EPE Min Corr flow for quantified edge placement error tolerances (after [9])

(i.e. largest pin capacitance) but gate delays would correspond to the tightest tolerance achievable.

Positive delay budgets correspond to increased gate length and positive EPE tolerances (EPE tolerance is a signed quantity in Mentor Calibre). Negative EPE tolerances (i.e., reduced gate length and signal delay) can also be obtained based on hold-time or leakage power constraints. For simplicity, one can assume equal positive and negative EPE tolerances for combinational benchmarks and timing, as below.

2.2.1.4 Calculation of CD and EPE Tolerances

To map delay budgets to CD tolerances, characterization of a standard-cell library with varying gate lengths, input slew and load capacitance values are required for every cell instance. For example, if an instance with specified load and input slew rate has a delay budget of 100 ps, we can select the longest gate that meets this delay. This largest allowable CD will lead to a more easily manufactured gate with less RET effort. Subtracting budgeted from nominal gate lengths yields CD tolerance for all design cells.

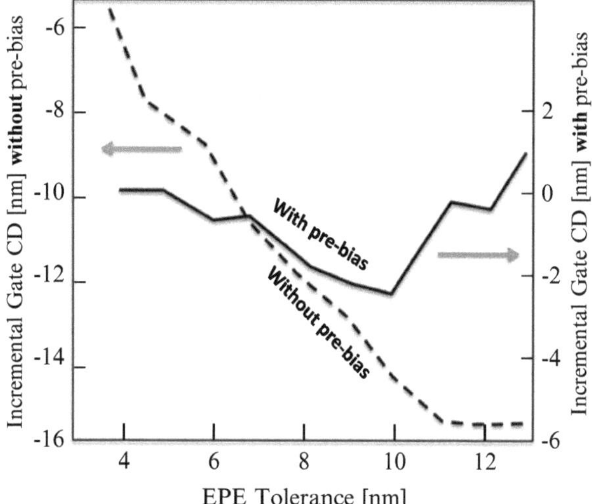

Fig. 2.22 Comparison of average printed gate CD with and without pre-bias (after [9])

The next step is to map CD tolerances to signed EPE tolerances for OPC insertion tools, assuming positive and negative EPE values to be the same. Since CD is determined by two edges of a geometry, max CD tolerance is twice the EPE tolerance.

In typical subwavelength lithography, gates shrink along their entire width such that the printed gate length is always smaller than the drawn gate length, except at the corners. OPC tool typically biases the gate length such that it is larger than the designer-drawn gate length. Thus, model-based OPC shifts edges outward ("positive" direction) until it meets the EPE tolerance specification. If the step size of each edge move is small enough, the EPE along the gate width will always be negative, because OPC tool is approaching the larger nominal gate length value starting from the smaller printed gate length value. As a result, the actual printed gate length will almost always be slightly smaller than the drawn gate length, leading to leakier but faster devices.

To achieve length deviation for an unbiased gate one can apply pre-biasing it, e.g. by its intended EPE tolerance. For a drawn 32 nm gate and EPE tolerance of 2 nm, the printed CD would typically lie between 30 and 32 nm (each edge shifts by 2 nm inward). If the gate length is biased by +2 nm so that the OPC tool views 34 nm as the target CD, the printed CD would lie between 30 and 34 nm, which amounts to a ±2 nm CD tolerance. In this way, pre-biasing achieves CD tolerances, as intended (Fig. 2.22). EPE tolerances can be enforced within a commercial OPC flow (Calibre [38]).

Mask cost reduction based on OPC optimization should rely more on the regularity of the layout than on reducing the count of high precision features. Even one critical spot in the whole chip may require an enhanced mask grade for lower variability, so design attention should be focused on delivering layout with equalized risk over the entire field.

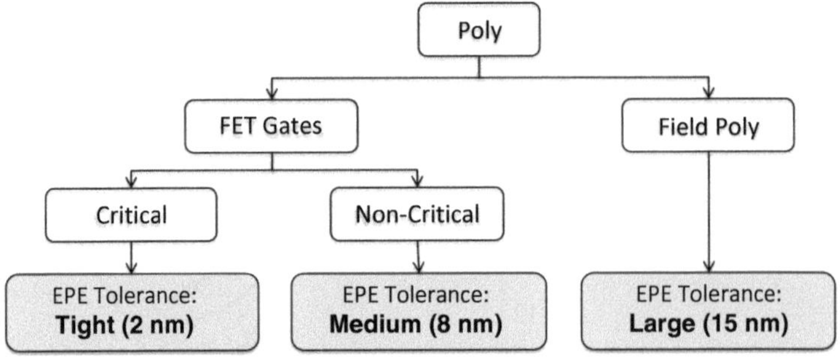

Fig. 2.23 Summary of EPE assignment for OPC level control (after [2])

2.2.1.5 Test Cases

Benchmarking can be performed e.g., on ISCAS85 suite of benchmarks and Opencores [39], synthesized, placed and routed in a restricted library containing cell macros. Multiple cell types (BUF, INV, NAND2, NAND3, NAND4, NOR2, NOR3 and NOR4, Table 3.2), should be characterized assuming total EPE tolerance levels e.g. ranging from ±1 to ±4 nm, with would correspond to the worst case gate lengths of 32 nm + EPE_Tolerance. Cell delays are mapped to EPE tolerance using circuit simulation but neglecting the dependence of delay on input slew.

Expected mask cost for each cell type can now be extracted as a function of EPE tolerance. Model-based OPC using Calibre on individual cells is followed by fracturing to obtain MEBES data volume for each pair (cell, tolerance). Though cell corrections depend on placement environment, standalone OPC is representative of data volume changes with changing EPE tolerance. The sensitivity of mask cost to delay change is close to a linear function (Fig. 2.20). A lib-like look-up table shows correction cost sensitivities (with respect to the tightest EPE tolerance of 4 nm). When slack is distributed, we extract the load capacitances to identify entries in the sensitivity table. Cost change is most sensitive to delay changes when the load capacitance is small, arriving at sensitivity numbers on the order of 1×–10× MEBES features per ps of delay reduction.

Model-based OPC flow may involve assist-feature insertion, which become an additional cost item. The EPE tolerance is tagged to each gate for the poly layer split into gate and field features (Fig. 2.23), e.g., with field-poly tolerance of ±4 nm and gate-poly tolerance from ±1 to ±4 nm tagged to cell names. Iterations are set to minimum values, beyond which the cost and CD distribution show little sensitivity to OPC. After model-based OPC, average gate CD and standard deviation are extracted from simulated wafer image. The corrected GDSII output is then fractured into MEBES to determine the total mask data volume.

With benchmark circuits synthesized (e.g. using Synopsys Design Compiler), place and route (Cadence Silicon Ensemble), slack report output of the top 500 critical paths STA can be run with a modified 34 nm EPE tolerance library with pin

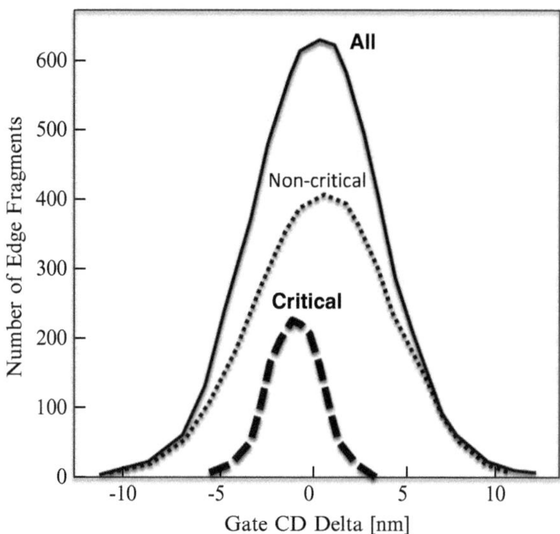

Fig. 2.24 Examples of gate CD distribution (after [9])

capacitances corresponding to 36 nm (loosest EPE tolerance) after slack budgeting (Synopsys Primetime). Two types of benchmarks are identified:

(i) Large designs with a "wall" of critical paths (Table 3.2), and
(ii) Circuits with small sizes, where a single iteration is sufficient to solve the budgeting problem.

For small EPE, more iterations may be necessary, because potentially critical paths not reported due to the constraint of maximum number may become critical later as they are not optimized. As a solution, one can selectively include those paths that may cause performance degradation as slack budgeting objects or increase the constraint of maximum number of critical paths in the slack report. After several iterations, performance degradation due to the selective OPC should be reduced several times.

The extracted CD distribution (Fig. 2.24) shows that Calibre consistently enforces assigned tolerances, with tighter CD distribution for critical gates. Table 3.3 compares OPC for minimum and standard correction. For small circuits, a single iteration ensures no timing degradation from the traditional to the reduced flow, and the budgeting runtimes are small (1–11 s). For large designs, more iterations avoid performance degradation with runtimes of several hours. Mask data for reduced flow has MEBES data volume lower by up to 20 % and OPC runtimes are improved by almost 40 % [2].

2.2.1.6 Optimizing RoI of OPC: Conclusions

Controlling mask cost by lowering computational complexity of OPC can be achieved by leveraging EPEs as the key parameter directing OPC tools to correct the drawn design to the levels required to meet timing specifications instead of minimizing printability distortions. Iterative linear slack budgeting for each gate is mapped to allowable critical dimensions in the standard cell. EPEs generated from the CD budget and tags placed on gates drive the OPC tool to the required level of correction, which should result in MEBES data volume and the runtime of the tool reduced by 20 % or better.

A constraint to setting CD tolerances may be the leakage power or field poly extensions for contact enclosures. One can also verify the impact of fragmentation and minimum jog length on mask accuracy and cost. An example of optimized OPC'd layout reported 25 % shot count reduction and up to 32 % reduction in mask write time [40]. The impact of these algorithms – based database reduction procedures would also depend on technology node. Its complexity makes it likely to be worthwhile mostly for the leading edge process rather than for the cost reduction retrofits.

2.2.2 *Post-OPC Timing Analysis*

Post-OPC embedded static timing analysis extracts residual OPC errors from a full-chip layout and derives MOSFET gate CD values (calibrated to silicon). Through a combination of layout back-annotation and selective extraction from the global netlist, this approach improves upon the design flow with ideal (drawn) values. For post-OPC values, timing analysis shows substantial differences in the order of speed path, critical and worst-case slack. This prevents from effort wasted for optimizing paths that are not critical.

This OPC/timing flow can be used to locate critical patterns in the die where OPC could not achieve the desired CD control. A calibrated algorithm could then be applied locally to save OPC run time e.g., in designs with matching FETs (mixed signal clock generation in processors), to identify cells with high occurrence in critical paths or with large pin slacks.

2.2.2.1 Variability Reduction

OPC, which compensates CD distortions, becomes itself a major source of variability [1, 2], causing speed degradation. Fixing speed paths based on ideal (not simulated) L_{gate} values may be misleading in the sub-wavelength lithography, and the alternative is process CD simulations on the entire chip post-OPC. It would be very time consuming, based on calibrated optical and resist models, and lithographic conditions for a variety of features. One can reduce it to gates on critical paths as predicted in the full-chip timing analysis, tagged with a critical layer ID and extracted along with their peripheral geometries, within a distance corresponding to the optical diameter, beyond which geometries have no proximity impact for the

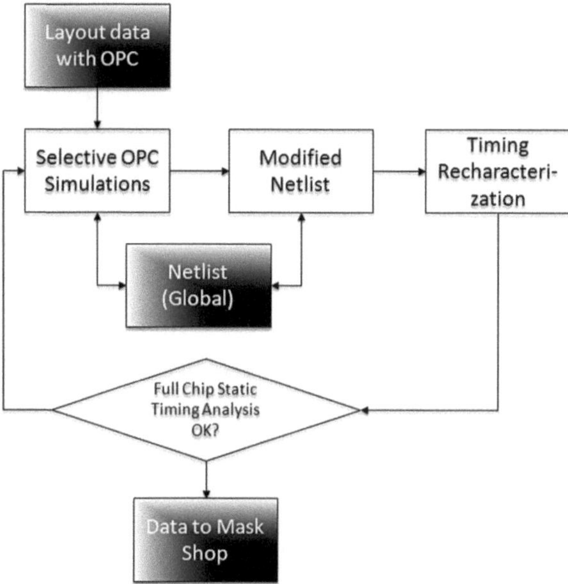

Fig. 2.25 STA – OPC flow (after [2])

lithography system, on a path-by-path basis. The Si-based CDs are extracted at the center point for each transistor. Each CD value is identified with the GDS coordinates of the corresponding transistor and mapped back to the circuit netlist, according to the following timing/OPC flow (Fig. 2.25):

- The cell timing library is updated with Si-based process CDs by creating a location-aware SPICE netlist for each cell. The process CDs for critical gates are back-annotated in the original library, with each transistor identified by its lower-left coordinates. A special LVS extracts the location of each transistor and the gate lengths in the SPICE netlists are modified with the process CDs. SPICE simulations account for the impact of systematic L_{gate} variations in timing re-characterizations.
- A full-chip timing analysis is performed on the global chip netlist and the top critical paths are reported. This may include cells not re-characterized in the previous step (i.e. on paths that the original pre-OPC timing analysis did not flag as critical), in which case re-characterization of newly critical cells will be necessary.
- Timing analysis based on process-simulated CDs are compared with ideal gate CDs at typical operating conditions (rather than the worst process corner to avoid the overly pessimistic worst-case) with CD extractions at best focus corresponding

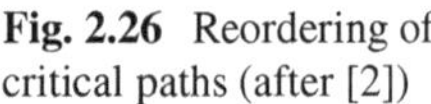

Fig. 2.26 Reordering of critical paths (after [2])

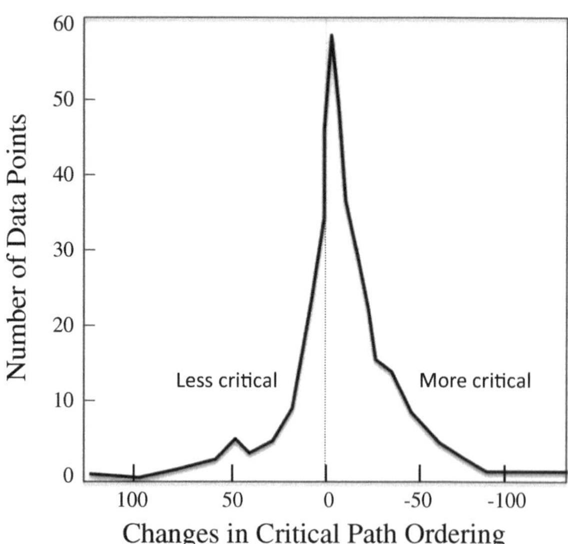

to a typical process, to determine whether the systematic L_{gate} variations introduced by RET/OPC have a significant impact on typical performance.

The optical diameter may be set e.g., at 4 μm based on experimental results. Due to adding new critical cells with Si-based process CDs, timing re-characterization runtime may be several hours per iteration. In the final speed path report, under 2 % of critical cells have not been re-characterized using their extracted CDs, in paths with lower critical ordering. Note that the majority of the layout geometries remain the same and there is no need to regenerate OPC patterns for the bulk of features. The risk of this approach depends on the risk of the process of tagging the critical nets.

2.2.2.2 Reordering of Critical Paths for OPC Optimization

Critical paths with slack ≤ 0 are rank ordered (most critical ones first) with I_{DS}. The Si-based timing report lists more paths as critical than the ideal CD- based timing report (Fig. 2.26). More negative Si-based slack paths are due to a number of paths with high critical ordering not reported in the traditional timing analysis (indicated as −1).

- For paths existing in both timing reports, the range of critical ordering difference indicates fewer paths becoming more critical and more paths less critical. The path slack changes (Fig. 2.27) show the slack increased by less than one third.
- When post-OPC CDs are used, new critical paths are identified, some of them highly critical with large slack violations and worsened path slack (Fig. 2.28). A traditional timing analysis for design optimization will be misleading: paths will be not be considered for resizing, although they will actually be critical post-fabrication.

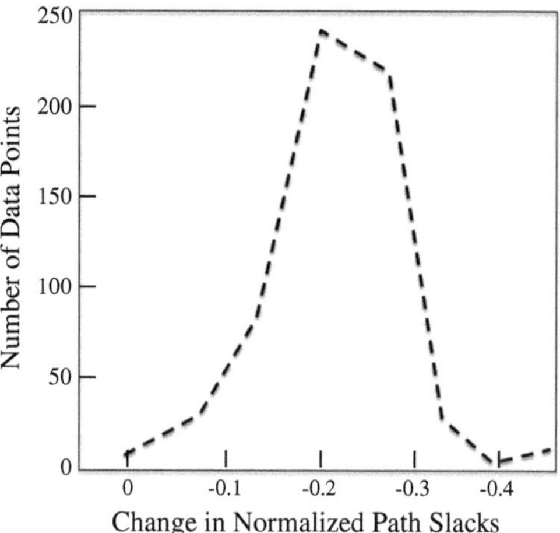

Fig. 2.27 Slack degradation for old critical paths (after [2])

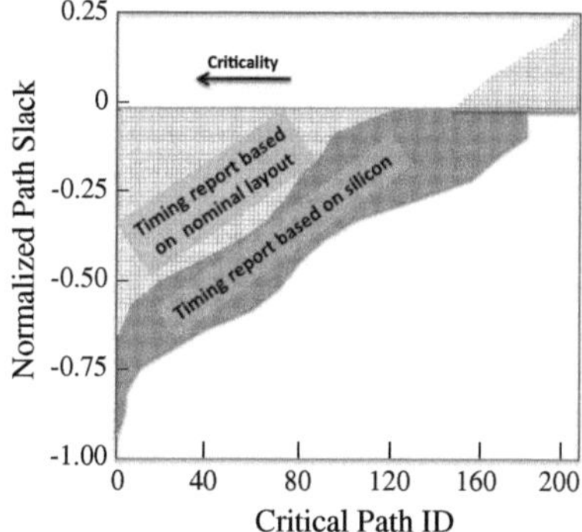

Fig. 2.28 Slacks for new critical paths (after [2])

Certain cells may consistently appear on critical paths and be good candidates for re-design and OPC optimization. The distribution of pin slack for critical cells is wider in the Si-based timing report (Fig. 2.29), with a shift to larger slack violations. The ordering of the top ten most frequently used critical cell types changes in the post-OPC analysis (Fig. 2.30). Therefore, design optimizations should be made post-OPC, increasing the complexity of design flow.

Model-based OPC corrects the layout on a point-by-point basis, considering all neighboring features as well as stepper and mask settings. OPC quality can be

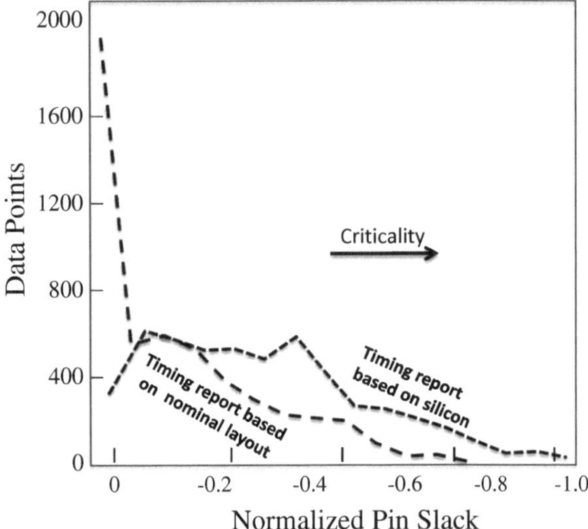

Fig. 2.29 The difference between (**a**) layout based timing report and (**b**) silicon-based timing report for critical cells

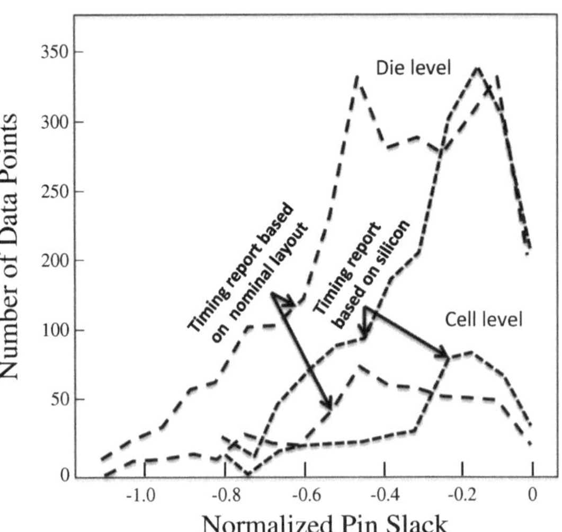

Fig. 2.30 Top IO most frequently used cells in ideal L_{gate} based timing report versus Si – based timing report

improved through fine-grained edge movements constrained by neighboring geometries, and the prioritization may limit the convergence of OPC correction. The number of iterations for edge movements required converging below the residual EPE error may increase. For cells with both large CD errors and a high frequency of occurrence (Fig. 2.30), customized OPC recipes can be created or cell layout patterns adjusted. As an example, using single pitch of the critical layer for a P-transistor in an inverter and avoidance of non-rectilinear shapes (e.g., L's or T's) reduces the normalized L_{gate} errors by several times. With the subsequent technology shrinks, only straight shapes are permitted on poly and active around the transistor areas.

2.2.2.3 Post-OPC Timing: Summary

Changes in the timing analysis after OPC demonstrate that traditional approach is no longer valid in nanometer-scale designs. A post-OPC performance verification can enable process variation-aware design optimization and drive tradeoffs when significant variability is unavoidable. With only slight modifications to the traditional flow put them in (gray boxes and no dashed lines (Fig. 2.31)), the post-OPC interconnect parasitics are extracted and back-annotated. In this way, the gate CD variations due to RET/OPC are taken into account during static timing analysis to improve performance prediction.

RETs, historically a post-layout procedure, now became part of a design flow with libraries and layouts optimized based on conflicts discovered by the RET tool. Integrating OPC step into the design flow will allow design-time optimizations to be aware of the manufacturing process and achieve improved performance and yields in the as-fabricated design.

Risk reduction of post-OPC performance verification requires the use of highly regular, restricted layout. This will be discussed in Chap. 3.

2.3 Incremental Improvements of IC Designs and Products: From 2D to 3D

In addition to optimizing the manufacturability of flat layout pattern as discussed in previous sections, DfM techniques are also dealing with design methodology development from to 2D to 3D [41]. Recent years have seen propagation of IC DfM into the multiple aspects of semiconductor IC products. The traditional DfM related to the corrections of circuit layout e.g. by pattern resolution enhancement techniques, recently broadened its scope by adding stacked die verification, 3D die packaging, floorplanning, and wiring, to help with the growing diversity of IC applications. These extended categories of DfM would be useful especially for Systems-on-Chip (SoC) as well as for the further pursuit of IC shrinkability, e.g. by double patterning. Many of the new concepts aimed at improving design efficiency have not developed a consistent, flow-based methodology at this stage. For that reason, they are published as isolated patent disclosures, which one can divide into the ones pertaining to DfM definition, DfM execution, and DfM verification (Fig. 2.32).

The focus of the first disclosure group (definition) is the correct-by-construction (CBC) methodology e.g., for layout (active devices, metal routing), die floorplan, or package architecture.

The second group (execution) pertains to process proximity correction (mostly OPC).

The third group (verification) concentrates on process model calibration and identifying the sources of process variability.

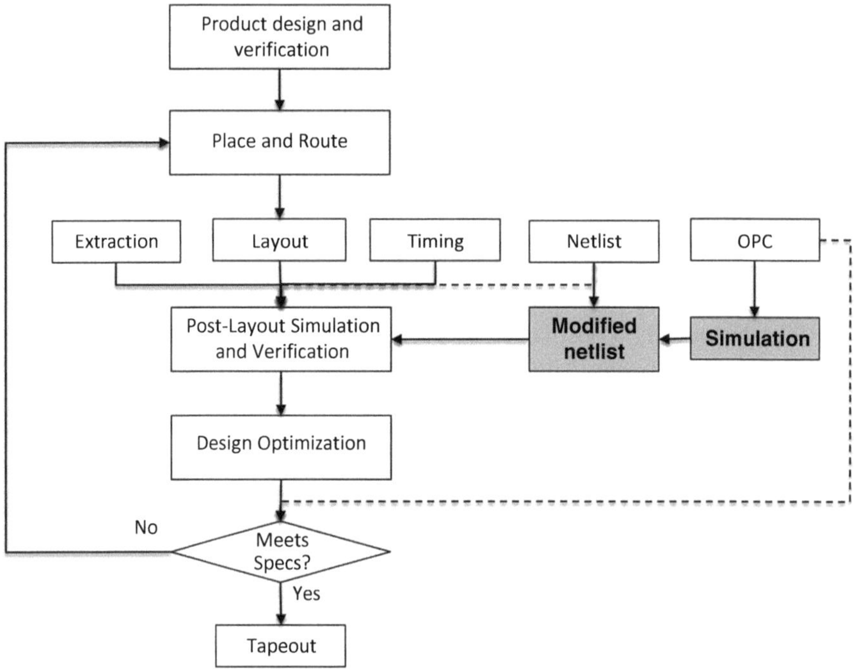

Fig. 2.31 Design flow with post – OPC verification Design flow with post – OPC verification: standard-with dashed lines, modified - no dashed line, with gray boxes

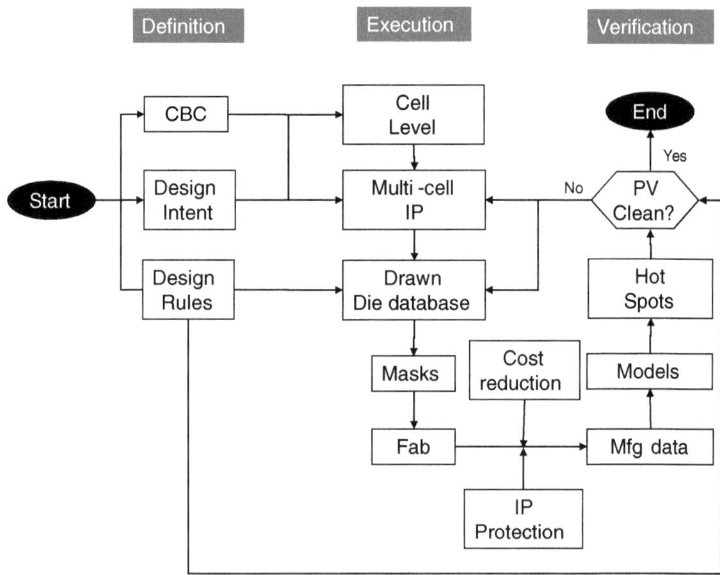

Fig. 2.32 Three domains of DfM and their implementation in design cycle (after [41])

The directions for engineering concepts of 2D ⇒ 3D DfM development, consistent over the recent years, show good alignment with the existing DfM approach to solve future design problems.

One can ask, why would package architecture or floorplanning become a DfM discipline. The answer is that DfM is a cross-disciplinary art of balance, based on the knowledge from multiple engineering areas, such as electrical, optical, and mechanical engineering. It has to accomplish product level goals such as high die yield, attractive performance, and zero reliability fallouts, at low cost. It is at the discretion of the design and manufacturing to decide which of the techniques belong or do not belong to DfM. By linking DfM with multiple disciplines, product engineering makes it easier to find synergies among the different aspects of product functionality. Also, new IC applications would cause such extended DfM approach to drive layout optimization.

It is difficult to quantify the value of many proposed DfM concepts. Return on investment (RoI) of engineering disclosures is to be judged by the reader. On the one hand, DfM can pull in the schedule of a new product yield ramp, which, when supported by the multi-million dollar quarterly revenues, would justify the cost of DfM including software tool licenses and manpower. On the other hand, delaying new product design in order to comply with stipulated but unproven DfM principles is undesired. Therefore, qualifying and implementing good DfM ideas as early in product definition as possible would not only help reduce delays but also improve cost efficiency.

For the purpose of this analysis, we can identify the following IC DfM implementation levels in product design (from the local to the global ones):

- Cell DfM – optimizing a self-contained set of simple layout geometries
- Block DfM – optimizing interactions among sets of cells performing common function at higher level
- Die DfM – optimizing medium and long-range process effects on a complete design of IC product
- Wafer DfM – reducing D2D variation on a manufacturing entity containing many dice
- Package DfM – managing mechanical stress for an IC product ready for customer implementation.

For these five levels, DfM concepts can be assigned to three product design stages.

At the earliest, Definition stage, DfM precedes IC design rules and guidelines, as correct-by-construction (CBC) approach to layout or product architecture, from cell to die and to package level. At the intermediate, Execution stage, DfM aligns with design or mask data verification for the IC layout, with emphasis on process proximity correction (OPC, PPC) from cell to wafer level. Finally, at the Verification stage, DfM improves manufacturing feedback to design/layout at all levels. Accordingly, examples of recent patent applications and publications can be grouped by the following categories:

(A) Definition – CBC (cell to package level):

1. Dialed – in design scaling [42]
2. Double patterning with hard mask [43]

3. Spacer based – fin-FETs [44]
4. Advanced die floorplanning [45]
5. Stress reduction by metal slotting [46]
6. Antenna ratio reduction [47]
7. Three-dimensional packaging [48]

(B) Execution – Validation (cell to die level):

8. Pixel-based OPC [49]
9. Cluster-based OPC [50]
10. Resolving OPC conflicts [51]
11. Auxiliary OPC patterns [52]
12. OPC preserving design intent [53]
13. A robust DfM flowchart [54]
14. Process compensation in IP libraries [55]

(C) Verification – Feedback to design (die to wafer level):

15. Etch process model [55]
16. Mismatch evaluation [57]
17. Process variation for on-chip sensor [58]
18. Planarity-related hot spots [59]
19. Verification of 3D devices [60, 61]
20. Table-based DfM [62]
21. Encryption of DfM data [63]

2.3.1 DFM Definition: Correct-by-Construction (CBC) Architectures (Cell to Package Level)

Because physical design verifications may take up to 80 % of design development effort, one should maximize the use of the correct – by – construction (CBC) architecture based on known good design intellectual property (IP) for cost reduction. When a CBC layout is first developed, one may assume that no design rules need to be checked to ensure the 100 % product yield. Restricting layout freedom up-front with CBC translated into RDR requirements should remove the need for corrections. However, even for CBC cells, manufacturing problems may occur due to their placement in blocks and in the die. Therefore, a design rule check (DRC) with complexity depending on the confidence in IP robustness to process and design variations, is still required. Three CBC robustness levels can be proposed, driving the restrictiveness of DRC rules. A proven, robust CBC IP needs only basic, geometric DRC check of width and space to prevent major printability violations. A layout not verified to DfM rules will require a more comprehensive rule system to address pattern fidelity. A fully random layout calls for a complex DRC to ensure pattern matching or fab transferability. These levels of DfM and CBC robustness should be considered when understanding new concepts of production – worthy IP, e.g., created by a dialed – in layout shrink, from a

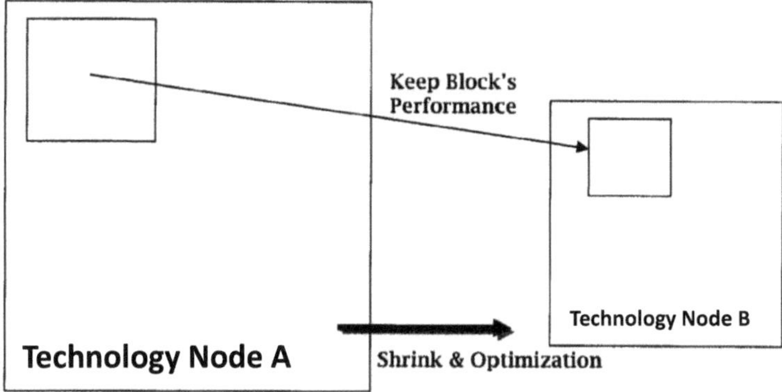

Fig. 2.33 The concept of shrinking design IP from larger to smaller technology node [42]

p-Cell implementing metal slotting, or suppressing antenna effects. New die-level concepts involving CBC layout such as die floorplanning or implementing 3D devices may require new design rule systems e.g., for System-on-Chip (SoC) products.

2.3.1.1 Dialed: In Design Scaling

One fast path to building CBC layout is scaling down known good design IP, to convert an IC product from the less advanced (A) to the more advanced (B) technology node (Fig. 2.33, [42]. Silicon proven IP kill, by definition should be DfM – compliant, if at the previous technology node. Geometric DRC for the target process may be expected to suffice for the scaled down but robust layout to perform without model redevelopment and resimulations. But the question is, how much effort is saved by the optical shrinking of the layout vs. designing it from scratch, and what opportunities are lost by skipping the relayout which may allow for product optimization.

As one approach to make sure that the shrunk design is comprehensively checked, the pre-shrunk design libraries need to be extended to include tables and formulae called electrical patterning (ePatterning) database [42]. The concept of ePattering reflects the correlation between wafer pattern and electrical parameters of the devices to drive layout corrections. When wafer pattern of a transistor is electrically simulated, the shapes of gate electrode and active region are simplified to rectangles, from which electrical parameters are determined based on the ePatterning tables.

IP migration between two technology nodes involves shrinking, gridding, and compaction (annealing) of hard IP blocks (Fig. 2.34), which then must meet timing and power constraints at the target technology node. This means, changes of litho process, materials, implants, SPICE models, due to different device and process targets need to be implemented, using the ePatterning IP database. To make sure that performance of an IP block for the target technology node is substantially the same as that for the original technology node, IP migrates in two phases: Phase 1 = DRC clean and Phase 2 = electrically matched. In the process, layout geometries are first divided, compared, and scanned for repetitive blocks, extracted and stored into a

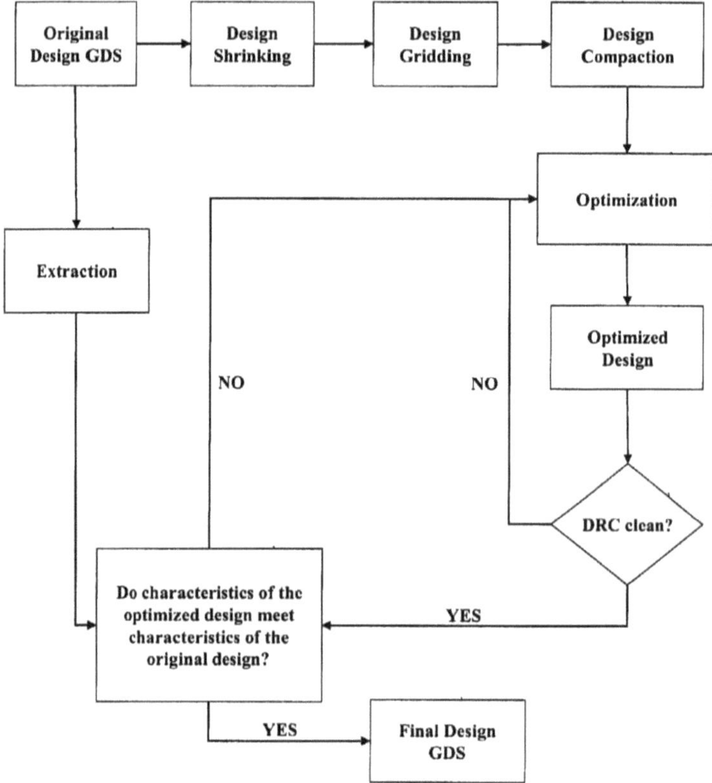

Fig. 2.34 IC layout migration flow using dialed – in shrinking [42]

library. Optical OPC (scattering bars, serifs, hammerheads) and electrical eOPC (drawn geometry extensions) are applied to achieve high resolution according to optical simulation and electrical properties equivalent to those of the existing silicon.

Next, rule-based or model-based ePatterning corrections are performed. First, a maximum rectangle inside the channel contour of each transistor is defined by overlaying it with the gate contours. The equivalent electrical width of the channel is then derived by electrical simulation (e.g., SPICE). Parametric correlations between the geometry expected on silicon and in ePatterning equations are saved in the ePatterning database. This way, when an IC product is to be transferred from one manufacturing technology to another, the designer does not need to be involved. The semiconductor manufacturer can independently redesign the IC by adjusting the IP.

The CBC concept of ePatterning based on optical and electrical simulations, self-adjusting OPC, and data tabularization has a number of risks. The apparent benefit of the low - cost IP scaling can be undone by months of product debugging on silicon if the product was launched without complete simulation. While layout migration may become a useful technique for design cost reduction, product design cannot typically be achieved without statistical corner models, for two reasons:

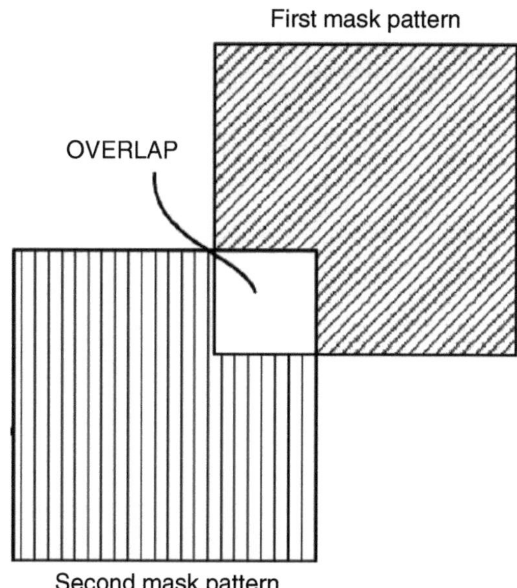

Fig. 2.35 Layout overlaps for first and second mask patterns [43]

- Complexity: the dialed-in IP scaling does not provide sufficient insight about the practicality of ePatterning equations,
- Lack of experience: a dialed-in full-chip layout scaling has not been the mainstream design activity. It is still preferred to design new products from scratch, based on new models, especially fprn the analog circuits.

In summary, comparing full product redesign to an algorithm based IP reuse for product scaling, one should note that layout footprint is not the only trade off. Sometimes, more important is the time to market. Layout translation based on fast but poor verification carries a risk to the project RoI that a lot of effort would be spent on debugging both correction algorithms and the suboptimal design.

2.3.1.2 Double Patterning with Hard Mask

Improving resolution of pattern transfer from design to wafer is a fundamental challenge for DfM. Masking structures and methodologies that enable printing layout geometries as small as 50 % of the critical dimension (CD) of the photolithography tool in the fab, are in high demand. The concept of multiple (double, triple, quadruple) patterning (DPT, TPT,...) by exposing different elements of the layout with a number of masks has enabled significant CD reduction compared to its single – patterning counterpart. But image distortions of the overlying patterns make multiple patterning a methodology more difficult from single patterning repeated several times. To reduce these distortions for DPT, the first pattern could be transfered from the photoresist to the hard mask layer, followed by the second photoresist layer patterned over it (Fig. 2.35 [43]). A CAD and photolithography operation splits layout geometries into the first and second pattern and creates overlap area where the

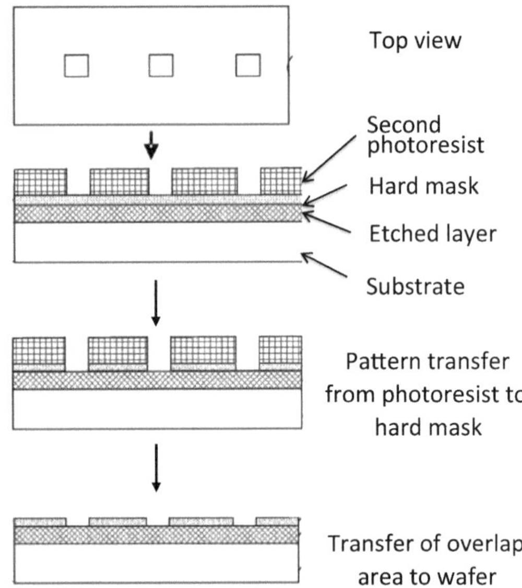

Fig. 2.36 Cross-sections of wafer pattern after first and second masking process [43]

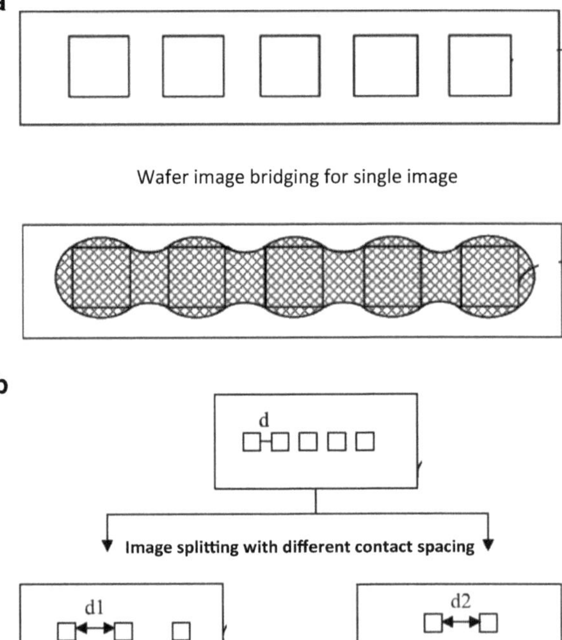

Fig. 2.37 (**a**) Bridging of wafer pattern due to single mask approach, (**b**) layout splitting between two masks [43]

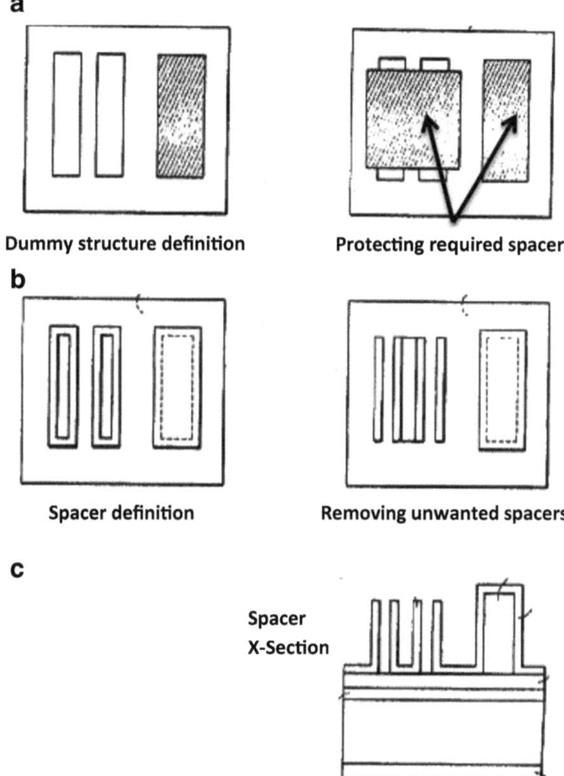

Fig. 2.38 (**a**) Dummy structures laid out to form spacer pattern, (**b**) top view of spacer pattern, (**c**) cross-sections of fin FETs [44]

patterns are projected on a common surface. First pattern is then etched into hard mask and a photoresist deposited over it transfers the pattern of the second mask (Fig. 2.36). Next, the overlap area is transferred to the wafer. Critical dimensions of both masks are larger than the resolution of a photolithography tool. Therefore, the manufacturability of process sequence is not compromised. Physical shrink is done without mask or wafer image bridging due to the conventional photolithography (Fig. 2.37).

In summary, the proposed concept of multiple patterning is already becoming the industry workhorse, due to the well defined path of reducing line CD's below the tool resolution limits. This concept may require DfM rule support for yield and reliability assurance.

2.3.1.3 Spacer Fin FETs

One flavor of multiple patterning is fabrication of spacer-based finFET devices to reduce mask overlay errors [44]. A spacer-based layout may be created on a global grid pattern of a dummy layer. A logical operation is then performed on the overlay

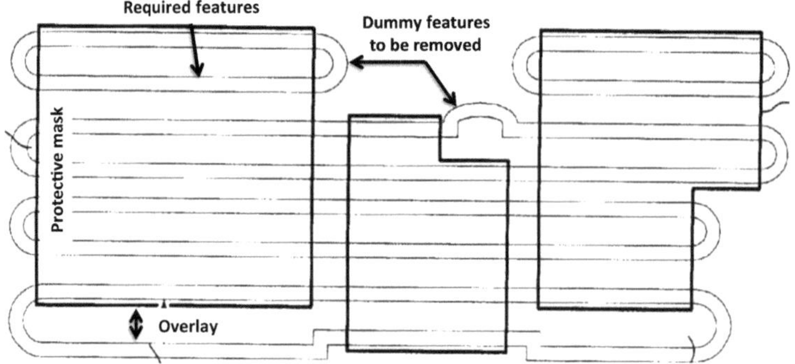

Fig. 2.39 Masking elements creating spacer pattern [44]

pattern to find critical margins and regions where bias may be added to the adjacent features, in order to protect geometries of the first pattern. This is followed by simulation, physical verification (e.g., DRC, LVS), parameter extraction, and layout place and route matching of paired elements, to verify that the placement would meet product requirements.

The first pattern defines dummy structures to form spacer elements abutting the pattern line (Fig. 2.38). The spacers define critical dimensions for the features, such as fins, on the substrate. The second pattern removes top spacer material and the third, cut pattern, defines the active area by removing unwanted spacer elements. Their width and pitch define finFET devices at less (~50 %) than minimum critical dimension of the photolithography.

The pattern formed from the overlay of the first, second, and third patterns are then used to etch the substrate to form a masking element. The hard mask is patterned as derivative dummy or a spacer layer and then removed from the substrate, to form a grid of masking elements (Fig. 2.39). The fins are created by etching the substrate.

In summary, the spacer – based methodology of extending the manufacturability of MOSFETs by defining them as 3D devices is one of the key drivers for further technology shrinks. This type of device architecture DfM has already been proven successful at product level.

2.3.1.4 Layout Auto-floorplanning for Reduced Routing

Another area to use the CBC design principles is automated layout floorplanning. It should be done right for the first time, as it cannot be easily corrected and it influences many parameters of the product including its manufacturability and RoI. While efficient floorplanning tools need significant machine power, parallel processing can reduce calculation time by 30 % compared to the prior art based on serial

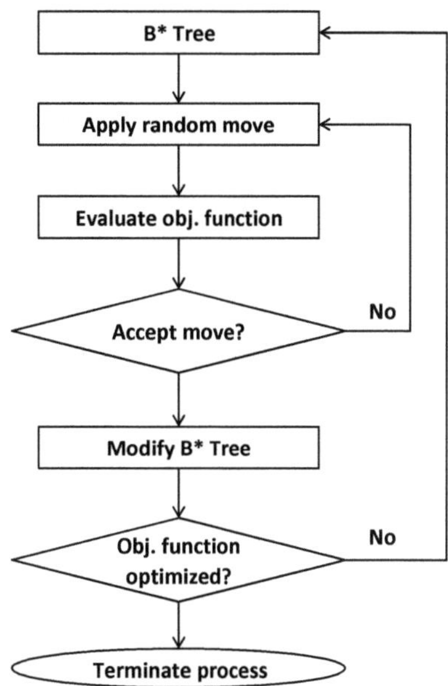

Fig. 2.40 Annealing algorithm using the B* tree [45]

computing architectures [45]. The floorplanning requires placement criteria (or objective), related to a DfM deliverable, such as wirelength area or pattern density.

Efficient block structure should represent non-slicing floorplans. One is an ordered binary tree (B* tree) in which the nodes are kept $^2/_3$ full by redistributing keys to fill two child nodes, then splitting them into three nodes. Ordered tree (O – tree) is a data structure where the children of every node are ordered: first, second, third child, etc. For any acceptable placement, B* tree inherits desirable properties from the O tree but it overcomes the irregularity typical in O trees and has a 1–1 correspondence with its compacted placement, where individual modules cannot move down or left. This adds certainty to the complexity involved with B* trees.

Typically, simulated annealing, i.e. probabilistic floorplanning optimization of IC layout within its allowed footprint, is applied to the B* tree. The random moves to explore the solution space are: (a) rotating a module; (b) moving a module to a new location; and (c) swapping two modules (Fig. 2.40). Starting with an initial floorplan, annealing it (i.e., shaking it up) randomly, applies one of the three moves to evaluate the newly generated B* tree. The move is accepted or rejected depending on whether it improves placement objective such as area or wirelength, and repeated until a satisfactory solution is obtained.

The sequential floorplanning relies on repeatedly applying a random move of a block and modifying the floorplan based on the acceptance of the move. A typical run will evaluate thousands of moves, creating a long chain of control and data dependencies, which must be broken to restructure the process for efficient mapping onto a GPU (Graphing Processing Unit), while preserving the solution quality.

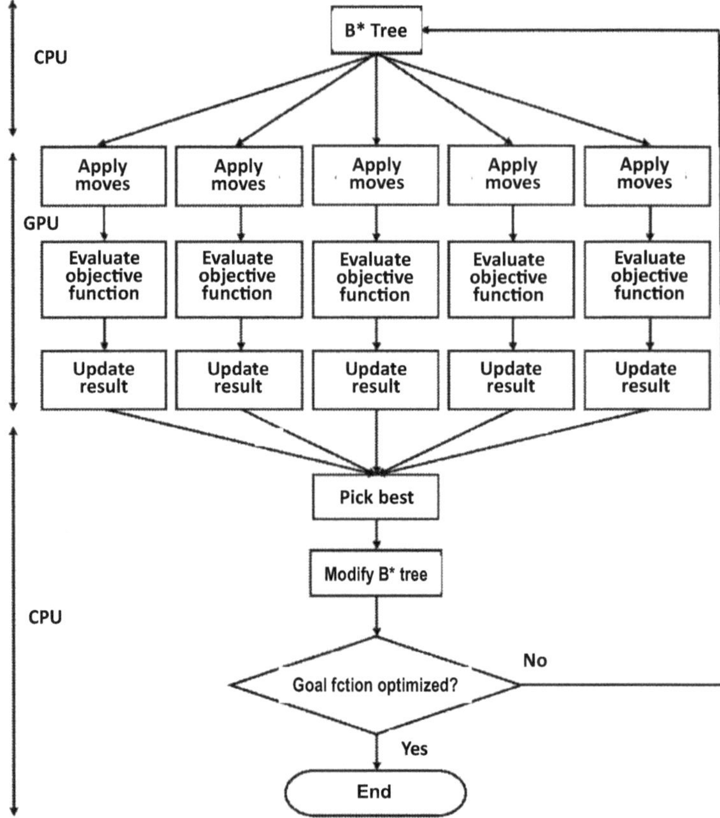

Fig. 2.41 Breaking the dependency chain in floor planning process by multiple concurrent moves [45]

At high-level, the CPU uses the GPU as an accelerator for a specific function (kernel function). All threads execute the same kernel function, but different threads may perform different data manipulation. To parallelize the floorplanning by breaking the dependency chain of sequential processes, one can apply multiple concurrent moves on a given floorplan (Fig. 2.41). For the initial floorplan B* tree selected in the CPU, several concurrent GPU threads are launched after copying the state from the CPU, with each thread applying a separate move and concurrently evaluating the objective function for the resulting floorplan. The CPU then inspects the evaluations and accepts one of the moves evaluated during the concurrent phase. The process repeats unless a stopping criteria is met, according to the flow:

Input

1. A circuit with blocks

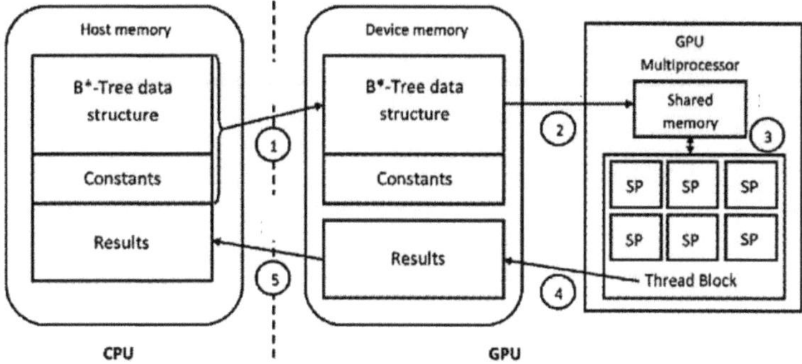

Fig. 2.42 Parallel floorplanning data flow between CPU and GPU [45]

2. A floorplan that optimizes the objective. Possible objectives include: area (small or matching another die) wirelength (short), pattern density distribution (within given range), etc.

Begin

1. Read input circuit
2. Construct initial floorplan B* tree
3. If stopping criteria (objective: area, wirelength, pattern distribution) not met, do:
4. Copy tree and attributes to GPU device memory
5. Launch B parallel thread blocks
6. Copy tree and attributes to the shared memory
7. Select and perform move
8. Modify tree (local copy in shared memory)
9. Evaluate objective matching to floorplan (pass or fail objective)
10. Write objective in GPU device memory
11. Copy B objectives from GPU device memory to host memory
12. Pick best move
13. Modify tree with best move
14. End while loop

In the dataflow between the CPU (host) and the GPU (device), the B* tree structure of a floorplan along with circuit related information (e.g., width, height) are copied to the device memory (Fig. 2.42). Subsequently, multiple thread blocks copy the tree to their own shared memories. Different moves are explored in different thread blocks, and the objective function is evaluated and stored. Finally, the objective results (i.e., whether the target parameter value such as the routing length is met by the floorplan) are copied back to the host memory.

This floorplanner shows about 30 % run time speedup compared to the sequential version of kernel execution based on four steps:

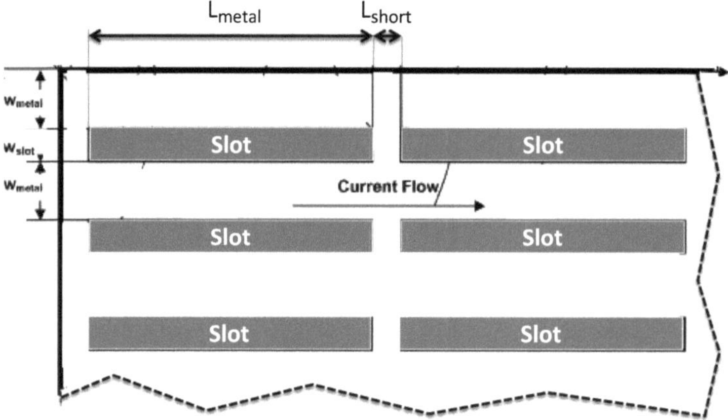

Fig. 2.43 Wide metal bus with slotting (after [46])

1. Copy device to shared memory
2. Move B* tree
3. Evaluate objective
4. Write to device memory

Interestingly, 97 % of the total run time in GPU takes copying data from the device memory to the shared memory and 3 % of the time is spent for the computation to evaluate the newly generated floorplan.

Restructured tree data and access pattern give dramatic speedup, e.g. by 20×. Larger circuits would show greater speedup, as the data copy cost is better amortized by concurrently computing costlier move operations.

In summary, the parallel floor planning of IC layouts claims two major advantages:

– Applicability to the most complex IC architectures, to optimize the objective function (e.g., die area or routing length)
– Processing time reduction over the serial floorplanning process, by 30 %.

That latter advantage could still be increased when data copying from CPU to GPU is improved.

2.3.1.5 Stress Reduction by Metal Slotting

To prevent the high resistance caused by dishing of copper wires during the polishing process, copper pattern density (PD) is required to be within a range, e.g. 85 %, checked every 50 μm of square area of an IC design. The maximal width of copper wire is also limited [46]. These limitations do not create significant layout challenge, as only a small subset of interconnect lines in the design are critical for modeling as transmission lines (T-lines) and are wide enough to disturb pattern density.

The width of an interconnect line (depending on the current it should carry) determines the number of elongated apertures (slots) to be arranged inside it to meet PD rules. If the wire is wider than the maximum defined by the design rules, it has to be slotted (Fig. 2.43).

Another aspect of wide bus slotting is stress reduction due to the mismatch of coefficient of thermal conductivity (CTE) between wide metal buses and the surrounding dielectric. Slotting breaks up stressing forces and prevents cracking.

While manual slotting is admissible, it is inefficient and inconsistent. A slotting algorithm proposed in [45] first defines the number of slots and determines if technology rules require any bridges or shorts in the middle of the slots along the length of the interconnect line (Fig. 2.43). The optimum interval between the shorts should be less than a tenth of the shortest signal wavelength and the highest speed signal dictates the spacing of approximately 50 μm. Slots should be large enough to keep OPC at minimum, especially that their CD accuracy is not critical.

Metal density in wide slotted wires should comply with pattern density rules, checked within windows of specified size. The slotting and tiling should preserve the capacitance and inductance effectively the same as for the one-piece copper line. In an automated DfM flow, the splitting of interconnect lines into parallel fingers symmetrical with respect to the conductor hole pattern, can be implemented as a parametric cell (Pcell), making it a CBC solution.

The change in electrical parameters of a transmission line due to the slotting is programmed into its model for time and frequency domain simulations. The low frequency resistance of a line with slots can be easily extracted from the transmission line models. The high frequency line parameters can be calculated due to the systematic 2D nature of the slotting approach compared to the 3D nature of the "isotropic" hole-generation process. The "anisotropic" length is equal to the interconnect line length. This is unlike for the "isotropic" slotting where the effective current path is always greater than the interconnect length, causing additional resistance. The finger shorting at every predetermined length causes no periodic interference and has a negligible effect.

In summary, the proposed metal slotting approach addresses two important aspects of IC DfM, device planarity and stress-related DfR. While IC manufacturers have readily implemented slotting solutions of their own, a novelty here is the use of pCells as DfM CBC solution, with parameters depending on the density and direction of the current flow.

2.3.1.6 Reduction of Antenna Ratio

Key aspect of Design for Reliability (DfR) [47] is preserving gate oxide integrity (GOI), which may be impacted by plasma process induced damage (PPID). Plasma damage depends on the density of the charging current through the gate insulating

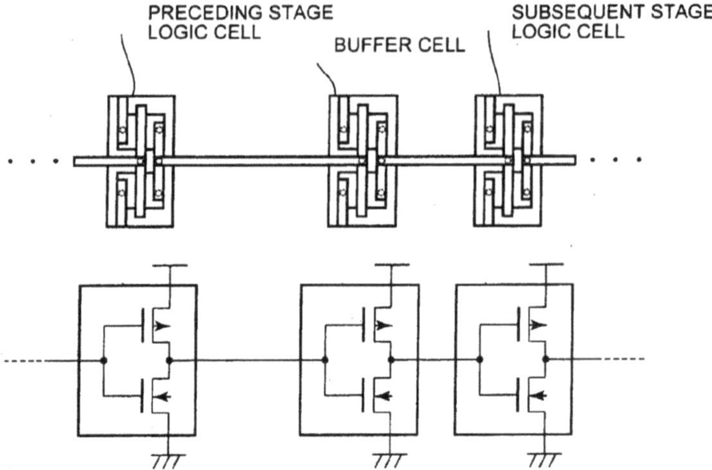

Fig. 2.44 Insertion of buffer cell for antenna ratio reduction [47]

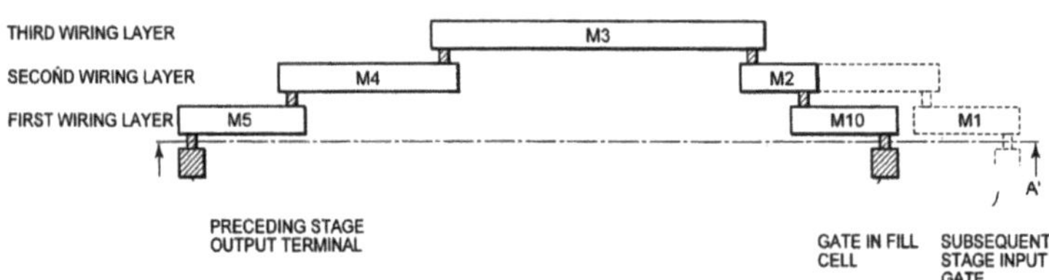

Fig. 2.45 Cross-sectional view showing buffer cell connectivity

film with charge carriers collected by the large area of the gate connectivity areas that function as antennae. An IC layout should have gate connector (antenna) no greater than a predetermined threshold (antenna criterion). The antenna ratio (AR) represents the ratio of an area of the signal wiring (usually the metallic wiring) connected to the gate, relative to gate area in the transistor. When the antenna criterion is exceeded (producing antenna error), the layout has to be corrected, e.g., by inserting a protective diode rerouting the charge current that concentrates in the gate oxide, directly into the silicon. But when the area coverage ratio of standard cells is high and no empty region in the die exists, the area of the circuit would have to increase to accommodate protecting diodes to reduce the AR. Also, if a protective diode is inserted inside the cell, it increases the cell input capacitance.

A preferred solution to reduce the AR is to increase gate area by inserting a buffer circuit [47]. However, since one has to do so in metal wiring, arrangement of other cells and wiring lengths may have to change, possibly causing a timing error.

The proposed method of AR reduction without adding die footprint or altering its timing requires computing the gate area to be added to avoid PPID. The layout is modified by arranging a buffer cell with a second gate electrode in an empty region

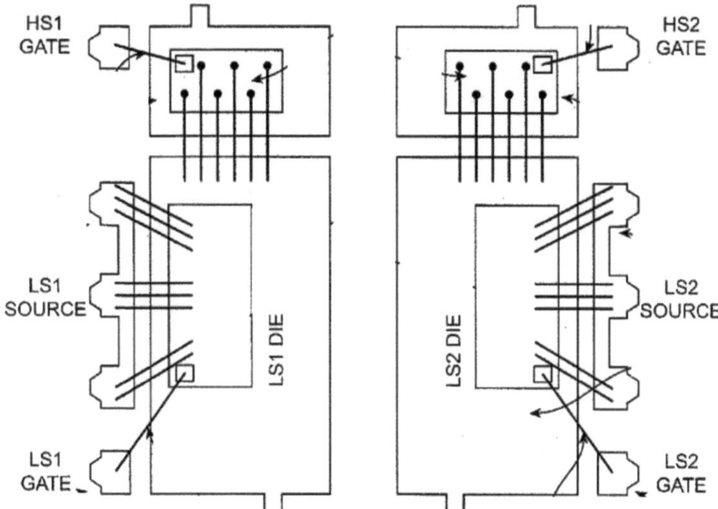

Fig. 2.46 High – and low – side components of a 3D package (after [48])

of the layout, and connecting the second gate electrode to the wiring without altering to the circuit logic. Since the gate area is increased by insertion of a cell that performs no logic operation, the antenna ratio can improve without affecting main circuit cell arrangement and wiring.

In a circuit consisting of three logic cells (Fig. 2.44), the gate electrode in the first and the second cell, and the second gate in the third logic cell are connected together. The third logic cell makes no contribution to the logic. The gate area connected to the wiring that acts as an antenna in the plasma process is enlarged by the second gate electrode (Fig. 2.45). Since the cell that performs no logic operations it is placed in the empty region of the die, improving the AR should not largely alter the layout and the timing due to the layout correction should not change either.

The design support system verifies the antenna ratio on the preceding stage logic (second logic cell) and the subsequent stage logic (first logic cell) connected together through metallic wiring. An equivalent circuit of the verification object circuit is the same.

Since a primitive cell can be inserted as the fill to eliminate antenna error, it is not necessary to design a dedicated cell (e.g., a diode). But one should prepare a bank of various buffer cells to find one consistent with a function of the circuit where the antenna error arises. During layout correction, the algorithm searches for empty regions on the chip nearest to the subsequent stage input gate, places the logic cell there, and connects the gate electrode within a shortest path.

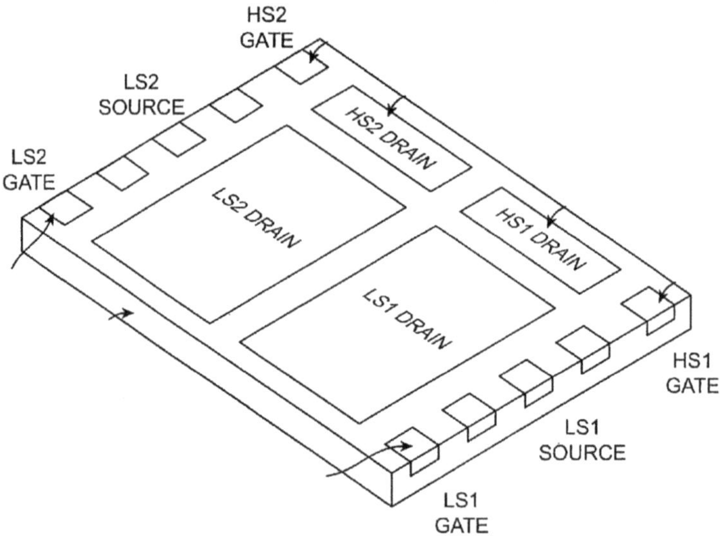

Fig. 2.47 Inside view of a 3D package (after [48])

In summary, the AR reduction algorithm addresses an important DfM/DfR concept and can be useful to define chip-level CBC layout for high GOI and reliability.

2.3.1.7 3D Packaging Techniques

Another aspect of DfM by CBC is related to SoC product applications requiring multi-purpose die packaging [48]. A new package design can contain up to four semiconductor dice, with one or more internally connected switch nodes, e.g., to form a dual output or phase synchronous buck converter. The package would have control leads at opposite sides, with semiconductor dice oriented perpendicularly to one another (Fig. 2.46).

Semiconductor die packages containing several devices with similar connections (e.g., inputs or outputs) separated from each other require longer or more complex electrical connectors to the rest of the circuit. Output connections may be on opposite sides of a package, such that the circuit board to support it would need to have IC design geometries (e.g., pads) similarly separated, making it difficult to design a circuit board.

An inside view of a package without molding material and its bottom perspective show the leadframe structure with multiple die attach pads and leads (Fig. 2.47). The first and the third control lead can be at the opposite sides allowing four dice within the package to be oriented perpendicularly to each other. The leads extend

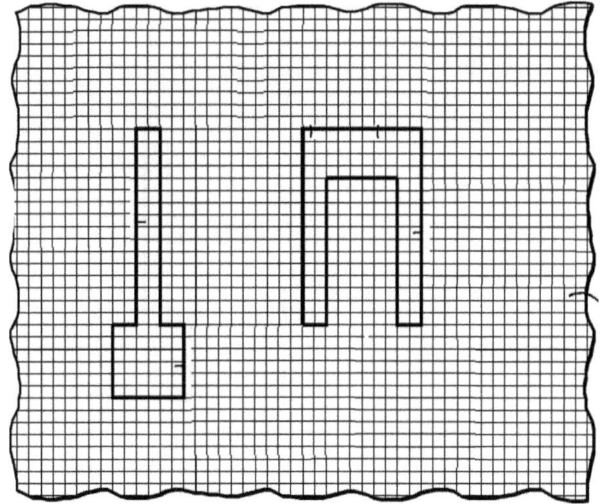

Fig. 2.48 OPC – standard method of creating a layout mesh (after [49])

past the lateral edges, and depending on whether the molding is used, a housing could be assembled around the dice and the leadframe structure. Vertical devices may include an input on the one side and an output on the other side of the die so that current can flow vertically. The pinout (i.e., location of the leads) allows for easy layout of a dual output e.g., of phase synchronous buck converter or other suitable vertical or horizontal device. A user can place power train components (inductors, capacitors, and transistors) to create dense power supplies, critical for reducing board area. The design enables PCB layouts to reduce parasitic loss due to routing of internally connected switch nodes with low inductance or other parasitic losses for the die package to be used at high operating frequencies.

In summary, package development is becoming an important aspect of DfM. Mechanical and electrical simulations may be required to confirm the robustness of the build (see Chap. 4).

2.3.2 Execution: Modifying Design Database

When CBC methods are not readily applied at the cell or block level, post-layout corrections at die level may be required to make the design DfM-compliant, typically by post-processing algorithms such as OPC. Such post-processing has to preserve both design intent and schedule. The new concepts related to OPC apply to its cycletime reduction (a common DfM goal) by focusing on the regions in the die requiring significant computing effort. OPC algorithms would need to resolve the tradeoff between the magnitude of edge placement errors (EPE) and device sensitivity. The proposed procedures take advantage of simple or redundant geometries to speed up the calculations.

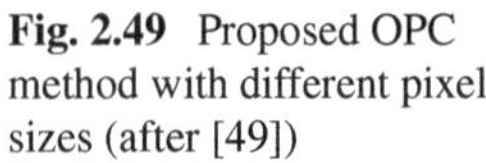

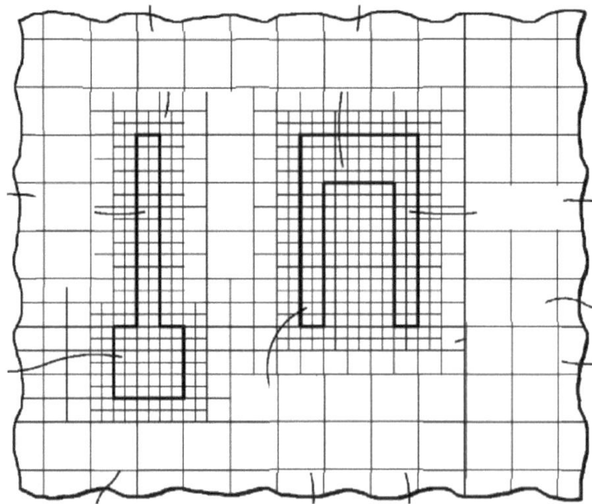

Fig. 2.49 Proposed OPC method with different pixel sizes (after [49])

2.3.2.1 Pixel-Based OPC

OPC corrections to drawn design layout reduce the effect of light refraction at the boundaries of small features on the photomasks, by adjusting their widths or lengths through corner serifs, line end extensions, etc. The OPC operation is costly as it may take many days for an entire device design to converge to a solution acceptable for all geometries in the die. The new methodology focused on optimizing OPC processing time and accuracy [49], replaces the regular OPC mesh typical for time-consuming methods (Fig. 2.48) with mesh of smaller pixels only in sensitive areas. In the first step, empty regions are found by scanning the die at a large pixel size. When a feature is detected, the pixel size is reduced for calculations (Fig. 2.49).

For conventional (sparse) OPC, chip layout is fragmented along the edges of features. Edge placement errors are then measured against a simulated image and the layout edges are moved to reduce them. Such “sparse OPC” is often not capable of making corrections to the accuracy target.

A more accurate approach referred to as “dense OPC” defines pixel size to analyze portions of the aerial image to determine the corrections, pixel by pixel. An OPC tool places a global grid over a design layout, calculates image parameters for each pixel, and compares them to the target image to determine the correction. A small pixel, e.g. 20 nm × 20 nm, is required to achieve high accuracy, but at the cost of extensive run times.

In the approach proposed in [49], portions of the calculated aerial image are compared to the target image for each pixel, to determine the correction. Then, the layout of the calculated aerial images is divided into an array of large pixels, e.g. 100–200 nm, and scanned by an OPC tool to determine whether a portion of a pattern resides within each pixel (Fig. 2.50). If pattern is found, that pixel location is stored in the memory of the OPC tool. Then, the algorithm defines smaller pixels

Fig. 2.50 Flow chart of OPC process using different pixels (after [49])

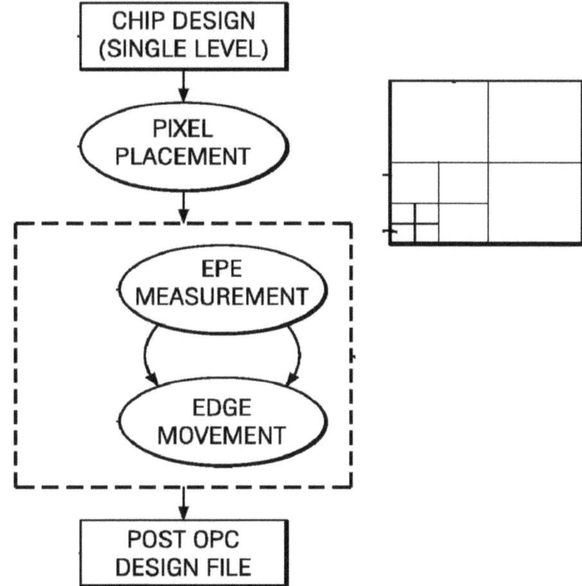

close to the edges of the patterns for detailed calculations. For higher accuracy, pixels may be divided into smaller ones, e.g. quarters or ninths. Small pixels close to the edges and corners of features and larger pixels in areas with no features present save computing time.

In summary, the accuracy – dependent pixel size, a concept similar to nonuniform mesh definition for mechanical or electrical simulation programs, should be taken advantage of for any simulations relaxed to DfM rules.

2.3.2.2 Cluster-Based OPC

As the complexity of IC layout becomes too high for efficient OPC implementation due to increasing optical interaction range, clusters of elements containing an increasing number of geometries instead of individual polygons can be extracted and then handled separately. A set of polygons forms a cluster if for any two polygons, the distance between them is less than or equal to a given threshold number. Rather than analyzing each and every polygon in the design, unique but repetitive polygon patterns are analyzed once and then replicated for all clusters with the same general architecture [50].

In a cluster, OPC algorithms search for polygons with a segment or a wall not near enough another polygon (Fig. 2.51). Light refraction in such region could result in a curved line segment on the wafer, which needs to be corrected to the acceptable error EPE value. As mitigation, a scattering bar, i.e., a geometry below the resolution limit of the lithography equipment, may be added – space permitting – parallel

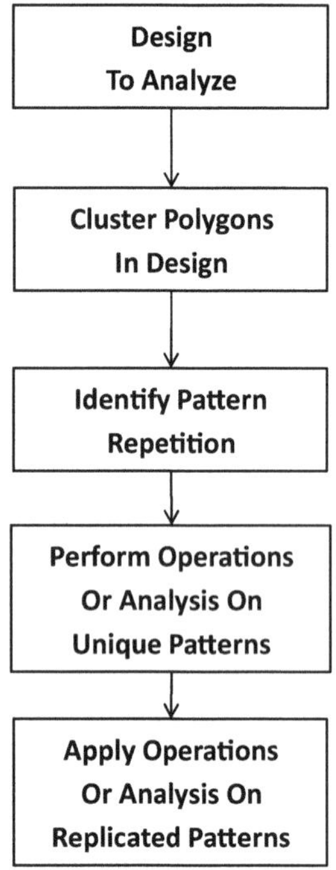

Fig. 2.51 Extraction of polygons for cluster OPC definition (after [50])

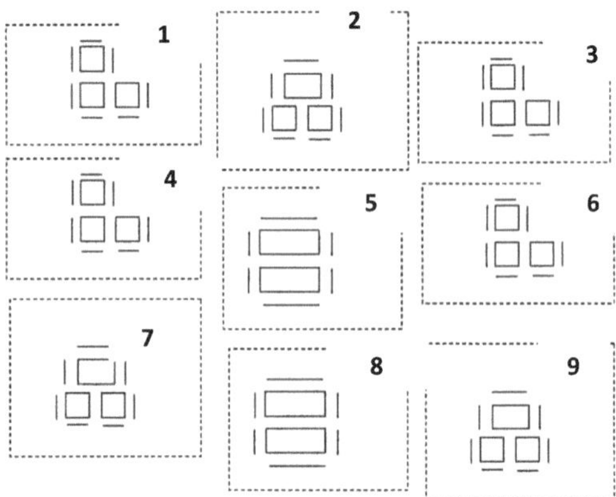

Fig. 2.52 Examples of polygon clusters (after [50])

to the segment of interest to produce a straighter line. Given the large numbers of components in a typical IC design, it takes time and resources to perform a full set of scatter bar analysis. Therefore, a cluster – based approach is a timesaving measure, but also adds consistency to pattern matching.

The algorithm recognizes all clusters in a given layer and forms a list of classes of equivalent clusters with repetitive patterns. Adding scattering bars is done for each polygon individually. The next action is pattern recognition to identify the set of repeatable, unique patterns. For clusters 1–9 from the example set, there are three

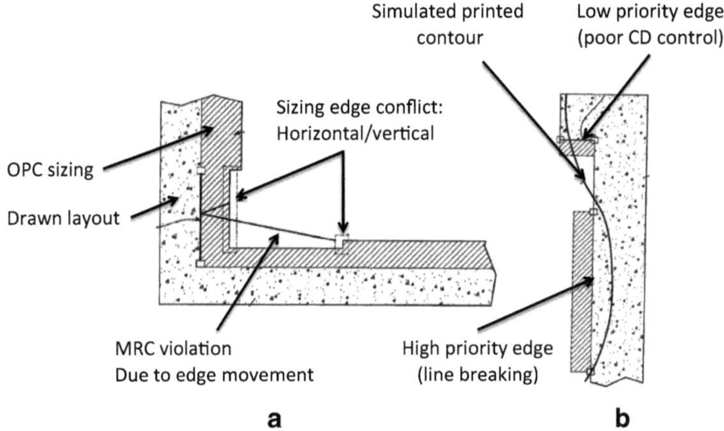

Fig. 2.53 Examples of edge shift conflicts for OPC corrections (after [51])

unique patterns: A, B, C. To replicate the processing for each instantiation of these patterns, the set of scattering bar for Pattern A is repeated for clusters 1, 3, 4, and 6, etc. (Fig. 2.52).

The distance between clusters is modifiable to optimize the clustering. If the distance is too large, there would be only a small number of clusters, each of them potentially having a large number of polygons. If it is too small, then there may be a large number of clusters, with only few polygons each. The process can be configured to iterate with different cluster spacings, optimally at 3–4 times the value of the minimum spacing design rule.

In summary, parallel analysis of similar layout elements for process corrections should simplify layout optimization. But the value of the cluster based approach may be compromised by the complexity of the code leading to the high cost of debugging, compared with the runtime reduction, which can be accomplished by a more powerful computation engine. Another alternative is to use cells from a standard library, pre-processed for layout corrections.

2.3.2.3 OPC Conflict Resolution

During OPC implementation, conflicts may arise between sequential edge corrections and mask manufacturing rules for minimum width, spacing, or notch reduction. One way to resolve these conflicts is by assigning edge correction priority [51]. Because a suboptimal wafer pattern may not be identified correctly by the OPC algorithm, a dedicated system may need to alert the user about it. As response, the user may change the layout or relax system constraints for critical corrections.

The proposed OPC tool assigns priority to edge segments of a feature to be corrected such that adjustment of a less important edge does not hinder the correction

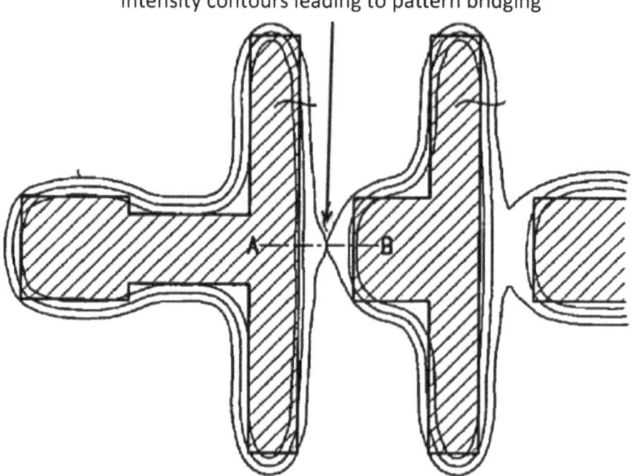

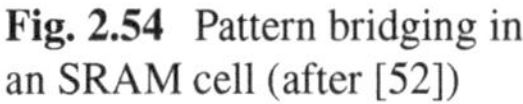

Fig. 2.54 Pattern bridging in an SRAM cell (after [52])

of a more important one (Fig. 2.53). In order to print properly on wafer, the OPC tool may suggest for a portion of a feature (stippled area) that its boundaries be extended outwards (dashed area). An edge segment of the feature is to be moved to avoid a conflict with an other edge segment that forms a jog. But if the OPC tool moves the edge segment further outwards, the distance between that segment and the jog violates a mask (MRC) rule. The designer would have to address the edge segment which was assigned a less than desired OPC correction. Either the area around the concave corner would need to be refragmented such that the distance between the final OPC corrected positions is greater than the minimum MRC, or the MRC would need to be relaxed in order to permit the correction, possibly at the expense of a mask of higher accuracy. The system would keep track of which edges cannot be moved to their desired OPC corrected position without causing a conflict.

Edge segments can be categorized based on their function in the layout or on their MRC constraint information, e.g. as being part of a gate, a line end, a signal wire, etc. That way, critical portions of layout are corrected first.

In summary, this approach intends to resolve conflicts between the complex OPC algorithms and mask manufacturability requirements. While tagging layout edges by a recognition system would prioritize the corrections, involving layout designer at chip level post-processing may require working the design back to IP block level. If, on the other hand, the conflict resolution algorithm is allowed to work at IP block level, decisions can be made more easily assuming that the designer is aware of the lithography setup.

2.3.2.4 Auxiliary OPC Features

Improving pattern resolution can be achieved by adding fine auxiliary features within a light-transmitting region of the mask. Overlapping one mask with another

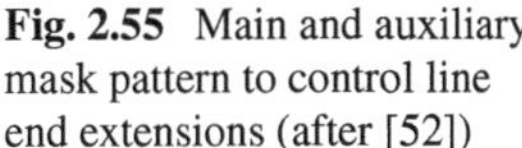

Fig. 2.55 Main and auxiliary mask pattern to control line end extensions (after [52])

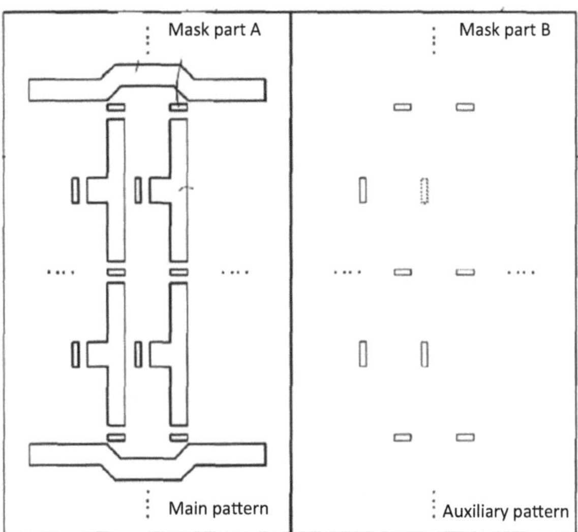

one, with the opposite tone over the desired pattern would remove areas of pattern bridging [53].

Pattern quality in the line end areas of an SRAM cell (Fig. 2.54) can be improved by auxiliary mask features, placed in order to extend the end portions of the lines, i.e. by a dedicated mask (Fig. 2.55). Since pattern bridges at line ends are difficult to correct, fine auxiliary patterns are not inserted there. If a fine pattern having a line width less than the obtained resolution is independently applied to a mask, a pattern where the light physically passes though the mask but does not appear to affect the photoresist material can be defined.

An overlapping exposing process may not need to use an additional mask plate, but instead a different mask pattern on the same mask, to easier align the layouts. An example mask may contain a pattern including light-blocking features and fine auxiliary patterns within a light-transmitting region. The second pattern may have the opposite tone of the first pattern. The light-blocking and the light-transmitting auxiliary patterns would be aligned at the same positions during an overlapping mask exposing process.

Dividing a mask into two regions, containing respectively main design and auxiliary patterns with opposite field tones, can help improve process margin for small gap printing. However, this DfM integration scheme may not fit well within the current state-of-the-art manufacturing, due to:

- The need to use double – patterning (DPT) which can accomplish much more to optimize layout features, than the proposal discussed here, if utilized to its full extent,
- Placing clear – and opaque – field patterns next to each other on one mask plate which would create ambiguous CD targeting,

– Reducing the usable printing field size by at least two times, due to the two adjacent patterns, thereby increasing stepper runtime.

It is possible that the concept of printing sub-resolution features using a separate mask layer would become of interest, but it would likely be an implementation different from the proposed one.

The approach discussed above indicates that directions of OPC development are not always driven by engineering disclosures. The ad-hoc OPC concepts do not necessarily reflect the actual process roadmaps for the involved companies.

2.3.2.5 OPC Preserving Design Intent over Process Window

The ultimate goal of OPC is to preserve design intent on wafer. MBOPC results in greater fidelity in the printed image and therefore better alignment with design intent, compared to RBOPC, but it would require significantly more computational resources [53]. This is due to iterative optimization of mask pattern that involves:

– Generating simulated contours of the initial mask layout, modified by basic RET
– Comparing the simulated contours to the wafer target
– Adjusting the RET layout to compensate for offsets between the simulated contours and the wafer target, thereby generating the first estimate of the mask pattern
– Repeating this process using the interim mask pattern from one iteration as the input for the next iteration.

This cycle is repeated until the offset between the simulated contour and the wafer target is acceptable, or until a maximum number of iterations is exhausted (leading to non-fully converging OPC solutions). The output of the final iteration becomes the actual mask layout, which is sent to the maskhouse.

A common assumption required by electrical simulations, is that the initial design input layout is assumed to be equal to the wafer target. With this in mind, the development of lithography and wafer etch processes and chip designs typically occurs concurrently over periods from several months to several years. This time frame makes it practically impossible to give designers accurate descriptions of the RET and OPC solutions as well as accurate process window models during the design of the chip. Having designers optimize layouts to inaccurate models and RET/OPC solutions while they are operating under the assumption that they have accurate insight into the patterning process, can lead to catastrophic failures and would make manufacturability worse, not better.

As additional complication, the primary customers for model-based layout optimization are fabless design houses, which design chips to be manufactured at outside foundries. A key requirement for these fabless design houses is to maintain foundry portability (i.e. the ability to move their business from one foundry to a competing one) or even to outsource their product to multiple foundries at the same time. The success of model-based layout optimization is based on a detailed, accurate model of a particular foundry's RET/OPC and imaging solution, which may fundamentally link the optimized layout to a specific foundry. Thus, performing a model-based layout optimization using the detailed process model for each individual foundry would be impractical. An alternative solution of the least common denominator model that describes the worst case printability failures for multiple foundries would be conservative and produce noncompetitive layout densities, which is of particular importance for multiple foundries collaborating or competing for fabless business.

Lastly, a designer adjusting the original layout based on simulation feedback is effectively introducing a new polygon set, i.e. the optimized layout no longer represents the original designer's intent. It represents what the designer had to do to the intended layout to make it pass the model-based optimization. If this manipulated layout is introduced as the input layout to the RET/OPC flow, the added polygon complexity and uncertainty over designer's intent will introduce manufacturability risk and could have the exact opposite effect of what DfM is intending to achieve.

Implementation of OPC results in process variations by replacing the design – given wafer target with a wafer target band and by replacing the simulated contours with simulated contour bands. The iterative optimization remains the same, but the wafer target bands need to be generated either by the designer based on an understanding of shape tolerances required for circuit yield, or by the OPC tool from the input layout by applying tolerances communicated in the design rule manual. Such modifications to OPC have been termed process window OPC (PWOPC) proposed as a key component of a strategic design for manufacturability.

However, implementation of PWOPC with DfM has difficulties. Wafer target bands generated by the designer are unaware of the process capability. The designers know what they would like, but can't tell what is reasonable to ask for in all layout situations. Wafer target bands generated by the OPC application are unaware of designer's needs, i.e., the process limitations are well known, but acceptable tolerances are not. The generation of the wafer target bands in either case is rule-based, i.e. a series of sizing operations and Booleans is performed to generate rectilinear approximations to the desired bands. Challenges in reliably manipulating layouts through complex rule sets drove the implementation of model-based OPC in the first place, and reestablishing a dependence on such rule-based operations would effectively be taking a step backwards and would introduce significant yield risk.

A DfM solution that avoids the misalignment with design intent, would provide a mask design that minimizes or avoids printability and/or manufacturability errors during mask verification at multiple foundries, and provides an efficient design process for fabless designs [53]. The layout is to be optimized using a process model until the design constraints are satisfied by the image contours. The process model supporting the design phase needs not be as accurate as the lithographic model required to prepare the lithographic mask. The initial image contours are included with the modified mask layout and later optimized using the lithographic model, for example, including RET and OPC. The mask layout optimization matches the images simulated by the lithographic process model with the image contours generated during the design phase, which ensured that the design and manufacturability constraints specified by the designer are satisfied by the optimized mask layout. The

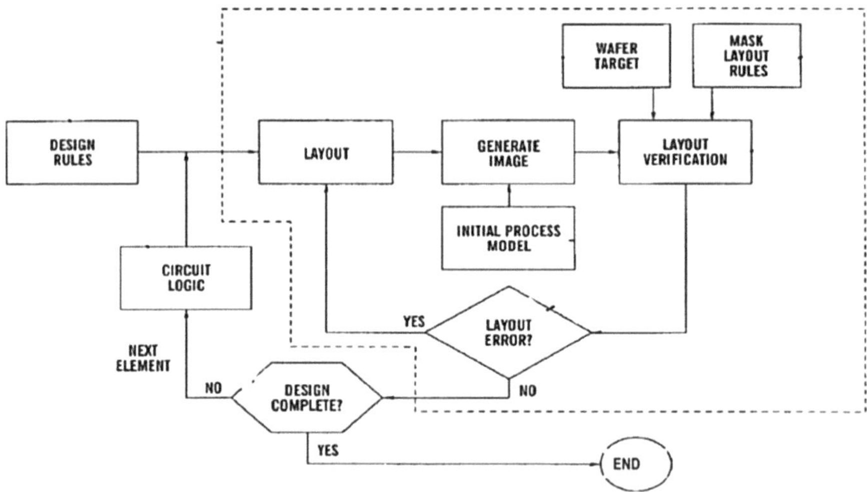

Fig. 2.56 Proposed DfM flow to resolve conflicting requirements for tolerance band definition and data preparation (after [53])

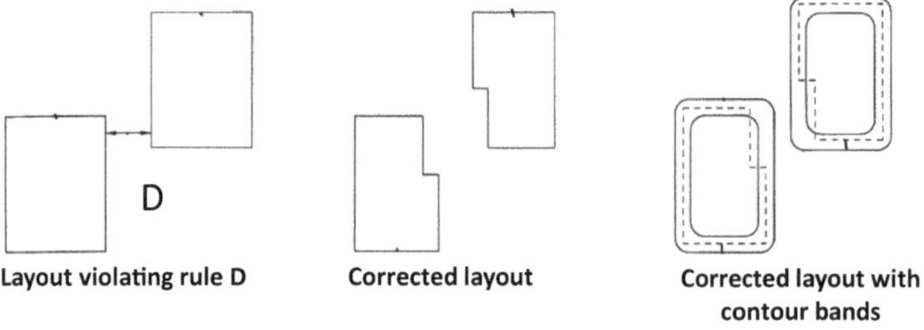

Fig. 2.57 Contour bands generated during layout optimization (after [53])

standard data prep and mask verification process is enhanced to produce the new DfM flow (Fig. 2.56). Layout shapes during the design phase of a DfM process may differ from contour bands generated during a layout optimization phase.

The process window model generates image contours, which simulate the printed image on the wafer with a certain degree of confidence (e.g. $\pm 3\sigma$) over a range of process conditions. The generated image and contour bands are verified by comparing to the wafer target and ensuring a match, within the provided tolerances. If the tolerances are violated, then the layout is modified by the designer until the deviations between simulated contours or bands and the wafer target are within tolerances. When the design is complete, the layout, by virtue of the verification optimized for manufacturability, is provided to the foundry for data prep.

For example, for initial circuit shapes if the spacing D between the shapes is less than a predetermined minimum spacing (Fig. 2.57), or if the simulated image using a mask based on the initial circuit shapes otherwise violates rules of a process model/tool, the shapes may be modified by the designer, or by an automated OPC tool.

The resulting shapes now satisfy the design rules and/or the process model/tool. In standard design flow, the resulting shapes would then be provided to the foundry as input (i.e. in the initial layout) to the data prep process with the wafer target set equal to the input layout, which the OPC tool will attempt to match, even though the target would include the shapes that no longer represent the designer's intended shapes. By contrast, in the new process using the process window model, simulated contour bands generated by the process tool satisfy the designer's rules and will meet the designer's tolerances over the range of expected process conditions. The contour bands, which preserve the designer's intended shapes (within tolerances), are now provided to the data prep process as input targets.

The data prep system receives the resulting manufacturable layout as input into a process window OPC or a RET layout tool. In addition, the contours or contour bands generated during the design layout optimization are provided as wafer target input to PW OPC. The litho process model for data prep is expected to be more accurate than the model used during the design layout optimization. If contours are provided as the target input, then the PW OPC may be configured to match the simulated contours generated by the current PW model within tolerances and mask layout rules, which may include manufacturability rules from the mask house, potentially not available during the design phase. Stated another way, the differences between the simulated contours and the target input (i.e., the contours or bands determined during the layout optimization in the design phase) must satisfy the tolerances. If contour bands are provided as the target input, then the PW OPC may be configured to ensure that the simulated contours fall within (i.e. substantially match) the provided target bands. Optionally, modified tolerances are provided to allow the simulated contours to deviate from the original target bands.

In summary, a key advantage of the design-oriented OPC method is that the mask layout will satisfy the designer's intent and manufacturability rules and thus the layout will not need to be sent back to the designer, as opposed to the standard method where the mask layout may produce shapes outside design intent. The methodology of closing the design loop for DfM during data prep allows for the mask layout to be optimized according to individual mask house rules and the original design rules without requiring a customized design.

2.3.2.6 Process Bias Compensation in IP Libraries

Pattern transfer from design to wafer is never 100 % accurate. Cell layout transfer can be optimized with process bias compensation techniques (PBCT) across the hierarchy, from leaf cell, through block, to die level, involving resolution enhancement techniques (RET), etch proximity, gap fill, pattern density adjustments for chemical mechanical planarization (CMP), etc. For accurate simulation, circuit extraction should be performed on aerial images of the PBCT adjusted layout before placing and routing [55]. Aerial images of the cells subject to PBCT based extraction would then be added to the library of IP attributes. Since the as-fabricated features on wafer do not exactly match the polygon shapes of the layout, the accuracy of geometry dependent parasitics (resistances and capacitances) would need to be improved.

Fig. 2.58 Defining litho buffers by duplicating the main cell (after [54])

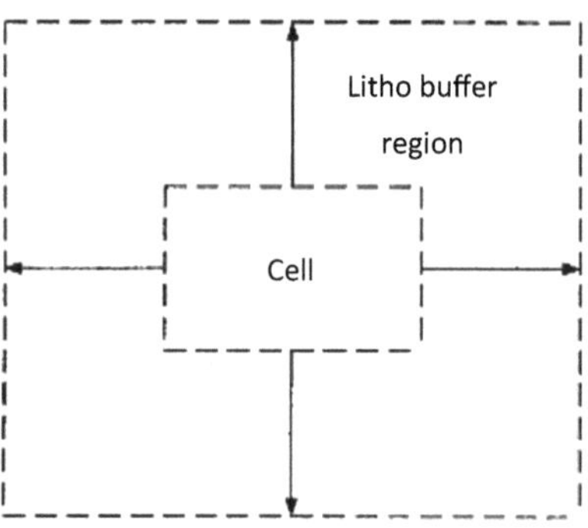

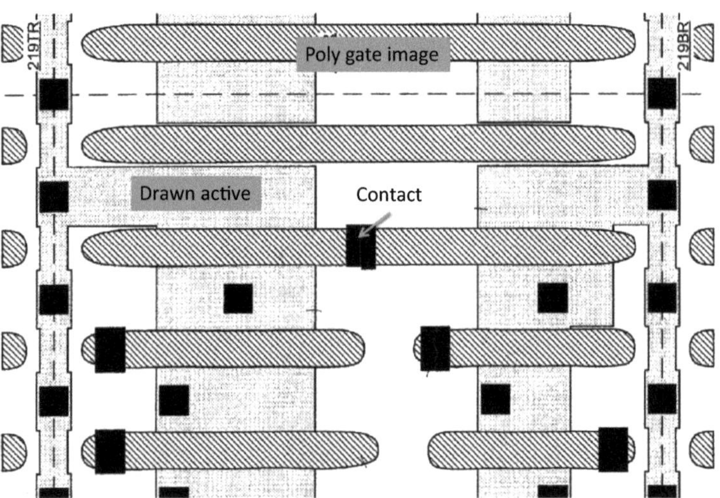

Fig. 2.59 Corner rounding and line end pullback of poly gates (after [54])

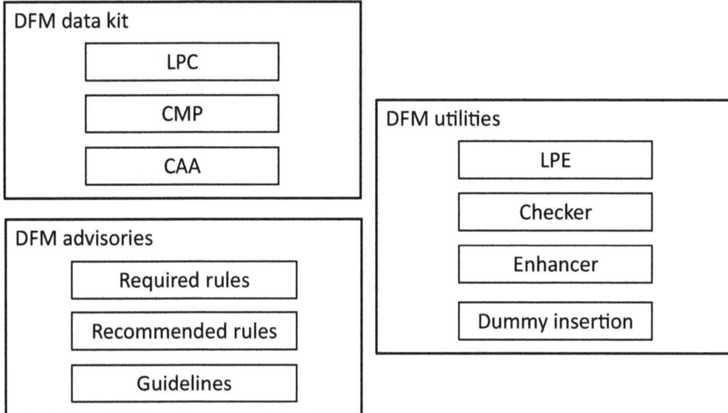

Fig. 2.60 Integrated DfM system (after [55])

At timing closure, extraction tools would combine layout information in the form of idealized polygon shapes with aerial images based on wafer fab technology data, to add parasitic components to the circuit schematics.

Process compensation at cell level is easier compared to the one at full chip level. Cell area may be surrounded by a large litho-buffer to reset its printability conditions (Fig. 2.58). The outward extent of this buffer (indicated by arrows) could be about 1 μm out from each side of the cell. If the difference between the contour of the simulated aerial image (i.e., the EPE) and the drawn layout of the cell prior to PBCT processing are larger than specified, further refinement of PBCT is required. If the EPE's are acceptable, the final PBCT version of the cell and its aerial image would be stored in the library.

Layout of transistors with gate layouts that extend over an active region have aerial images not matching their polygon representation (Fig. 2.59). This causes MOSFET characteristics to vary along the length of the gate electrodes. For accurate modeling, each transistor can be sliced into segments along the length of its gate, such that each segment is modeled separately [27]. Consequently, one may include PBCT extraction methodology into IC design hierarchical analysis.

In summary, while agreeing with the concept and assumptions of PBCT extraction, it is difficult to evaluate its RoI without reducing it to practice. Cell level PBCT may be readily built into many simulation engines. The potential savings would depend on the type of the circuitry and the required accuracy.

2.3.2.7 DfM Ecosystem and Ownership Distribution

DfM as a methodology to translate design intent to silicon should also improve communication between designer and manufacturer. Manufacturing data is created, quantified, and integrated to reduce design time and cost. EDA tool and intellectual

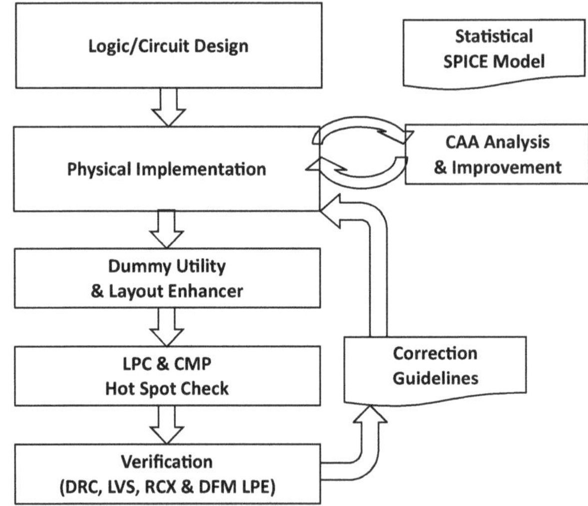

Fig. 2.61 Flowchart to implement integrated DfM system (after [55])

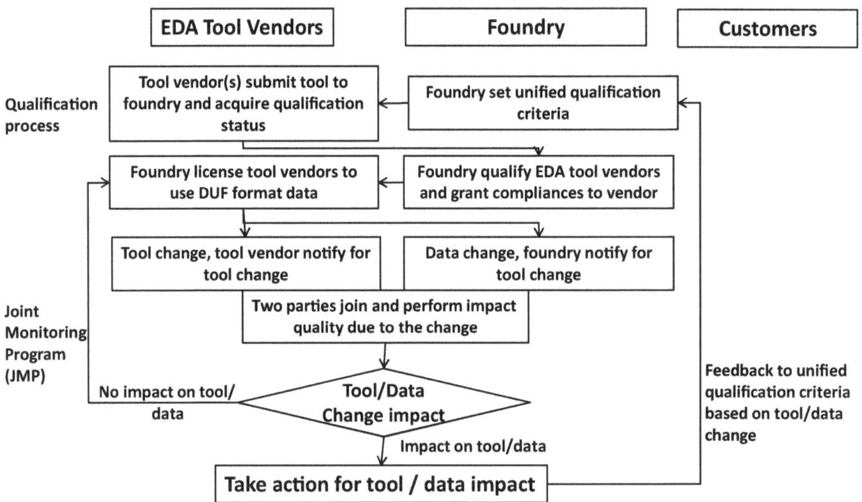

Fig. 2.62 Flowchart to implement JMP (after [55])

property library vendors, as well as the customers may involuntarily duplicate the effort, if they are not sharing resources, potentially providing inconsistent results [54].

Implementing DfM to benefit semiconductor manufacturing, design tool vendors, IP library (IPLib) partners, and customers requires an integrated system (Figs. 2.60 and 2.61). A joint motoring program (JMP) is recommended to unify qualification criteria of design tools utilizing partnerships among the interested parties (Fig. 2.62).

DfM unified format (DUF) categories help share manufacturing/DfM data in a tool and data kit (DDK) containing processing recipes, production statistical information, etc., compiled for manufacturing simulation. DDK can be provided to an IC designer, either readily integrated into a design tool, or distributed directly with DfM rules and guidelines extracted from the manufacturing information.

In such enhanced design flow, DfM would provide model-based utilities from simulations and rule-based utilities from guidelines. LPE (layout parasitic extraction) deck may be implemented to eliminate hotspots at extraction or at timing simulation, before or after the tape-out. After logic design with input of a statistical design model (SPICE), the flow proceeds to physical design implementation including CAA analysis, layout enhancer, dummy insertion, layout tuning, LPC and CMP hotspot check, and then to design verification: layout vs. schematic (LVS), design rule check (DRC), resistance and capacitance extraction (RCX), and DfM LPE.

An IC manufacturer can license DfM modules integrated with data in a unified format (DUF) to IC design suppliers to be built into an EDA design tool. Similarly, the IC manufacturer may license DfM modules to IPLib suppliers to be built into an IPLib package. Thus the design tools and IPLib are integrated for DfM enhanced design. The joint motoring program JMP (Fig. 2.62) unifies qualification criteria of design tools. The design tool can be submitted with criteria to the manufacturer for qualification and the tool supplier may grant compliances and licenses to utilize DfM information in the unified format. The manufacturer may change the manufacturing data and notify the tool supplier, to change tool configuration, database, etc.

In summary, DUF as one unified format to distribute manufacturing data among design tool supplier, manufacturer, and customers for DfM associated integrated circuit (IC) design, would include at least three categories of information: lithography process control, chemical mechanical polishing, and critical area analysis. The unified DfM data may be plugged into a manufacturer certified DfM design tool to validate the associated design process. This methodology may become important for DfM standardization across the different technology nodes.

2.3.3 DfM Verification: Defect Reduction on Silicon

One key task of DfM verification is defect reduction on silicon, to resolve any quality issues left behind by CBC and process execution. Defects are identified thorough modeling and sensing of process variations.

2.3.3.1 Integrating Etch Process with OPC Models

Not all semiconductor process steps have mathematical models. One critical missing link affecting pattern printability is the gap between the photolithography and the

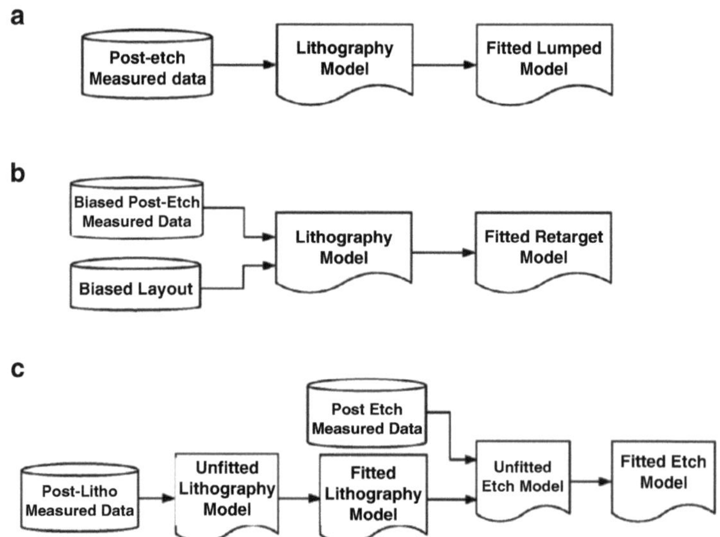

Fig. 2.63 Derivation of lithography and etch models: (**a**) lumped model, (**b**) retarget model, (**c**) staged model (after [56])

etch process. Conventional (staged) OPC model can be enhanced by a rule table for correcting CD's based on photolithography and etch bias [56]. Large variations in feature sizes can lead to micro-loading and aperture effects, which should be captured in OPC model. A key inaccuracy arises from modeling of the non-uniform and non-linear etch bias i.e., the difference between a resist contour after the photolithography and an etch contour after etch. That inaccuracy can be mitigated by calibrating a model for the combined photolithography and etch process, with etch bias data for critical dimension (CD) difference between the values measured separately after these two steps [56].

Three types of models can include the etch effects to calibrate an OPC model: the lumped model, the retarget model, and the staged model (Fig. 2.63).

(A) For the lumped model, post-etch data should directly fit into a lithography model, based on the optical and resist parameters. Unfortunately, the lumped model ignores the intermediate measurement after the photoresist development, which loses the lithography process window information.
(B) For the retarget model, both the layout and the post-etch data are biased by the values stored in a rule table to account for the etch effects. The modified data, i.e., the biased post-etch data and the layout biased to compensate for etch effects, are then used to fit a lithography model. The retarget model has the advantages of using one stage correction while maintaining the lithographic model properties. However, the rule table based on previous measurement data is required for the retargeting and therefore the OPC model accuracy is limited by the previously measured pattern structure.

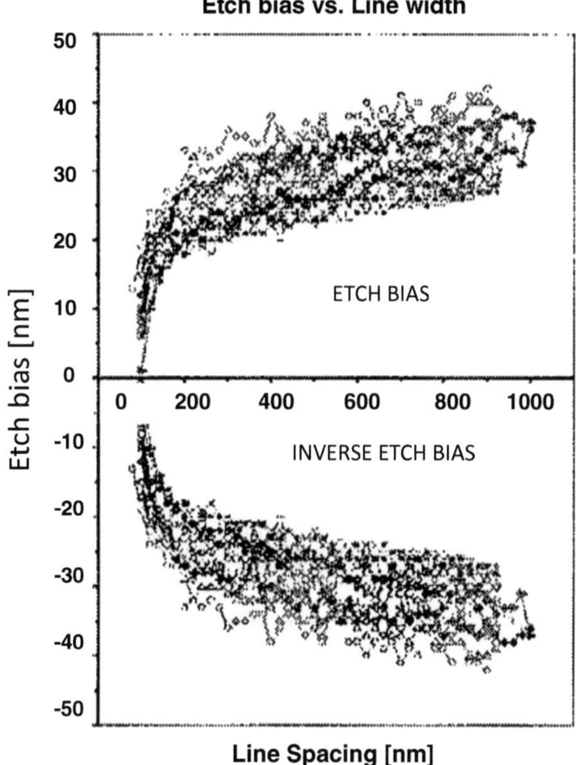

Fig. 2.64 Etch bias as a function of line width and spacing (after [56])

(C) For the staged model, the lithography and etch are modeled separately. The staged model is an unfitted lithography model, calibrated by post-litho data. The etch model is calibrated with both post-etch data and the output of fitted lithography model. Unfortunately, staged models calibrate the lithography independently from etch. Because the post-litho measurement data can be noisy and the lithography model output is an input to the etch model, the staged model accuracy is limited by the accuracy of the lithography model.

The etch bias and its inverse depending on the line width and spacing, can be expressed as:

$$Etch\_bias = AEI_{wafer} - ADI_{wafer} = AEI_{wafer} - ADI_{model} \qquad (2.7)$$

ADI_{model}	Simulated lithography contour
ADI_{wafer}	After-photoresist-development inspection, ADI, on wafer
AEI_{wafer}	Post-etch measured data as (after-etch inspection AEI), the etch bias

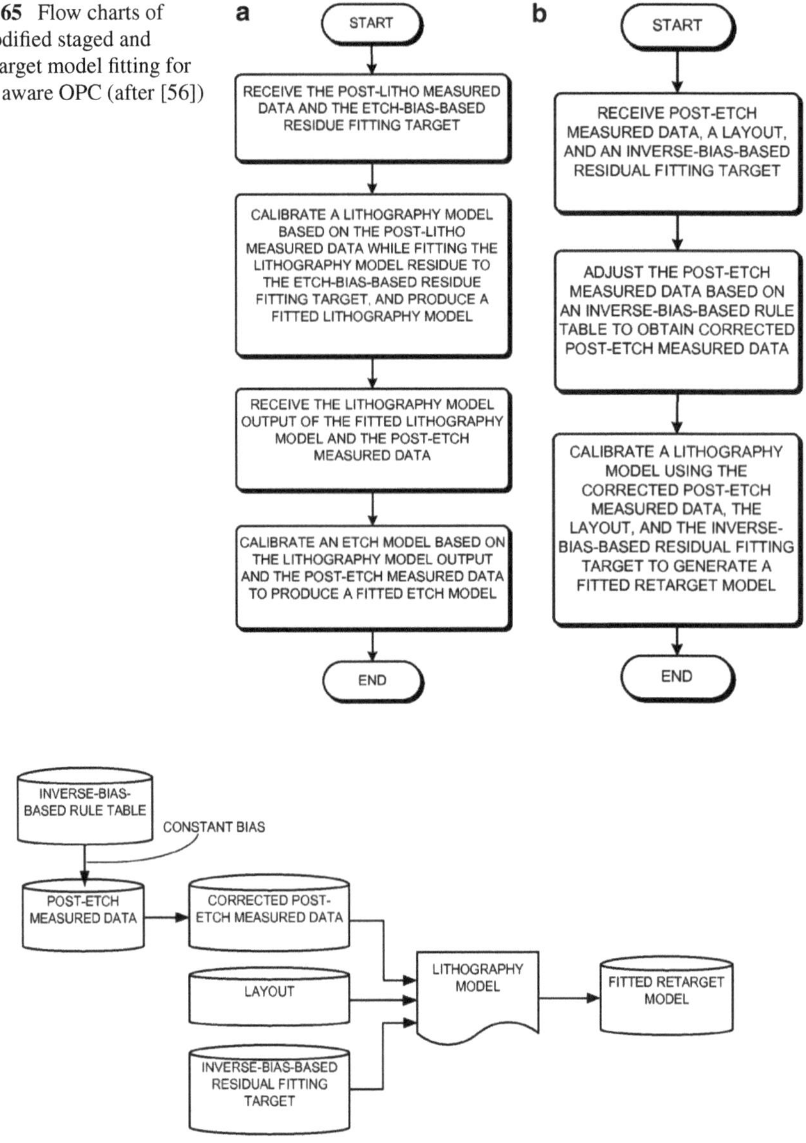

Fig. 2.65 Flow charts of (**a**) modified staged and (**b**) retarget model fitting for etch – aware OPC (after [56])

Fig. 2.66 OPC calibration with modified retarget model (after [56])

Calibrating the lithography model should cause ADI_model to match ADI_wafer for a minimized model residue. However, when ADI_wafer cannot be accurately measured, this calibration does not produce an accurate lithography model by simply fitting ADI_model to ADI_wafer.

Etch bias, which should be determined where the residue is computed, also depending on trench width, trench density, polygon width, and pattern density, displays an increasing trend with respect to an increasing line-spacing and to a decreasing line-width (Fig. 2.64). Each curve represents the etch bias (Y-axis) at a constant line-width as a function of line-spacing (X-axis).

The coefficients can be determined by fitting the set of functions with measured etch bias. In order to trace the lithography behavior and maintain the accuracy of the model, one may fit the residue to the reversed etch bias with a mean value substantially equal to zero. For example, if the first linewidth corresponds to an inverse bias between (−20 and −40 nm), the system can choose the constant to be 30 nm, so that the shifted fitting target for the first linewidth is between (−10 and 10 nm).

At the fitting of a modified staged model, the lithography model receives both post-litho measured data and an etch-bias-based fitting target as inputs (Fig. 2.65). During the calibration, the system fits lithography model so that its residue is equal to the etch-bias-based fitting target. The calibration flow (Fig. 2.66) may be configured to receive an output of the photolithography process model and post-etch CD data.

In summary, the proposed modeling approach goes a long way towards including the etch process into the overall model of pattern transfer from design to wafer. For the upcoming technology generations, the accuracy of the aggregate model would be of increasing importance. One may recommend algorithmic coverage of the proposed process, by adding user interfaces to EDA tools, in the model flow chart.

2.3.3.2 Evaluation of Pattern Mismatch

Analog circuitry requires matching of layout geometries and their environments to ensure identical performance of identical devices in different locations of the wafer [57]. Verification flow based on the coordinates of paired pattern elements would identify desired geometries for comparison to determine a mismatch depending on the contextual differences (Fig. 2.67). Layout verification (e.g., by an exclusive OR operation) can detect whether there is a match between paired elements and their surrounding patterns. The mismatch is determined at a distance between the paired element and the surrounding pattern either calculated as the shortest of the X-axis and Y-axis Manhattan distances or a radial distance. The process to verify whether that distance, called the length of characteristic influence, is acceptable, consists of:

- Determining mismatched patterns
- Dividing verification region into comparable sections
- Calculating mismatched pattern distance and area
- Layout modification.

When characteristic influences cancel each other, layout modification is not required (Fig. 2.68).

In summary, the method improves long-range layout ordering, critical for both IC manufacturing and device performance.

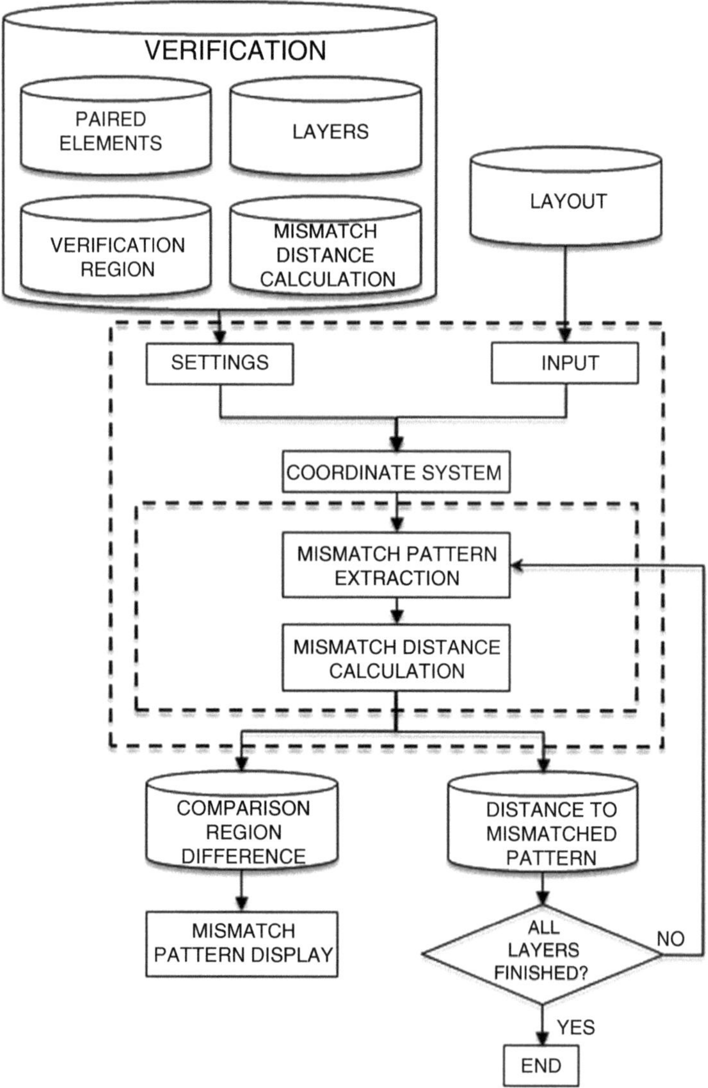

Fig. 2.67 Mismatch pattern determination and verification for interconnection layers (after [57])

2.3.3.3 On-Chip Sensor of Process Variation

As IC scaling down continues, the impact of process variations (PV) on circuit performance and robustness is increasingly more undesired. It is becoming critical to design circuits and systems with low sensitivity to process fluctuations. Process

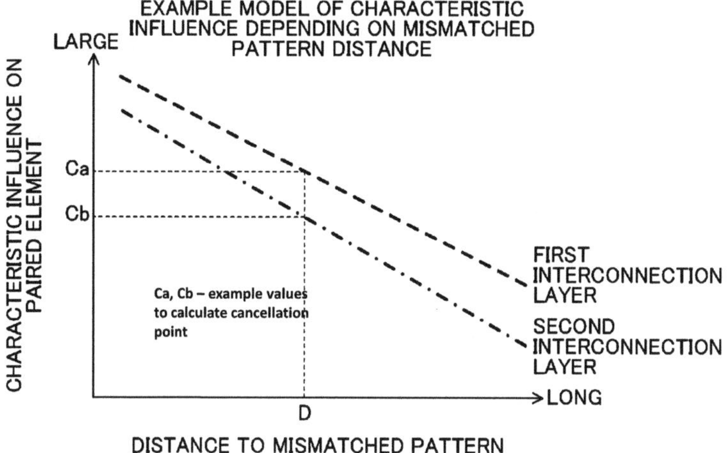

Fig. 2.68 Characteristic influence as function of distance (after [57])

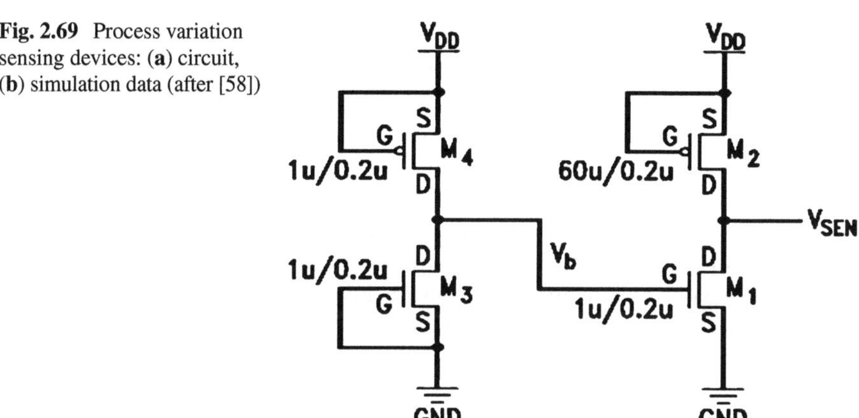

Fig. 2.69 Process variation sensing devices: (**a**) circuit, (**b**) simulation data (after [58])

variation sensors identify both global and local variations associated with transistors in the circuit layout. The first step in designing such sensors is to provide calibration data to compensate their variations, and understand what is going wrong during manufacturing [58].

One way of detecting process variations is by measuring MOSFET leakage current due to its high sensitivity. Sensor circuits usually suffer from low gain and area overhead. Ring oscillators (RO) as process variation sensors occupy too much room to be placed in large numbers for better resolution. Since the impact of process variations is averaged out through the inverter chain in ROs, local variations cannot be detected.

A circuit to sense global and local process variations able to detect systematic components and random dopant fluctuations relies on transistors operating in the subthreshold region for high sensitivity and low power dissipation (Fig. 2.69). There is no need for external circuits since all transistors are directly connected to ground and supply voltage. The variations can be linked to shifts of threshold voltage.

Under subthreshold conditions, transistors act like resistors in series and the substrate bias, V_b is determined after voltage division between them. As the threshold voltage of each transistor changes, the resistances of PMOS and NMOS transistors change at a different rate due to the difference in DIBL (Drain Induced Barrier Lowering) coefficients. The value of V_b changes with the threshold voltage, due to the difference in DIBL.

With decreasing channel lengths, random doping in the channel would affect the threshold voltage of neighboring transistors. The circuit topology for sensing local NMOS variations is the same as for their global variations, except that the transistors are sized differently to sense the differences in threshold voltages, if placed closely to each other. Their sizes are then kept at a minimum to magnify random doping fluctuations. Other transistors in the circuit are kept relatively large, to guarantee sensing only the difference between two parallel transistors since all the rest of the circuit is common for all measurements. The individual transistors in the array are typically connected to the load and amplification circuit through a hierarchical switch network.

In summary, a methodology is proposed to close the feedback loop from on – wafer PV distribution to design modifications aimed at reducing process sensitivities, based on MOSFETs with various dimensions, operating in subthreshold range. The layout architecture must ensure good compensation of the sensing elements.

2.3.3.4 Topography: Induced Hot Spots

Defect reduction in the layout areas problematic for lithography requires modeling of wafer surface topography to determine best planes of exposure focus [59]. A three-dimensional surface height map of a chip generated from a lithography tool data prior to exposure of an upper metal level (Fig. 2.70) shows peaks (surface irregularity), which could become bad focus areas depending on the focal planes used by the lithography tool (scanner).

Lithography tools either measure surface topography in real time (during the scan) or pre-measure the entire wafer surface at the “idle” stage first, followed by choosing best average focal plane to expose. That plane can be moved up, down, and rotated, to achieve best average focus adjusted as the slit scans.

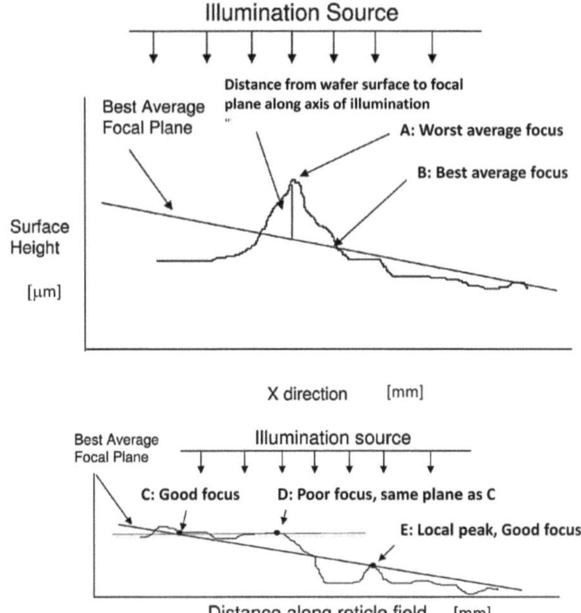

Fig. 2.70 Example of topography of a processed wafer surface. Y-axis in μm (after [59])

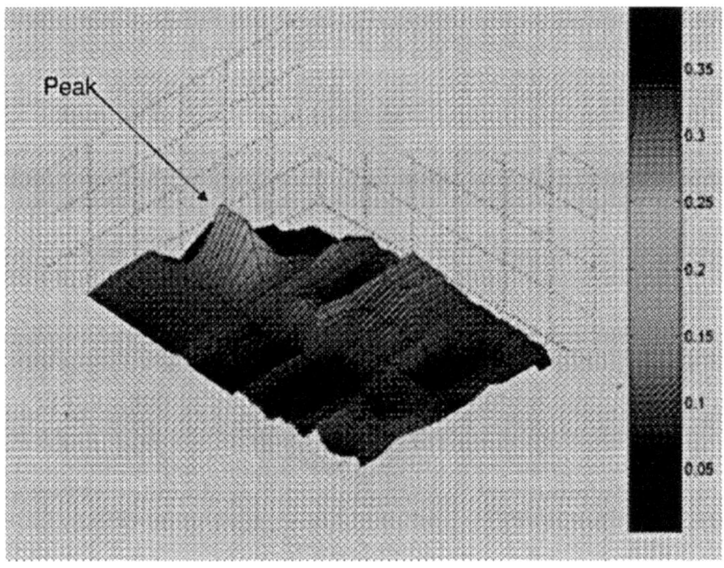

Fig. 2.71 2-D wafer planarity profile [59]

The lithography tool might choose two-dimensional surface profiles in the direction over the large surface "peak" as the best average focal plane (Fig. 2.71). The areas of the photoresist closest to that plane will be in better focus than others. The surface topography can be modeled by chemical mechanical polishing (CMP) simulation, which takes into account the details of the pattern. The surface height is the weighted average of the copper and dielectric height mapped by square wafer regions of a specific size (tiles). Still, modeling cannot accurately predict all areas problematic for lithography, by looking for high or low points within the CMP modeling surface, without understanding the way in which the lithography tool decides which focal planes to use. The points, which have the same height above the reference plane might be considered to be "high" along the surface profile, at risk of bad focus. But because of the way that the lithography tool must select the best average focal plane, one point would be exposed with better focus than another. Therefore, surface height alone is not an indicator of quality.

Problematic areas for lithography can be predicted based on surface heights of tiles on a wafer using a model of a lithographic tool to determine best planes of focus [59]. The average distance of the surface heights for each tile is its 3D representation at a location on the modeled wafer. The method predicts that distance calculating a predetermined number of focal planes for each tile in a reticle field, to ensure an equal percentage of exposure dose for the entire reticle. The focal planes are used to measure surface irregularity in 3D and to find an average focus offset. This is done by calculating the average distance along an axis of illumination from the best average focal plane. A plane which best fits CMP surface height data model has to be calculated for a predetermined number of values within a slit. The process predicts areas of poor focus (hot spots) the way in which a stepper decides the planes of best focus based on the surface heights of the wafer (Fig. 2.76). For example, the tool calculated ten focal planes for each tile in the reticle field. Each point (focal plane) would provide 10 % of the exposure dose for the tile. These focal planes are a reflection of a 3D surface texture. Once each plane of exposure for each desired point in the reticle field is calculated, the tool finds an average focus offset by calculating the average distance from the planes to the wafer surface.

In summary, the analysis of exposure field for wafer topography may become an enabling factor to ensure high fidelity of pattern transfer from design to wafer, especially for the short wavelength of the EUV. The proposed methodology helps reducing the risk related to wafer planarity and it should be of interest to see results of experimental studies in this area, especially for designs using multiple patterning.

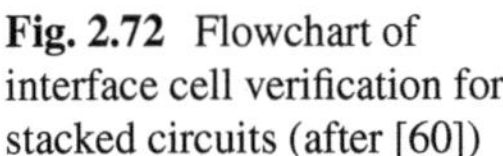

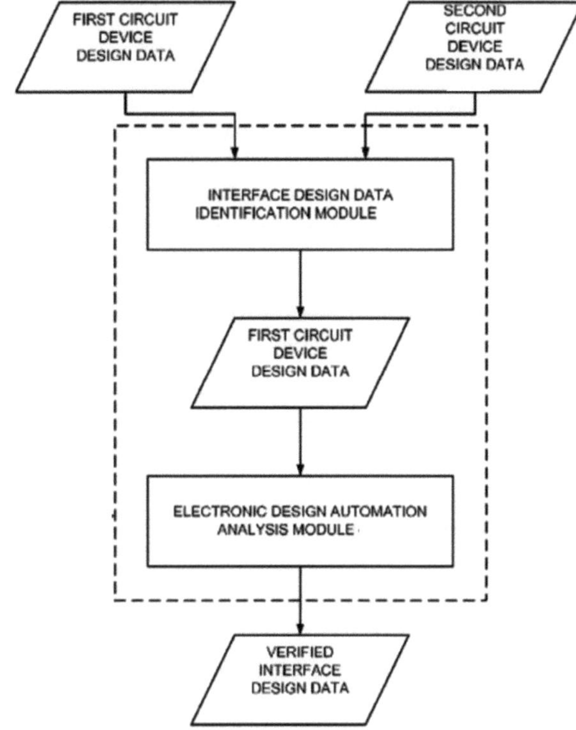

Fig. 2.72 Flowchart of interface cell verification for stacked circuits (after [60])

2.3.3.5 Physical Verification of Stacked IC's

A standard two-dimensional integrated circuit aimed to be a part of a stacked device is typically formed by a planar IC process and has metal lines on one side of a substrate ("front"). If a two-dimensional IC is to be stacked on top of another IC, it requires metal layers on the opposite side of the substrate ("back" side), connected to the circuit and metal layers on the front side by the vias that pass through the substrate ("through-silicon vias", TSVs). There are three basic techniques for manufacturing a stacked IC: wafer-to-wafer stacking, die to wafer stacking, and die to die stacking. EDA verification routines for stacked IC's would have to address these options [60].

One technique for verifying a stacked IP is to combine all designs of the component devices together for a multichip physical verification (PV). However, this may shift the focus from verification of TSV designs back to the verification of the component dice. Conceptually simple, such "mega merge" PV may alter the integrity of the initial design and of the standard verification rule file. A better method for merging two-dimensional IC databases into a single three-dimensional design is

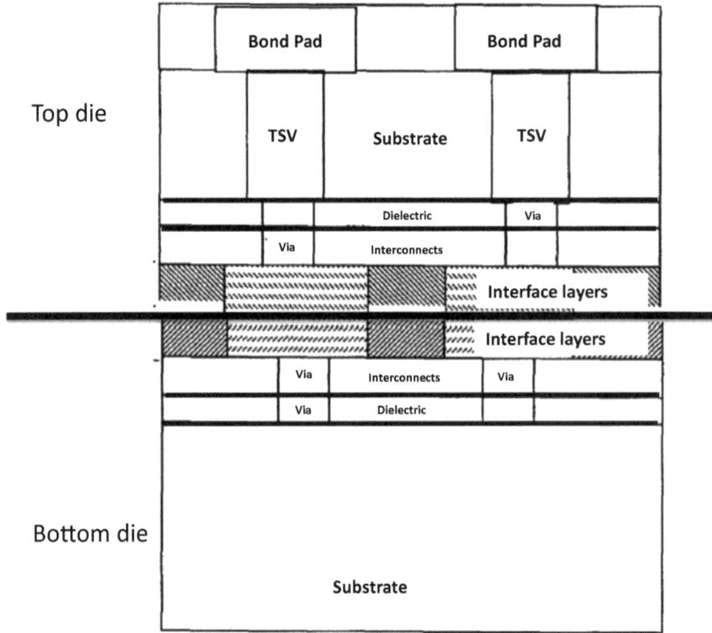

Fig. 2.73 A cross-sectional view illustrating an example of 3D IC (after [60])

"shifting" all the design layers and PV rule decks twin to incorporate them without collision [60]. For a stacked device, interface layers between 2 two-dimensional circuits that will be electrically connected is first identified. These layers are then combined and physically verified as a single set of design data. Once the interface, the first, and the second IC design data have been physically verified, the design can be recombined to form verified data of stacked device.

As an option, an interposer structure may be placed between the first and the second 2D device, to include interconnects that route signals from contacts from the first to the second circuit. If interface data includes the interposer, routing analysis can verify the interposer architecture.

When verifying interface design, data for first and second device are provided to the data identification unit, to identify their interface portions as the ones closest to the other device, combine them for physical verification and perform a routing analysis (LVS, Fig. 2.72).

In summary, the process flow of combining design data from two or more die IP's to take advantage of the TSV process is critical for the extension of silicon technology into three dimensions. The challenge is to find a concept generic enough for different types of devices to help create common 3D PV standards.

Table 2.6 Design flow for dummy feature insertion into integrated layout of interface layers [61]

1	Identify physical layout of first and second device
2	Extract physical layout of interface layer of first and second devices
3	Merge physical layout of interface layers
4	Run dummy feature utility to generate dummy features for merged interface layer
5	Insert dummy feature layout into interface layer of first device physical layout
6	Insert dummy feature layout reflected by y-axis into interface layer of second device physical layout

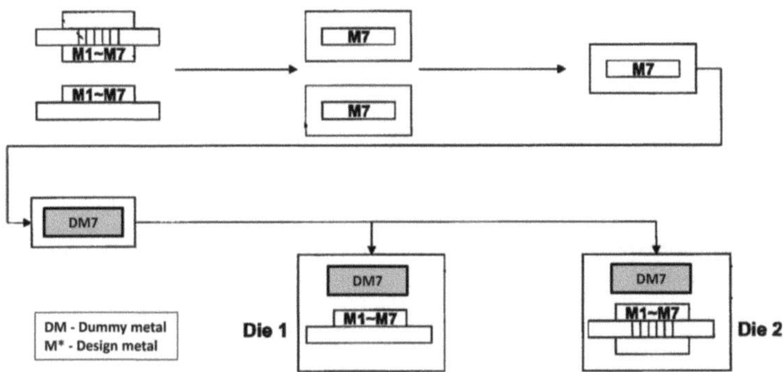

Fig. 2.74 A flowchart illustrating a method of providing a dummy feature pattern (after [61])

2.3.3.6 Dummy Pattern and Comprehensive PVs for 3D Integrated Circuits

Assembling a 3D integrated circuit (3D IC) from two or more layers of active electronic components (e.g., vertically stacked and connected) and with two or more interface layers, to be verified by DRC or layout-versus-schematic (LVS) would require stacking rules pertaining to the contents of the stacked dice [61]. In 3D IC technology, these components are built on two or more substrates and packaged to form a single circuit (Fig. 2.73). They are aligned and bonded together, either after dicing into singulated die or while in wafer form which may then be subsequently diced. The interaction within the stacked die creates design challenges, which often have not been addressed by designers or CAD tool developers. Both the physical and the electrical connections between stacked devices must be accurate and robust using a reliable method of connection and verification of such a connection.

One challenge is to place dummy pattern on the stacked dice such that it won't interfere with through-silicon-vias (TSV) through the substrate to enable a bond pad on a bonding layer to be operably coupled to the TSV. The bond pad includes an I/O pad providing connection to the component die.

The interface layer may include dielectric regions, conductive traces and pads (probe pads and bonding pads to provide connection to another die). The bottom die includes a substrate and all the other IC layers.

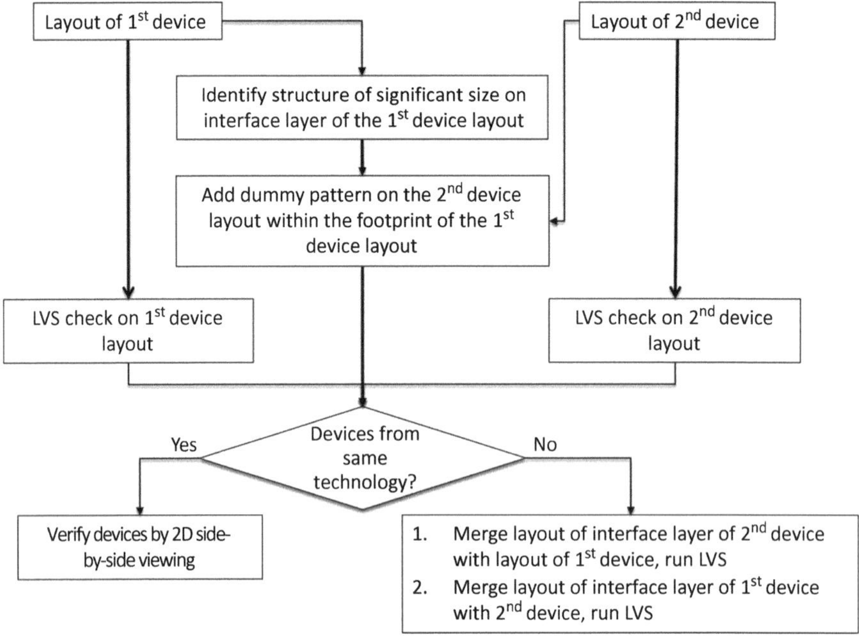

Fig. 2.75 Verification methodology of dummy pattern for a stacked device (after [61])

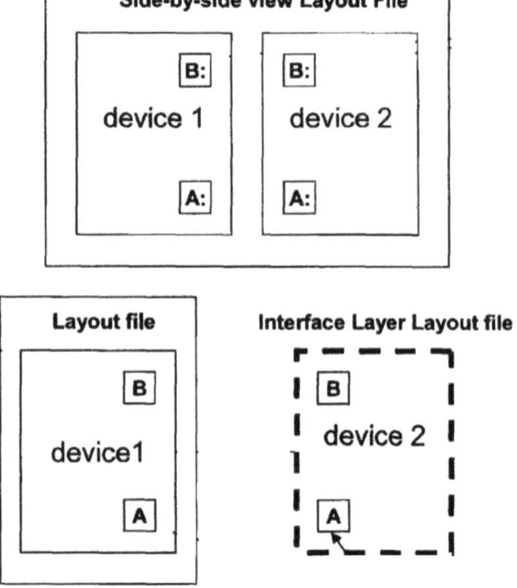

Fig. 2.76 Top-views illustrating implementation of design files with the 3D verification method (after [61])

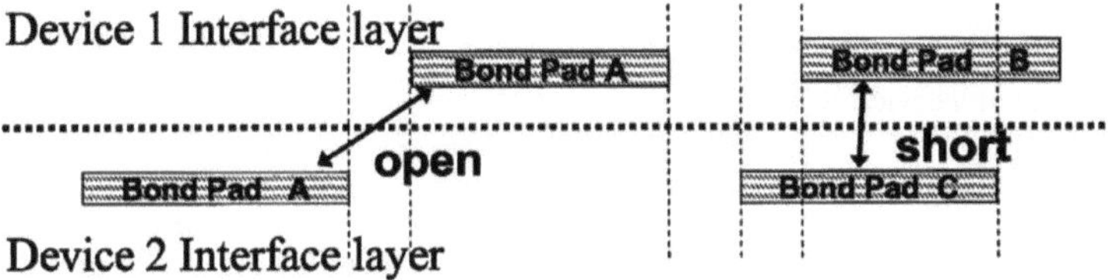

Fig. 2.77 A cross-sectional view of the alignment of interface layers of a 3D IC (after [61])

From the design verification standpoint, the layout of the interface layers of the first and second tier devices is first extracted. Two layout design files including only the interface layer of the first tier and the second tier devices are generated (Table 2.6). The layers are merged to form one layout file, which may include a single layer with the structures from the first and the second tier device interface layer. One file may be reflected on an axis (e.g., y-axis) such that the layouts are properly aligned when merged (e.g., to account for the orientation of the die in a 3D IC). Figure 2.74 illustrates a design file representative of the merged interface layer (M7) which includes the layouts of both design files, and reflecting about the y-axis accounts for the alignment of the layouts when die is "flipped" and positioned on die in a 3D IC. This verification needs to be sequenced with adding dummy pattern.

Therefore, a dummy feature tool is ran on the merged file. Dummy pattern is added in the first and the second tier device layout and in the interface layer of the first tier device layout (Fig. 2.74, DM7). The dummy pattern in the interface layer of the second tier device layout may be reflected upon its y-axis first. Its mirror image would be provided to the second tier device layout, to ensure proper alignment when the latter device is "flipped" and stacked on the first tier device. This way, dummy features are added to both the first and second tier devices taking into account the layout of the other device, and including the presence of device structures (e.g., interconnect traces, contacts, pads, etc.) on both of them. Dummy structures may be added to the interface layer only where there is open space in both the first and second tier devices, to prevent a short caused by their placement in a location on one device that may come into contact with the opposing device when the 3D IC is fabricated.

This method would improve the density of an otherwise lower density area of a 3D IC circuit. For example, if a large probe pad is found on a first device, no dummy pattern will be added in that area in the merged design file. Therefore, when the dummy feature pattern layout is extracted and inserted into the second die, a large vacant area (e.g., of the probe pad outline) may be required on the second device. This low density area can impact pattern fidelity. In the fabrication of photomasks and wafers, the method has to include LVS verification separately on the first and second layout, including the correlation of the design from netlist to the physical layout (e.g., a GDS II file, Fig. 2.75).

One also has to determine if the first and second devices are built in the same technology (e.g., 28 nm, 90 mm). Verification including a side-by-side view in 2D for LVS is performed, to check the logic and physical connectivity of the layout

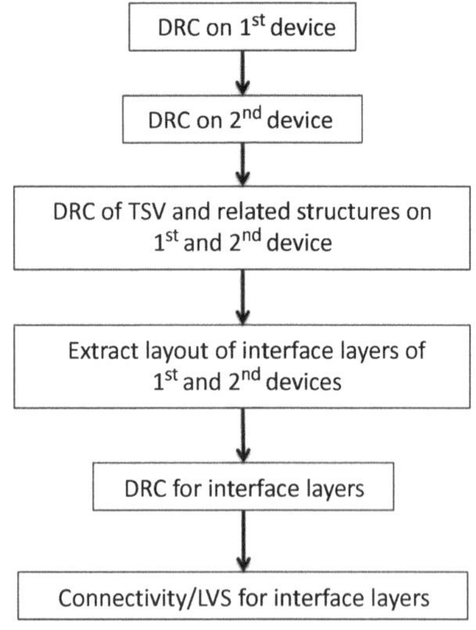

Fig. 2.78 A flowchart of a method of designing a physical layout of a device of a 3D IC (after [61])

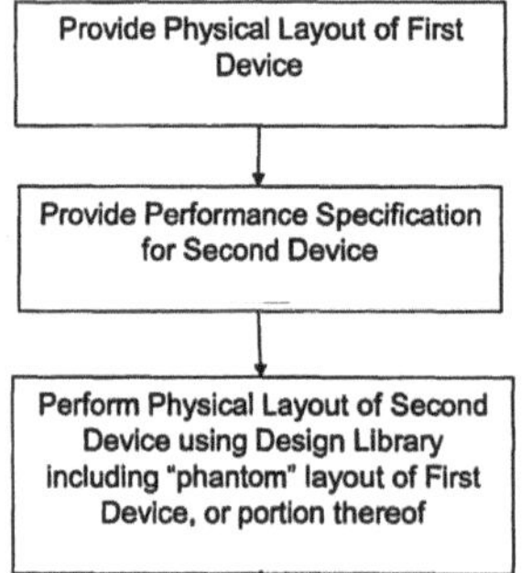

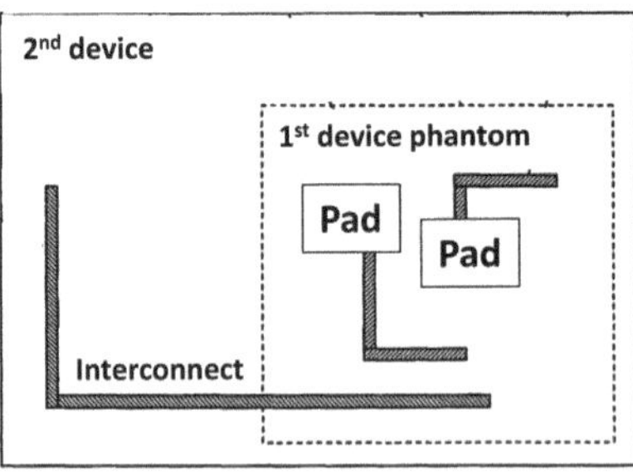

Fig. 2.79 A top-view of a physical layout design of a 3D IC device in the verification method (after [61])

design files. The side-by-side view may emulate the 3D IC layout (Fig. 2.76). The verification may be performed with the virtual connectivity feature of the LVS tool (e.g., providing for connection of the bonds between the first and second devices incl. their interface layers, such as illustrated by conductive bonding area, Fig. 2.77). Internal pin text may also be provided.

If the first and second devices are associated with different technologies, more complex verification procedures are required. The layout of the interface layer of the first device is merged into the layout of the first and second device and LVS verification is run. Only the interface layers (e.g., their connectivity to one another) are verified but the layout may include all the layers of the device. A layout including only the interface layer of a second device is reflected about its y-axis to simulate

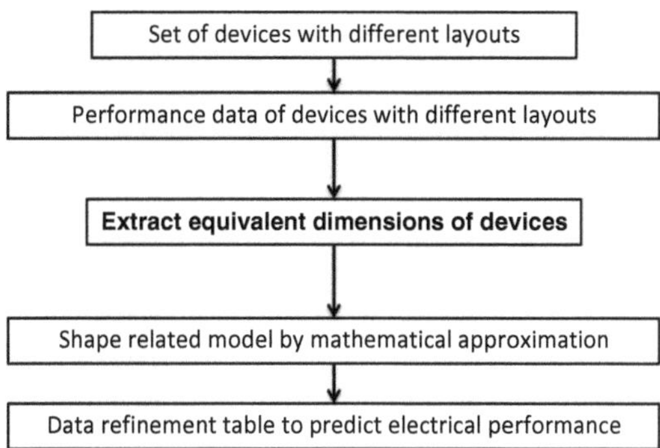

Fig. 2.80 Table based DfM flow (after [62])

the positioning as stacked in a 3D IC, and merged with the physical design layout file. LVS verification may then be run on the merged layout.

The output of the verification can be analyzed on a cross-section of the alignment of interface layers of the first and second device of a 3D IC: the first device interface layer including bond A and bond B, the second device interface layer including bond A and bond C (Fig. 2.77). The cross-section shows an open where bond A of the interface layers are not properly aligned in the 3D IC, and a short where bond B of the interface of the first die is coupled to bond C of the interface layer of second die (e.g., where bond B is to connect with bond B and bond C is to connect with bond C). This cross-section would be recognized by the LVS verification. Physical layout corrections may be then made. The representative cross-section and verification tools to generate it, should take into account alignment tolerances of the layers.

Design rule check (DRC) would verify where the layout of the interface layer of the first and second devices is extracted (Fig. 2.78). Separate GDS II files for the interface layer of the first device and the second device would be generated, followed by a DRC including physical rules associated with the interface layers and their connection to one another, including mutual spacing and pattern density. The spacing rules must take into account the alignment of the top and bottom device and different potentials that may be found on them. A connectivity check may determine an electrical open or short in the connection of the interface layers due to alignment tolerances.

Physical layout is then generated. CAD tools convert the netlist into a physical layout of the second device. In contrast to conventional methods, the second device layout is created taking into account the layout of the first device e.g. as a "phantom" view. This view may include a footprint (or the physical information) of the circuit including the locations of the gates, interconnections, isolation regions and information about connections (e.g., pin layouts) of the first device. A specific layer of the first device may be imported into the library for designing the second device (Fig. 2.79) including an interconnect structure. A phantom layout associated with the first device including interconnects and pads, provide numerous advantages.

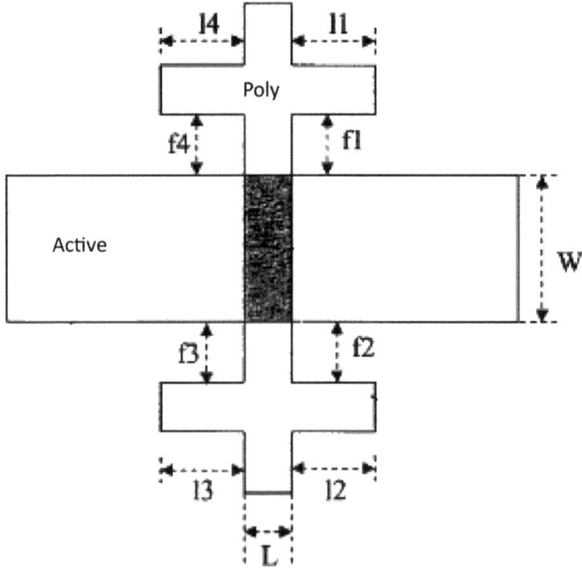

Fig. 2.81 Top view of device subject to table – based analysis (after [62])

In the earlier technology, designers were unable to check both GDS II interface layer rules correctly or to verify the connectivity between two GDS II files. The placement of dummy features in the interface layers may cause a short between the two stacked die (GDS II files) if the layout of both interface layers are not considered.

In summary, for physical verifications of two devices forming a 3D integrated circuit (3D IC), a third layout may need to be generated to include a portion of the first layout and a portion of the second layout, i.e., an interface layer. Dummy features inserted into the third layout are merged into the first and second layout and may be reflected about its y-axis such that it is properly aligned, when the devices are stacked.

2.3.3.7 Table-Based DFM

Circuit performance can be linked to device geometries by a shape model translating dimensions to electrical parameters [62]. DfM tools use equation-based solutions for layout parasitic extraction (LPE) to predict device behavior on chip. The equations are obtained by best fitting test patterns to the baseline silicon data. But electrical drift induced by process variation cannot be accurately predicted by an equation. Neither can equation-based approach handle abrupt layout transitions without costly high-order approximation and the risk of introducing singular points.

A table-based post-layout analysis helps estimate electrical performance of MOSFETs which lost their ideal, rectangular shape due to process effects, to investigate their manufacturability during patterning. For the distorted layout areas, such as rounding comers of the image contour, the table-based strategy provides a cost-effective and more accurate prediction of electrical behavior.

For a table-based IC DfM flow (Fig. 2.80) performed for a top view of an exemplary IC device (Fig. 2.81), in addition to "W" and "L", the device with the rounding H – shaped active region can be described by various heights "h1", "h2", "h3", and "h4", and spaces "s1", "s2", "s3", and "s4".

Accordingly, a set of IC devices with different dimensions are first designed followed by silicon based extraction of their equivalent dimensions. First, an IC contour is generated based on the layout and channel regions are defined as overlapping areas of the active and gate contours. Then, an effective rectangle from the layout contour is generated, to be simulated for electrical performance by a SPICE tool. A maximum rectangle inside the layout contour is defined followed by the width and the length correction to that rectangle.

The method proceeds to correlating the dimensional parameters to the electrical performance of the device, depending on the equivalent width "We" and length "Le" of the channel instead of the design width W or length L. The shape related model can be generated using a multiple regression, response surface approximation, etc. A refinement table is then generated to improve electrical performance using the collected data, the dimensional parameters, and the shape related model. It can be used for post layout analysis including tuning the layout and identifying the hot spots.

In summary, the table-based approach is aimed at adding another level of corrections over the model-based DfM approach. Though extraction of real geometries is important for accurate modeling, it may be logistically difficult to define such tabulated database for a variety of devices.

2.3.3.8 Encrypting DfM Data

IP preservation in DfM flow requires interactions among the multiple design and manufacturing entities. It is important that confidentiality of information is not compromised. To that end, DfM support organizations offer tools to encrypt DfM related information.

An encryption and decryption interface in DfM flow includes a decryption module embedded in the design tool [63]. An encrypted DfM data is provided to an authorized IC designer. A private key is provided for decrypting DfM data to manufacturing limits. For example, IC design flow may consider ideal or simplified models for metal thickness distribution. Signal analysis and design performance evaluation cannot reflect the variations in metal thickness during fabrication. The IC design layouts have no proper way to incorporate data from a chemical mechanical polishing (CMP) process. However, variations of metal thickness due to CMP seriously impact the signal wire characteristics, IC design functionality, and performance. For various layout environments, the same metal wire may have different thickness, which results in variations of its electrical properties.

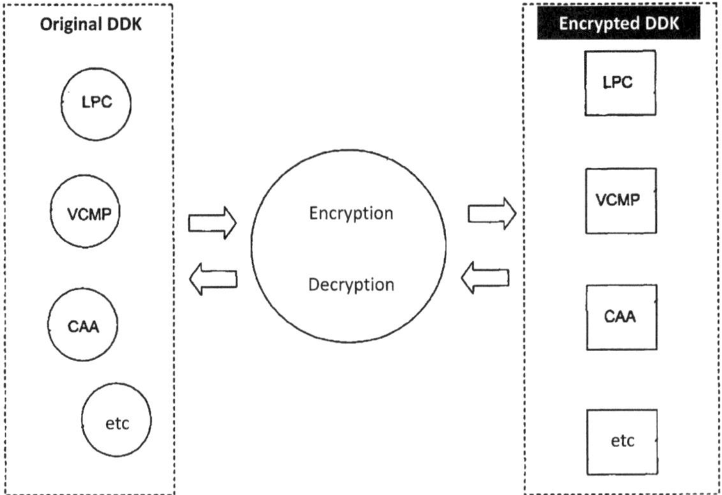

Fig. 2.82 Components of PDK flow (after [63])

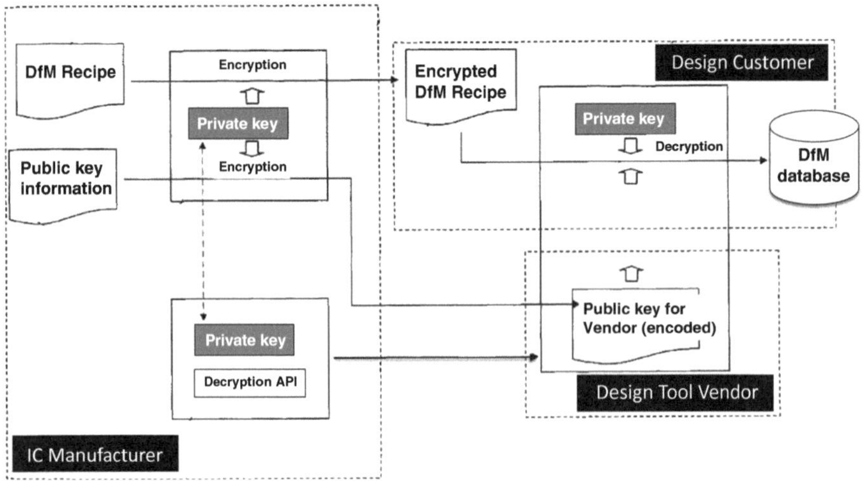

Fig. 2.83 Block diagram of an encryption/decryption flow (after [63])

A DfM tool kit may include data from various process modules. Manufacturing information, such as processing recipes, tool characterization, manufacturing environment, production and processing statistical data, and IC testing and measurement results, are compiled, accumulated, and formulated to form the DDK (DFM Data Kit) and provide a manufacturing simulation setup for lithography process check (LPC) chemical mechanical polishing (CMP) or critical area analysis (CAA), etc. In LPC, lithography process can be simulated for a design layout by

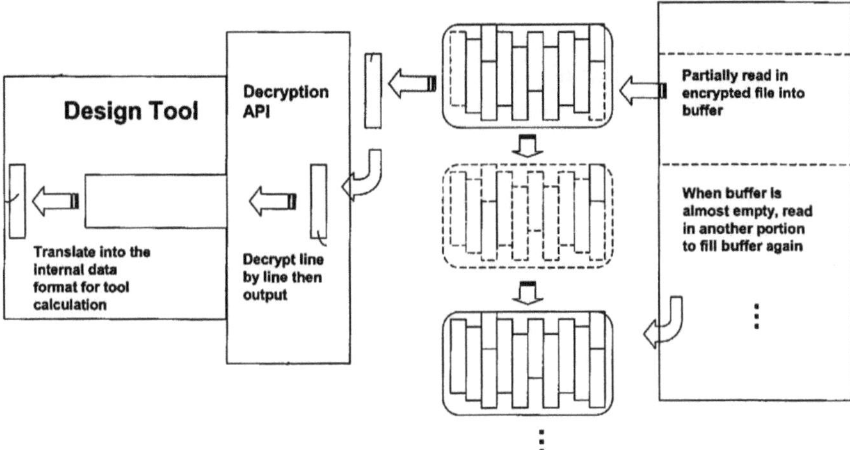

Fig. 2.84 A block diagram illustrating an API decryption with buffered decryption function (after [63])

implementing DDK. Various defect or weak areas associated with the manufacturing process, referred to as hotspots, can be identified for design tuning. To that end, in the CMP simulation with DDK, the design layout is supplemented with material thickness data. CAA simulation utilizes DDK for critical area identification and design improvement. DfM data may be provided in a unified format (DUF).

DfM tool kit also includes DfM advisories extracted from the manufacturing information for an IC design tool or a designer. The advisories may include "action required" rules (i.e. dispositions requiring further actions to eliminate the associated violation). Recommended rules are not binding and are suggested for design improvement. Guidelines provided for the designer to follow in implementing an IC design procedure may need a different level of encryption.

DfM tool kit provides model-based utilities from various simulations and rule-based utilities from DfM advisories. It can be implemented at various designing stages and predetermined manufacturing stages. For example, dummy insertion may be implemented at place-and-route step such that the dummy features are included in a layout at early design stage. LPE deck may be implemented at extraction and at timing simulation. LPC may be implemented before or after the tape-out of drawn database, to eliminate hotspots identified by LPC before fabricating a mask.

In a block diagram of data encryption/decryption flow, a data kit (DDK) includes lithography process check (LPC), chemical mechanical polishing (CMP) simulation, and critical area analysis (CAA) (Fig. 2.82).

An encryption and decryption interface (EDI) is provided for data and sharing to protect manufacturing organization and provide a confidential format for data communication among the fab, design customer (or designer), and design tool vendor. The DDK with its subsets, separately or collectively, is encrypted through the EDI such that each subset such as LPC, CMP, CAA and other DDK modules are encrypted. The DDK presented in a way that its information is scrambled and unreadable, and can only be decrypted with the authorized key, can be passed to a design customer (a design house) or can be loaded into a design tool through the design tool vendor. When an authorized designer is to implement the encrypted DDK in a design tool, the DDK or its subsets are decrypted, based on a private key.

In a block diagram illustrating data flow, encryption method, and system, the DfM data flows between an IC manufacturer (semiconductor foundry), a design tool vendor (EDA), and a design customer (fab-less design house). The encryption/decryption methodology includes an encryption utility to enable recipe release for the IC manufacturer internal use. Therefore, a DfM data package (or recipe) such as the DDK (Fig. 2.83) or a portion thereof, is encrypted by the encryption utility to be released to the design customer. As the manufacturing data are encrypted, they can be protected with additional methods.

The encrypt/decrypt methodology also creates a static library of description to be delivered to the design tool vendor and embedded into a DfM design tool, or a design tool integrated with DfM tool kit. The DfM tool is able to decrypt the recipe dynamically depending on the standard for commercial software and hardware encryption or other security features (NIST may provide an example of advanced encryption standard). The system utilizes an asymmetric cryptography with a public key for encryption and a private key for decryption, the information of public key is distributed to the design tool vendor for encoding design information and it may be further encrypted for safety consideration. The static library with decryption program and the private key are synchronized through version control.

A block diagram illustrating a decryption application program interface (API) with buffered decryption function (Fig. 2.84) shows the API picks up a single line, from the buffer, then decrypts and sends it to the design tool. In order to prevent the key in binary code from leaking or tracking, further expansion, interleave, and scramble of the encrypted key information are adopted. Expiration date, tool features, technology or design tool vendor name are implicitly checked through the encrypted information carried by the public key.

In summary, a data encryption/decryption system is becoming necessary for efficient but safe exchange of DfM information.

2.3.4 Conclusions and Roadmapping

In this Chapter, we discussed several approaches of classical DfM – at – large:

- How to launch DfM effort at design stage,
- How to optimize the layout based on best effort from design,
- How to comprehend the many design – related disciplines in product development which can be improved by DfM.

The diversification of the discussed concepts shows that classical DfM continues to be a discipline of growing complexity. Its domain already expands outside of the fields of design, layout, process integration, yield management, optics and lithography into the floorplanning and packaging, due to the emerging IC applications including Systems-on-Chip. Compared to the status from several years ago, DfM started to meaningfully improve the design or layout via tool optimization as well as by employing best practices to arrive at the results well beyond its original RoI expectations. With many different designs and companies adhering to the different standards, foundry design rules and DfM recommendations should continue to be conservative, also because, as stated in the Introduction, DfM will continue to be an art of compromise. Fabless companies respond to the DfM challenge by providing designers with actionable information, how to build standard cells, custom IP blocks, memories, etc., so that they can minimize the impact of the most critical defects, as defined by the foundries' design rule decks and datakits.

One reason for which DfM implementation requires careful RoI considerations is the need to compromise in the integrated approach. Restricted or local CBC solutions can simplify DfM by addressing selected yield loss areas. But they also hide DfM complexity instead of managing it and may be too rigid to play well with other DfM aspects, such as hot spot fixes. The real answer is to give the IC designer a complete, integrated solution.

Future work should continue to support developing tools for integrated and automated DfM design kit ensuring cost and complexity reduction combined with improved performance. Point solutions pertaining to floorplan, OPC, or package optimization may be attractive for the short term but IC design and manufacturing are mature enough to depend on systematic feedback rather than on being developed in isolation from each other. Every aspect of manufacturing flow has now to be modeled to account for the growing field of applications, equipment upgrades, and process complexity. Design will be required to follow the ever-more-comprehensive rule decks combined with iterative approach on product tapeouts to realize full potential of DfM.

References

1. Balasinski, A.: Semiconductors: Integrated Circuit Design for Manufacturability. CRC Press/Taylor and Francis, Boca Raton (2011)
2. Yang, J.: Manufacturability aware design. Ph.D. Thesis, University of Michigan (2007)
3. Schellenberg, F.M.: Sub-wavelength lithography using OPC. In: Semiconductor Fabtech (edn. 9), Semiconductor Media Ltd, London, UK, pp. 205–209 (1999)
4. International Technology Roadmap for Semiconductors no 7, http://public.itrs.net/
5. Nassif, S.R.: Delay variability: Sources, impacts and trends. pp. 638–639. ISSCC, (2000)
6. Optical Lithography Cost of Ownership – Final Report, http://www.sematedi.org/docubase/document/4014atr.pdf (2000)
7. Liebmann, W.: Resolution enhancement techniques in optical lithography: It's not just a mask problem. Proc. SPIE. **4409**, 23–32 (2001)
8. Levenson, M.D.: Wavefront engineering from 500 nm to 100 nm CD. Optical microlithography X. Proc. SPIE. **3051**, 2–13 (1997)

9. Gupta, P., Kahng, A.B., Sylvester, D., Yang, J., Performance-driven optical proximity correction for mask cost reduction, J. Micro/Nanolithography MEMS MOEMS (6) 3 031005 (Jul 2007)
10. Lin, B.: Phase shifting masks gain an edge. IEEE Circuits. Dev. **9**, 28–35 (1993)
11. Nassif, S.R.: Modeling and forecasting of manufacturing variations. In: Proceedings of the Fifth International Workshop on Statistical Metrology, pp. 3–10 (2000)
12. Ma, S.T., Keshavarz, A., De, V., Brews, J.R.: A statistical model for extracting geometry sources of transistor performance variation. IEEE Trans. Electron. Dev. **51**(1), 36–41 (2004)
13. Orshansky, M., Milor, L., Hu, C.: Characterization of spatial intrafield gate CD variability, its impact on circuit performance, and spatial mask-level correction. IEEE Trans. Semiconduct. Manuf. **17**(1), 2–11 (2004)
14. Visweswariah, C.: Death, taxes and failing chips. In: Proceedings of the Design Automation Conference, pp. 343–347 (2003)
15. Agrawal, A.B., Blaauw, D., Zolotov, V., Vrudhula, S.: Statistical timing analysis using bounds and selective enumeration. In: Proceedings of the Design Automation Conference, pp. 348–353 (2003)
16. Orshansky, M., Keutzer, K.: A general probabilistic framework for worst case timing analysis. In: Proceedings of the Design Automation Conference, pp. 556–561 (2002)
17. Postnikove, S., Hector, S.: ITRS CD error budgets: Proposed simulation study methodology. (2003)
18. Chen, L., Milor, L., Ouyang, C., Maly, W., Peng, Y.: Analysis of the impact of proximity correction algorithms on circuit performance. IEEE Trans. Semiconduct. Manuf. **12**(3), 313–322 (1999)
19. Stine, B., Boning, D., Chung, J., Ciplickas, D., Kibarian, J.: Simulating the impact of poly-CD wafer-level and die-level variation on circuit performance. In: Proceedings of the Second International Workshop on Statistical Metrology, pp. 24–27 (1997)
20. Orshansky, M., Milor, L., Chen, P., Keutzer, K., Hu, C.: Impact of spatial intrachip gate length variability on the performance of high-speed digital circuits. IEEE Trans. Comput. Aided. Des. Integr. Circuits. Syst. **21**, 544–553 (2002)
21. Gupta, P., Heng, F.L.: Toward a systematic-variation aware timing methodology. In: Proceedings of the Design Automation Conference, pp. 321–326 (2004)
22. Liebmann, L.W.: Layout impact of resolution enhancement techniques: Impediment or opportunity. In: Proceedings of the ACM/IEEE International Symposium on Physical Design, pp. 110–117 (2003)
23. Bowman, K.A,. Meindl, J.D.: Impact of within-die parameter fluctuations on future maximum clock frequency distributions. CICC, San Diego, CA, pp. 229–232 (2001)
24. Berkeley Predictive Technology Model, http://www-device.eecs.berkeley.edu/ptm
25. Kahng, A.B.: The road ahead: Shared red bricks. IEEE Des. Test. Del, Mar, CA (2002)
26. Bowman, K.A., Duvall, S.G., Meindl, J.D.: Impact of die-to-die and within-die parameter fluctuations on the maximum clock frequency distribution. In: IEEE International Solid-State Circuits Conference, San Francisco, CA, pp. 278–279 (2001)
27. Zhang, W., Yang, Z.: A new threshold voltage model for deep-submicron MOSFET's with nonuniform substrate dopings. In: Proceedings of the Electron Devices Meeting, pp. 39–41 (1997)
28. Boning, D., Nassif, S.: Design of high-performance μP circuits, chapter 6: Models of process variation in device and interconnect, pp. 98–116 (1998)
29. Park, T., et al.: Pattern and process dependencies in copper damascene chemical mechanical polishing processes. In: VLSI Multilevel Interconnect Conference, Santa Clara (1998)
30. Hymes, S., Smekalin, K., Brown, T., Yeung, H., Joffe, M., Banet, M., Park, T., Tugbawa, T., Boning, D., Nguyen, J., West, T., Sands W: Determination of the planarization distance for copper CMP process. In: Materials Research Society 1999 Spring Meeting, San Francisco (1999)

31. Orshansky, M., Milor, L., Chen, P., Keutzer K., Hu, C.: Impact of systematic spatial intra-chip gate length variability on performance of high-speed digital circuits. IEEE Int. Conf. Comput-Aided. Des, New York, NY, pp. 62–67 (2000)
32. Wong, S.-C., Lee, G.-Y., Ma, D.-J.: Modeling of interconnect capacitance, delay and crosstalk in VLSI. IEEE Trans. Semiconduct. Manuf. **40**(1), 108–111 (2000)
33. Chen, K., Hu, C., Fangm, P., Lin, M.R., Wollesen, D.L.: Predicting CMOS speed with gate oxide and voltage scaling and interconnect loading effects. IEEE Trans. Electron. Dev. **44**(11), 1951–1957 (1997)
34. Gupta, P., Heng, F.-L., Lavin, M.: Merits of cellwise model-based OPC. In: Proceedings SPIE International Symposium on Microlithography (2004)
35. Spence, C., et al.: Mask data volume – Historical perspective and future requirements. In: Proceedings SPIE 22nd European Mask and Lithography Conference, vol. 6281, (2006)
36. Nair, R., Berman, C.L., Hauge, P.S., Yoffa, E.J.: Generation of performance constraints for layout. IEEE Trans. Comput. Aided. Des. **8**(8), 860–874 (1989)
37. Chen, C., Bozorgzadeh, E., Srivastava, A., Sarrafzadeh, M.: Budget management with applications. Algorithmica. 261–275 (2002)
38. http://www.mentor.com
39. http://www.opencores.org
40. Zhang, Y., Gray, R., Nagakawa, O.S., Gupta, P., Kamberian, H., Xiao, G., Cottle, R., Progler, C.: Interaction and balance of mask write time and design RET strategies. In: Proceedings of the SPIE Photomask, Japan (2005)
41. Balasinski, A.: Optimizing IC design for manufacturability. Rec. Patents. Electr. Eng. Bentham. Sci. **1**(3), 209–213 (2008)
42. Cheng, Y.-C., Ou, T.-H., Feng, J.J.H., Tsai, J.J.H., Liu, R.-G., Huang, W.-C., Xiang, X.-G.: Practical approach to layout migration. US Patent Application 0161907, 30 June 2011
43. Park, S.: Photolithography method. US Patent Application 0171585, 14 July 2011
44. Shieh, M.-F., Yu, S.-S., Yen, A., Yu, S.-M., Chang, C.-Y., Xu, J.J., Wann, C.H.: Integrated circuit layout design. US Patent Application 0151359, 23 June 2011
45. Roy, S., Chakraborty, K., Han, Y.: System and method for circuit design floorplanning. US Patent Application 0185328, 28 July 2011
46. Gordin, R., Goren, D., Strang, S.E., Tallman, K.A., Tretiakov, Y.V.: Layout determining for wide wire on-chip interconnect lines. US Patent Application 0179392, 21 July 2011
47. Yoda, K.: Method for designing semiconductor integrated circuit which includes metallic wiring connected to gate electrode and satisfies antenna criterion. US Patent Application 0165737, 7 July 2011
48. Moreno, T.: Die package including multiple dies and lead orientation. US Patent Application 0169144, 14 July 2011
49. Herold, K.: Methods of optical proximity correction. US Patent Application 0197169, 11 Aug 2011
50. Irmatov, A., Belousov, A., Cadouri, E., Gratchev, A., Ryjov, A., Thenie, L.: Method and mechanism for extraction and recognition of polygons in an IC design. US Patent Application 0167400, 7 July 2011
51. Lippincott, G.P.: OPC conflict identification and edge priority system. US Patent Application 0167394, 7 July 2011
52. Lee, J.S.: Masks of semiconductor devices and methods of forming mask patterns. US Patent 7,655,362, 2 Feb 2010
53. Gaur, I., et al.: Closed loop design for manufacturability process. US Patent Application 0127029, 29 May 2008
54. Smayling, M.C.: Integrated circuit cell library with cell-level process compensation technique (PCT) application and associated methods. US Patent 7,979,829, 12 July 2011
55. Kuo, H.W.: Method of design for manufacturing. US Patent 8,136,067, 13 Mar 2012
56. Xue, J., Huang, J.S.: Etch-aware OPC model calibration by using and etch bias filter. US Patent Application 0179393, 21 July 2011

57. Kojima, S., Toyama, M., Yoshidome, H., Ito, M.: Method for layout verification of semiconductor integrated circuit. US Patent Application 0088006, 14 Apr 2011
58. Meterelliyoz, M., Song, P., Stellari, F.: Process variation on-chip sensor. US Patent 7,868,606, 11 Jan 2011
59. Brunner, T.A., Greco, S.E., Liegl, B.R., Xiang, H.: System and method of predicting problematic areas for lithography in a circuit design. US Patent Application 0184715, 28 July 2011
60. Hogan, W.M., Petranovic, D., Aslyan, A.: Stacked integrated circuit verification. US Patent Application 0185323, 28 July 2011
61. Lin, K.-Y., Lin, Y.-L., Tsai, C.S., Wang, C.-H.: Design and verification of 3D integrated circuits. US Patent Application US20090319968 Al
62. Hou, Y.-C., Cheng, Y.-C., Liu, R.-G., Lai, C-M., Cheng, Y.-K., Lin, C.-K., Chao, H.-S., Yeh, P.-H., Wu, M.-H., Ku, U-C., Ou, T.-H.: Table-based DFM for accurate post-layout analysis. US Patent 8,001,494, 16 Aug 2011
63. Cheng, Y.-K., Chang, G.S., Liu, J., Chiao, H.-S.: System and method for design – for – manufacturability data encryption. US Patent 8,136,168, 13 Mar 2012

Chapter 3
DfM at 28 nm and Beyond

Introduction of more advanced technology nodes carries two key risks:

- Increased process variability which impacts both performance and yield.
- New physical phenomena, which were not visible at larger geometries.

These risks cannot be mitigated by "classic" DfM, i.e., by layout optimization concepts working very well, but addressing concerns of prior generations.

New DfM approach is needed, to not only reduce variations of product performance, but to prevent them at the start of design. One such approach is restrictive design, which limits the type and placement of features and enforces gridding of critical layers. Restrictive design is being pursued to reduce the process risk by reducing degrees of freedom in the total physical design space available. It may rely on the predefined layout architecture and new flavors of design rules, such as "DRC+" or Restrictive Design Rules (RDRs), as part of CBC design methodology.

The root cause of restrictive design is that simple dimensional scaling of geometries on any layer is no longer sufficient to achieve electrical performance targets at 28 nm technology node and below. A well – controlled, CBC design space is required. Device, interconnect, and process innovations add enough unpredictability of layout sensitive detractors to yield ramp to make random layout a non-viable option (Table 3.1). At the same time, it is not possible to develop a new process and thoroughly characterize its layout sensitivities before starting the respective node's design work. Layout has to proceed in parallel to the process and device implementation. "Classic" DfM has to be enhanced to design-technology co-optimization practiced previously for memory arrays (SRAM, etc.), based on a highly simplified layout environment, and applied to all logic features. Layout density, patterning, electrical yield improvements, and layout simplicity should be quantified to enable a template-based design with an increased focus on interconnect redundancy and driven by the responses from the process side of DfM [1].

A. Balasinski, *Design for Manufacturability: From 1D to 4D for 90–22 nm Technology Nodes*,
DOI 10.1007/978-1-4614-1761-3_3,

Table 3.1 CMOS scaling challenges to lithography, device, interconnect options, and new process steps required to stay on a 2 year node-to-node cycle, bringing about new risks to yield and performance

Year	2003	2006	2007	2010	2011	2013
Node	90 nm	65 nm	45 nm	33 nm	22 nm	14 nm
Pitch	250 nm	200 nm	140 nm	100 nm	~70 nm	~50 nm
λ	193 nm					13,5 nm
NA	0.75	0.94	1.2	1.35		
k_1	0.60	0.44	0.44	0.30	0.25	
Common patterning solutions	Optical Proximity Correction					
		Off axis illumination with assist features				
			Water immersion			
				Double pattering		
					Source mask optimization	
						Sidewall
Common process solutions		Strained silicon				
			Air gap metallization			
				High-k metal-gate		
					Innovative interconnect	
						Fin-FET

3.1 Design Setup

Advanced technologies typically require advanced design tools, additional ID layers, etc. Because of the increasing role of the CBC approach based on the many restrictions, more emphasis may be put on parameterizing the layout. However, even with CBC, the critical aspect of DfM continues to be related to design verification.

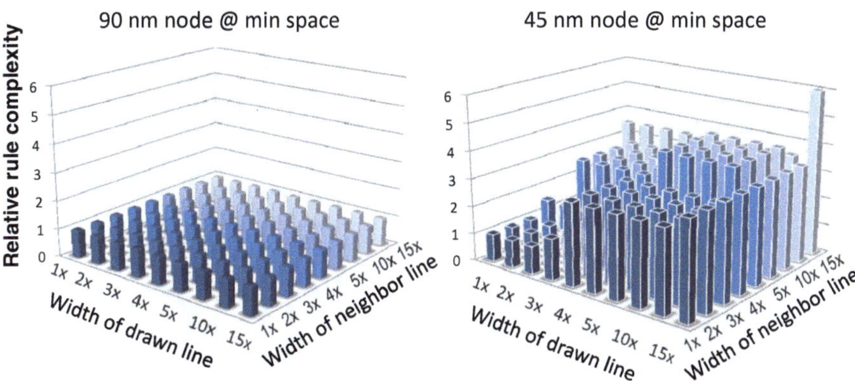

Fig. 3.1 Escalation in design rule complexity as response to layout sensitivities driven by CD's. An example of minimum allowed metal spacing as a function of drawn and neighboring widths (1× to 4×), a simple pass-fail limit has turned into a complex multi-body problem. Different shades correspond to the different rule values (after [1])

3.1.1 Design Verification

3.1.1.1 Design Rules: Where They Fall Short

Traditional design rules (DRC) provide simple solution for process and design development to proceed concurrently rather than sequentially. Design rules are first derived from competitive scaling of their precursors in older, reference technology, with simulation and optimization at the foundation of the new node. They establish control targets for process updates and to initiate layout work, and should be published early in the technology development to allow designers to begin product definition while process engineers drive the process to the control limits. On rare occasion, the committed control limits could not be achieved, adding a highly undesirable cycle of learning. This approach worked until non-monotonic layout sensitivities drove rapid escalation in design rule complexity (Fig. 3.1).

The shortfall of traditional DRC started when forbidden pitches in the patterning based on off-axis illumination required very complex multi-feature width-dependent spacing rules. DRC did not provide assurance that asymmetric width-space combinations or two-dimensional constructs would not cause yield loss while passing the check. It was also possible that some layout geometries fail DRC but would be adequately manufacturable. Rule-based design has to make compromises between being too conservative on one hand and too complex, on the other.

3.1.1.2 "Classic" DfM

Two "classic" DfM solutions respond to the tradeoff of DRC becoming either too complex or too conservative: model-based optimization and restrictive design rules.

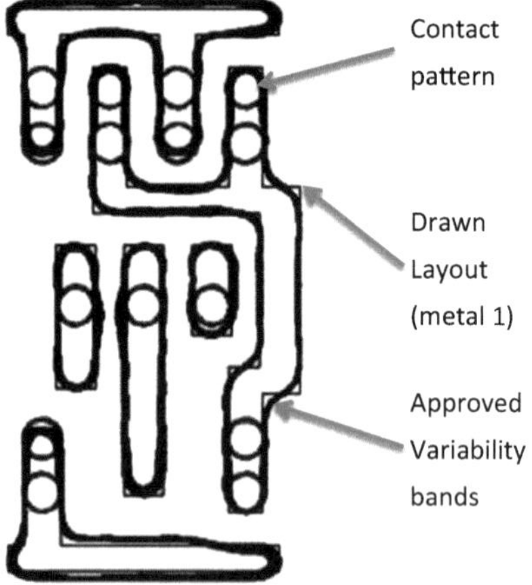

Fig. 3.2 Example of lithography variability bands of a metal 1 layout as they may be presented to a designer in interactive model-based DfM layout signoff (after [1])

Model-based layout optimization seeks to eliminate design conservatism by letting designers or design tools interactively and iteratively eliminate layout sensitivities by modeling the process response of layout constructs. Lithography response to a local wiring (metal 1) layout in Fig. 3.2 shows localized line and space narrowing. In theory, model-based layout legalization may seem like a good idea but it is infeasible for all except a few niche applications. Of the large number of process steps with unique layout sensitivities, only lithography and chemical mechanical polishing (CMP) have reasonably accurate process models. But even then, only the aerial image component of lithography (Fig. 3.2) not the chemical component of resist exposure, is based on predictive simulations of reasonable accuracy across the range of CD's and exposure conditions. Other physical and mechanical effects are typically captured in semi-empirical models that require extensive calibration. A partial set of such models, calibrated to a rapidly changing process, cannot provide an accurate layout legalization early in a technology node. Integrating model-based layout verification into a design flow would require substantial reengineering of the entire IP generation, synthesis, placement, and routing, and likely prefer escalating design complexity over increasing conservatism [2]. Mask pattern legalization for lithography and CMP modeling late in the design flow may increase design cycle due to the iterative convergence it may require [3, 4].

The alternative to model-based DfM, Restricted Design Rule (RDR) system, focuses on enhancing design efficiency while ensuring yield and performance by eliminating layout sensitivities [5]. RDR, which focuses on 'prescriptive' (rather than the traditional 'prohibitive') design rules, enables clarity in the design-process handoff by comprehensively defining all allowed feature placements (Fig. 3.3). RDRs minimize the complexity of width-dependent spacing rules. It first defines discrete placement options for narrow lines, followed by a more traditional, continuous design space with a single conservative design rule for intermediate width lines, and complete elimination of all lines of extremely large dimensions (with limited design value). In a process environment, where yield and performance no

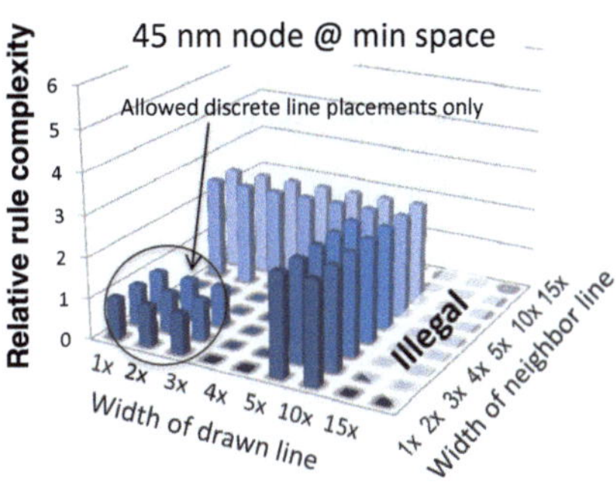

Fig. 3.3 RDR definition of the width-dependent-spacing relationship. *Dots* represent the allowed line placements (after [1])

longer improve monotonically as the rule values increase, RDRs provide targets for process optimization. Key to successful implementation of RDR-based DfM is a collaboration between the design and process teams, to ensure process development for a limited set of design rules that is actually useful to the designers [6].

The common element of "classic" DfM is that they focus on the 'layout space' after the actual design and before process optimization. This does not allow for taking full advantage of neither design nor process capabilities.

3.1.1.3 Design-Process Co-optimization

To get the best results from the high – cost IC design and manufacturing system, the desired approach is design-technology co-optimization [1]. It is achieved through technology options, starting from resolution enhancement techniques and extending to all aspects of process, device, and interconnect development, fine tuned to the unique requirements of the layout. This is not enabled by converging all designs onto simple one dimensional gratings represented by design rule values. Instead, design rules are pushed to the limit for each unique and complex layout configuration, such as SRAM cell (Fig. 3.4) to the point where the design rule is replaced by the actual construct being scaled. Parameterized variants of the cells are qualified using dedicated test vehicles to ensure understanding of yield implications. Design space allows competitive designs to be synthesized from a small set of predictably composable but limited logic constructs.

Design-technology co-optimization feasibility and benefits need now to become a part of logic designs in advanced technology nodes. This depends on:

- A competitive design generated from constrained set of logic constructs (templates)
- Predictably composable logic constructs subject to process optimization and characterization.

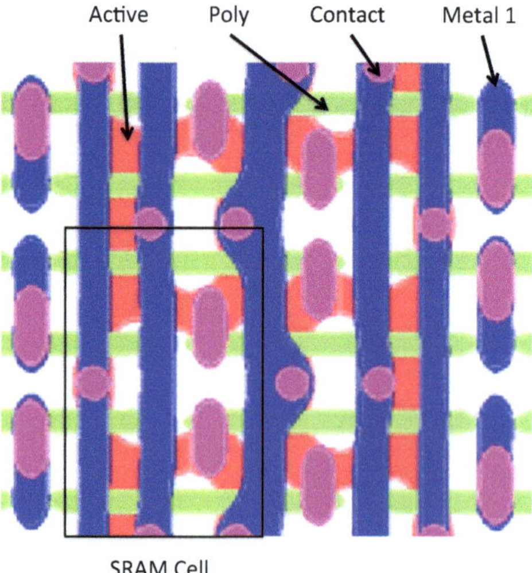

Fig. 3.4 A SRAM layout showing pattern complexity on multiple layers, subject to design-technology co-optimization (after [1])

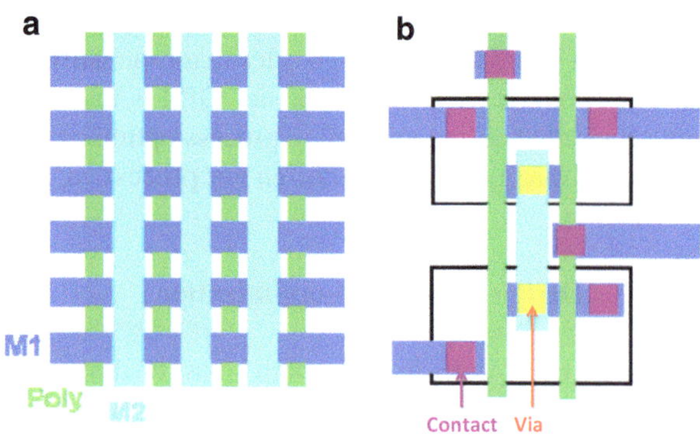

Fig. 3.5 (**a**) Layout fabric showing uni- directional fixed pitch pattern, (**b**) Simple logic cell mapped onto the fabric, showing selective track use (after [1])

One example of assessing a competitive design from a simplified set of layout constructs, is collaboration of manufacturing and DfM in the redesign of a PowerPC 405s core in 65 nm technology node [7, 8]). Starting with the original netlist, the redesign was constrained to maintain the same:

- Macro footprint and area (including reused memory blocks)
- Performance (Clock Period)
- Power (RTV/HVT transistor breakdown = 90/10 %)

The design flow began with the definition of an ultra-regular cell image, referred to as the 'fabric', onto which all logic was mapped (Fig. 3.5a) showing the constrained design grid for poly and metal shapes. A logic topology was then mapped onto design grid (Fig. 3.5b). While all features fall onto the prescribed grid, not all grid lines had to be occupied by layout features. The combination of this coarse layout grid and simple design rules yielded a fabric with:

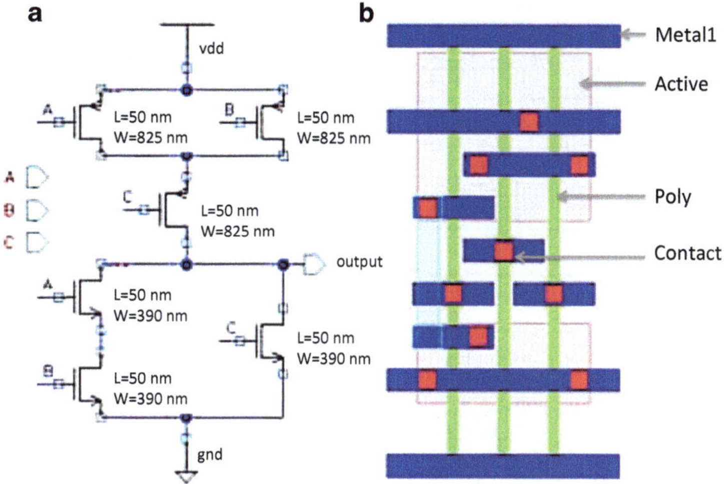

Fig. 3.6 (**a**) Schematic of a sample logic primitive, !(AB + C), representing a template, (**b**) Layout of the template (after [1])

- Limited diffusion corners
- Fixed-pitch, unidirectional Poly (vertical)
- Unidirectional metal (first metal, M1, horizontal; second metal, M2, vertical)
- Relaxed pitch for M1 (17 %) and M2 (25 %)
- Contacts and vias on grid.

The relaxed metal pitch and the predictable nature of the gridded layout enabled the use of tighter enclosures of vias by metals. Arbitrary schematic could have then been mapped onto the fabric to produce template layouts (Fig. 3.6).

3.1.2 Design Defect Reduction

The goals of design – process co-optimization are defect reduction and design RoI improvement, in several categories.

(A) Hotspot reduction
Lithography benefits of ultra regular layouts are due to simple frequency of line space for the fixed pitch, single-orientation layout without two-dimensional constructs that could lead to yield concerns. To confirm this, the conventional and the template – based layouts were scaled from 65 to 45 nm dimensions and run through process-window lithography simulations. Figure 3.7 shows the simulated lithographic process window bands for the conventional layout

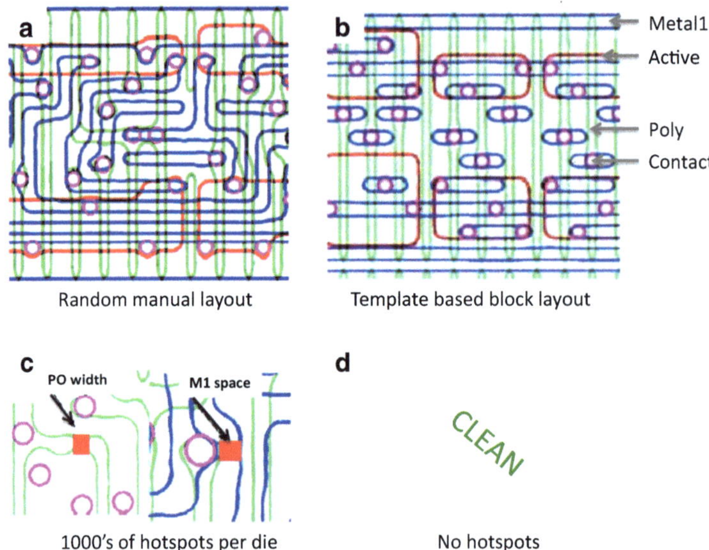

Fig. 3.7 Lithography simulations for: diffusion (*red*) poly (*green*), first metal (*blue*), and contact (*purple*) including dose, focus, and mask error: (**a**) standard layout, (**b**) template layout, (**c**) litho layout sensitivities: pinching and bridging hotspots occurring around the constructs for standard layout, (**d**) no hotspots for the template layout (after [1])

examples and an optimized block. The hotspot count reduced from over a thousand to zero testified to the improved patterning of the optimized block design.

(B) Variability improvement
In addition to process-limited yield (PLY), devices may suffer from circuit-limited yield (CLY) which reduces design RoI. The devices would not have any hard fails but may not meet timing or performance requirements. A primary factor in controlling CLY is the reduction of electrical variability. With the regularized layout, the lithography variability-bands (contours that capture deviations in exposure dose, defocus, and mask size (Fig. 3.8), showed extracted equivalent channel length reduced from 8.1 to 6.5 nm, corresponding to 10–15 mV reduction in threshold voltage variation, a very significant improvement in CLY.

(C) Area improvement
The restricted design of PowerPC microprocessor had the same footprint as the original layout. The total area occupied by sequential logic was identical in the two layout styles and the area occupied by combinatorial logic decreased by the total of 25 %, due to the following contributing factors:

- Construct-specific design rules, design specific complex gates synthesized from the primitive logic functions in layout templates,

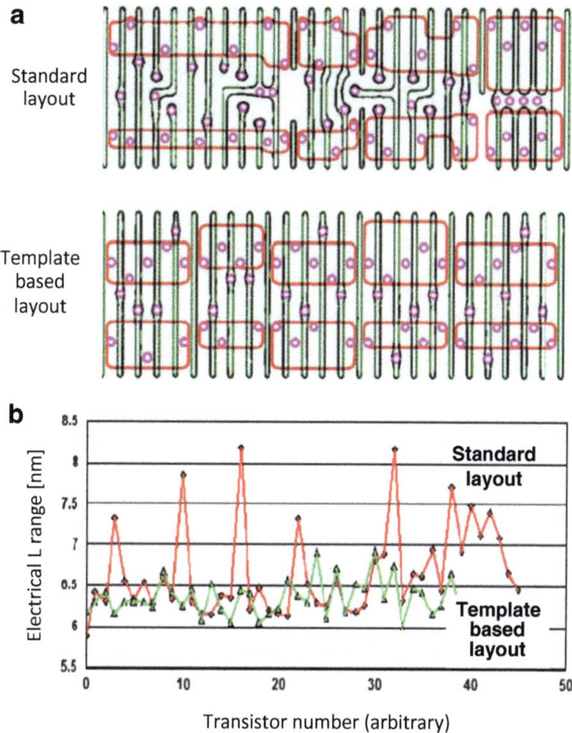

Fig. 3.8 (**a**) Diffusion, poly, and contact lithography variability-bands for random layout (*top*) and template layout (*bottom*), (**b**) Range of electrical channel length extracted for each transistor in the two layouts of (**a**). Increased electrical variability due to lithography variation of dose, focus, and mask size is visible in the standard layout (after [1])

- Optimal construction of application-specific logic cell functions. This was due to synthesizing an application specific library from a technology node to a specific set of templates, rather than using generic set of standard cells designed to cover all possible power/performance needs.

The core value proposition of regularized layout hinges on demonstrating substantially simplified set of layout primitives, i.e., the number of unique logic cells in the standard microprocessor core design versus the freeform version of those designs, as a function of the number of logical inputs to the operation (Fig. 3.9). The methodology for reduced variability shifted the design towards more complex and compact logic functions. The small number of logic cells common to both design styles further emphasizes that the reduction in cell count is enabled by the optimal use of application specific logic cells, not a constrained physical synthesis.

(D) Improved returns on litho-friendly design
Challenges in reticle/resolution enhancement technologies (RETs) to nanometer-scale designs: mask rule constraints, fragmentation, modelling, metrology errors, etc. increase silicon failure rates. Because each RET has its own layout limitations, some configurations are not easily combined in a specific RET recipe. The 'unfriendly' configurations are not only more vulnerable to process

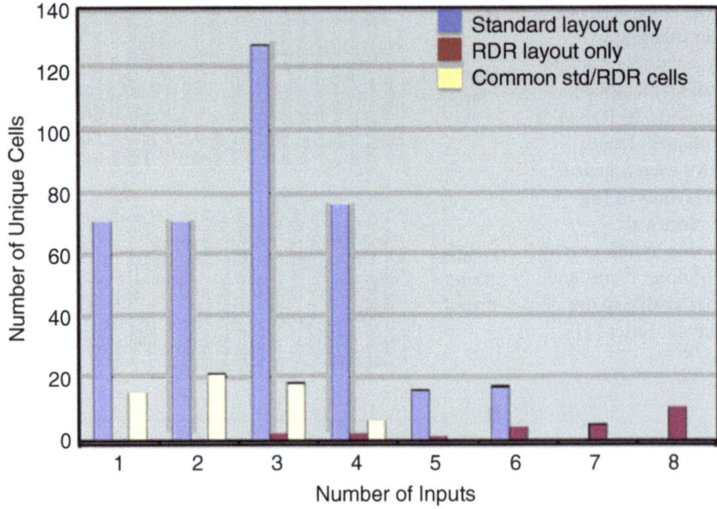

Fig. 3.9 Simplification of the design space as measured by the number of unique logic cells used in the standard versus the template design (after [1])

variation, but also slow down product yield ramp because they require diagnostic analysis and redesign.

Recommended design rules based on lithography knowledge help designers avoid sensitive configurations, and special tools let the fab perform post-OPC verification to identify printability issues before sending the layout to a mask house. But, despite these aids, designers need specific feature definitions.

3.1.3 Fabric Adaptability

Three DfM issues should be considered, when aligning the less advanced fabric with product specific internal design objectives at more advanced technology, to port the design between two nodes:

(A) Running uni-directional first metal perpendicular to the poly gates eliminates multiple diffusion contacts (feature A in Fig. 3.10) preferred for device performance and yield. The unidirectional metal adds vias for all 'wrong' way connections, which is not ideal for processes with via yield challenges.
(B) Forcing all metal to be equal width (feature B in Fig. 3.10) eliminates wide wires for power distribution, which impacts reliability on high current carrying constructs.
(C) Linking the contacted pitch (i.e. minimum poly pitch) to the minimum contacted tip-to-tip spacing of first metal (feature C in Fig. 3.10) creates a scaling problem. While pitch scaling is driven to 30 % reduction per node, tip-to-tip

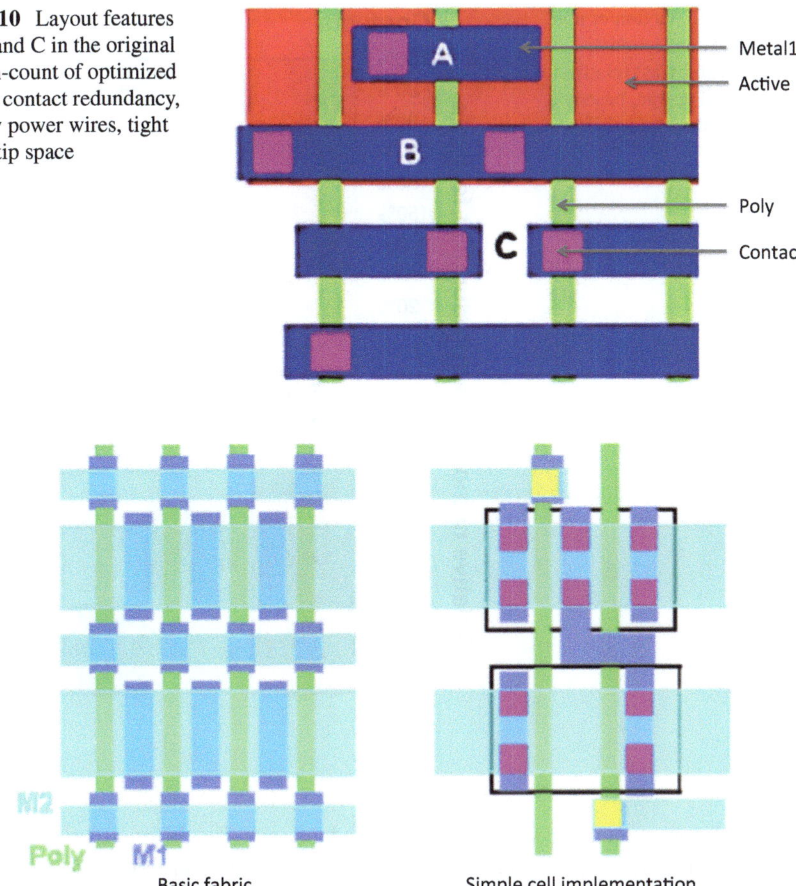

Fig. 3.10 Layout features A, B, and C in the original pattern-count of optimized fabric: contact redundancy, narrow power wires, tight tip-to-tip space

Fig. 3.11 Fabric preferred to optimize redundancy, reduce necessary connections, and allow wide power wire (after [1])

spacing has been scaling at roughly 20 % per node, forcing a decoupling of these two constructs in the overall density scaling.

A new fabric with different optimization priorities could be defined instead (Fig. 3.11), characterized by:

- Limited diffusion corners,
- Fixed-pitch unidirectional poly (vertical),
- Preferred orientation metal (M1 now vertical, M2 horizontal),
- Contacts and vias on grid,
- Relaxed pitch for M1 (25 %) and M2 (5 %),
- Limited wrong-way metal (on-grid)
- Wider power rails.

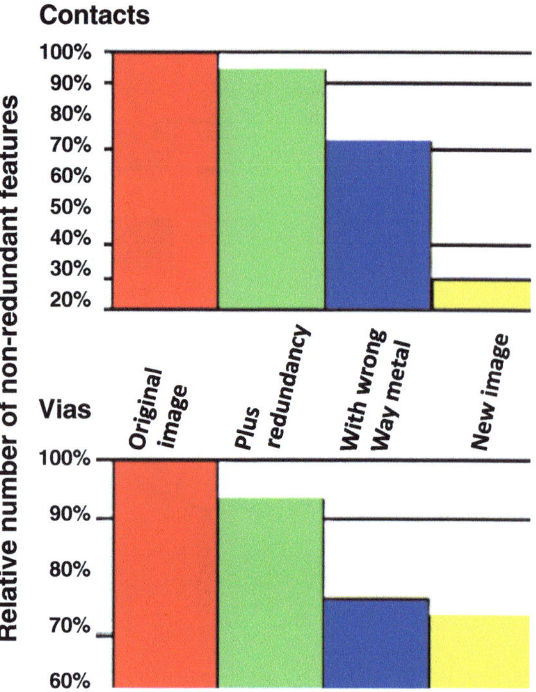

Fig. 3.12 Reduction of non redundant connections, (contacts and vias) relative to the original design, based on different fabric trade-offs (after [1])

The improvement in via and contact redundancy for different layout options (Fig. 3.12) is achieved by inserting redundant contacts and vias into the original image ('plus redundancy'), and limited use of wrong-way metal for a new cell architecture. Both 'eliminating connections' and 'adding redundancy to connections' fulfills the design intent.

Two fabric styles, one optimized for overall polygon count, the other for redundancy and wide power rails (Fig. 3.13) represent the same logic gate (NAND2) mapped onto the two fabrics. Defining the fabric with selective wrong-way metal does not affect layout density of logic simplicity. To ensure predictable circuit performance, layout sensitivities of active device parameters have to be minimized, affected by long range process interactions such as etch, stress, or rapid thermal anneal. These effects are best controlled through the regularity of the die architecture; i.e., global pattern uniformity on diffusion and poly (Fig. 3.14a). The first metal level acting as local interconnect is not a major contributor to parametric variability, making patterning hotspots the primary yield concern. The limited number of logic construct cells allows for the safe use of different layout configurations, as long as boundary conditions ensure that no new complex layout arrangement can be formed at cell edges (abutting areas). The more complicated metal patterns

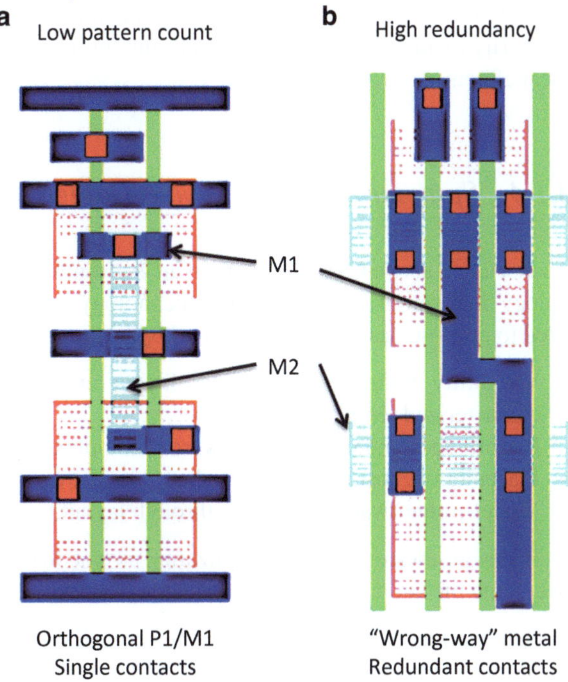

Fig. 3.13 A nand2 layout supported by 'low pattern count' (**a**) and 'high redundancy' (**b**) fabrics (after [1])

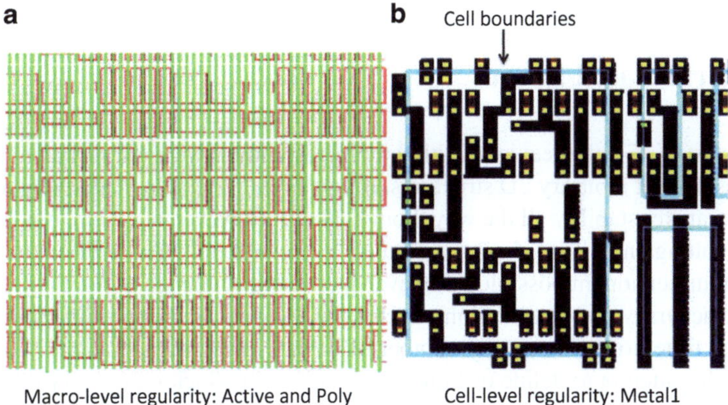

Fig. 3.14 Long range layout sensitivities on diffusion and poly minimized through macro regularity (**a**), local hotspots on metal1 prevented through control of boundary conditions (**b**) (after [1])

(i.e. staircases or dense T-shaped line-ends, right layout of Fig. 3.14b) are constrained to the center of the cell, as are 'belt buckle' constructs that require tight dimensional control for connectivity to other layers. Layout constructs within the boundary region of the cell (i.e., outside the outlines in Fig. 3.14) need to be simpler and more conservative (e.g. line-end connectivity in boundary region belt buckles is not allowed).

Predictable layout sensitivities at different length scales are ensured, through macro regularity to address long range effects and through boundary conditions to address local effects.

3.1.4 Summary of Logic Layout Guidelines Beyond 28 nm

As discussed above CMOS scaling in an environment of continuous disruptive technology innovation requires design-technology co-optimization. EDA companies need to propose design studies verified by test runs, demonstrate:

- Variability reduction by regular layout environment, incl. predictable, pre-characterized templates
- Performance and power benefits from complex logic gates
- Placement agnostic delay through regular die architecture
- Model-to-hardware correlation through layout templates.

If design-aware process optimization is not feasible, the key requirements to overcome the shortcomings of physical layout are:

- Late binding of logic to layout to preserve design creativity and performance
- Simplified layout to allow construct-driven process development
- Characterization and qualification test vehicles to drive yield ramp.

3.2 Grid-Based Layout

At the start of 45 nm, preexisting lithography techniques were not sufficient to print many useful but arbitrary 2D structures. Immersion lithography tentatively restored pattern transfer stability all the way though 28 nm. However, at 22 nm, it is going to be difficult again to print shapes accurately even with immersion lithography (and without immersion, impossible). The ever-increasing challenge of printing with reasonable accuracy a representation of what the designer is trying to implement has led, over time, to an increasing number of design rules and operations. The additional rules attempt to define or limit the situations in which it gets very difficult to print accurately. The number of operations that must be performed to implement the rules are increasing faster than the rule count itself, because also the existing rules are becoming more and more complex. A fundamentally different approach is needed to provide more ergonomic technology and design cooperation [9].

From EDA, startpoint, rather than trying to define everything the designer cannot do, the rules could prescribe what the designer can do (prescriptive design rules), and make the lithography problem better defined to solve. An intermediate approach of restricted design rules (RDRs) leads to more regular layout, based on manifested as pitch and grid restrictions. Physical verification will need to manage checking such restricted design constructs, which may be conceived in the process of technology – design co-optimization [1] but would also require an adequate DRC rule deck to verify.

Early designs allowed designer a lot of architectural flexibility. Layouts needed not be regular – designers were free to use a wide assortment of configurations and shapes and routing types, and to be very creative in reducing die floorplanning as long as they met some very simple, basic design rules. However, at lithography nodes not yielding without wavefront engineering, variability in device performance reduced the available physical design space. Especially horizontal-to-vertical transitions causing curvatures combined with alignment variability, are not printability friendly (due to the inability to print accurate rectangles) and create devices outside of the model space.

The basic concept behind restricted layout is to limit what the designer is allowed to do outside of the set of – pre-optimized cells. Designers want an assortment of features or shapes or constructs that they are allowed to use. The fab would like to limit that set – ideally, everything would be perfectly regular and repeatable, making designs much more simple to process and more robust against manufacturing aspects of variability. The restricted design compromise is to have an assortment of features or constructs, much smaller and more controlled than what has been allowed in the past. With restrictive design, future IC's will be much more regular to become manufacturable and provide the same robustness that we have today.

The traditional design regularity, allowing for a wide assortment of features and constructs being used, keeps the gates on a regular pitch. However, there is a fair amount of variability in the metal routing and the active layout.

If the same design is laid out in a more regular style, the performance variability can be decreased. The gates would not only be vertical on a given pitch, but there would be no horizontal/routing in the polysilicon whatsoever. All the active areas need to be perfect rectangles (no L's, T's, or H's), and all the metal lines need to be unidirectional going only horizontally or vertically (no jogs and no bends). Implementing these restrictions may at first constrain the design options and impact density. By technology enablement, until design is ready to switch to fully regular matrices, there would be a middle ground and transitional period where designability and design options are not too restricted, but the physical design space is more regular. This is attained through careful gridding of the layers.

A layout is gridded when the vertices of all drawn shapes within a given area are restricted in placement to coarse grid points. The shapes in a gridded layout are drawn as rectangles, paths and polygons, in two stages. In the first stage, each shape is checked for size legality (e.g., widths) and placement (gridding). In the second stage, additional rules that apply between geometries (spacing, overlap, etc.) are checked. Checking in the second stage is not performed if the layout objects do not pass the checking in the first stage.

One can define three basic types of layout objects (or shapes):

1. Line objects (polysilicon, metals): should be linear and on-pitch
2. Point objects (contacts, vias): should all be same size and preferably on pitch
3. Block objects (diffusions, implants): should be rectangular

In a gridded design, the vertices of all layout objects will be on a coarse grid which we may call the Layout Base Unit (LBU) grid. The anchors of all shapes – centerlines of line objects, centers of point objects, and edges of block objects – lie on a coarser grid called the placement grid, which is both layer-specific and orientation-specific. For example, a horizontal metal line will have a different placement grid as compared to a vertical metal line or a horizontal polysilicon line.

The LBU grid size may be arbitrary for a given technology. The contacted polysilicon pitch (CPP), wire pitches and placement grids should be multiples of the LBU grid. The coarser the LBU, the lower the number of design configurations, better for manufacturability but not for design density. A robust gridded design should effectively determine the values of the different grids by design-technology tradeoff analysis [9].

Each layer in the design will have a legal set of X and Y coordinates, e.g.:

- Poly PC X-grid – centerlines 0, 2*CPP, 3*CPP, etc.
- Metal M1 X-grid – centerlines 0, CPP/2, CPP, 3/2*CPP, etc.

However, placement periodicity is only a part of grid definition. Each layer will also have a set of valid offsets, or sub-grids. The general form for grid definition is as follows:

$$\text{<level>:<D>} = \text{O1[O2, O3...]} + \text{PG*n} \tag{3.1}$$

Parameter	Explanation
<level>:<D>	Level and orientation (direction) for the grid
O_1 $[0_2, O_3]$	Normalized offset with respect to a global origin, and O_2, O_3 ... are optional additional offsets(sub-grids)
PG	Poly grid
n	An integer (DRC code would check to determine if n is an integer or not)

Values are specified in the design rule manual for Os and PGs for all layers in both X and Y orientations. The top and bottom edges of diffusion may have different offsets (CPP ± x), but the same placement grid (CPP) (Fig. 3.15). In the diffusion grid equation, (−CPP/2-x, CPP/2 + x) + CPP*n x represents half contact size. Contact on active has a single offset of CPP/2, and a placement grid of CPP.

In the RDR approach, the rules may allow for some geometries to be only roughly specified, with retargeting to set the exact edge positions (Fig. 3.16). The valid minima for restrictive design rules are determined by contextual process restrictions or criteria (PR1/2/3). Typically, two or more PG criteria may define a single RDR (many to one relationship), e.g.:

Criterion 1. No Necking of the vertical Metal1
Criterion 2. No Metal1 tip-to-line shorting
Criterion 3. No Metal1 covering Via1 opens (overlap must be sufficient).

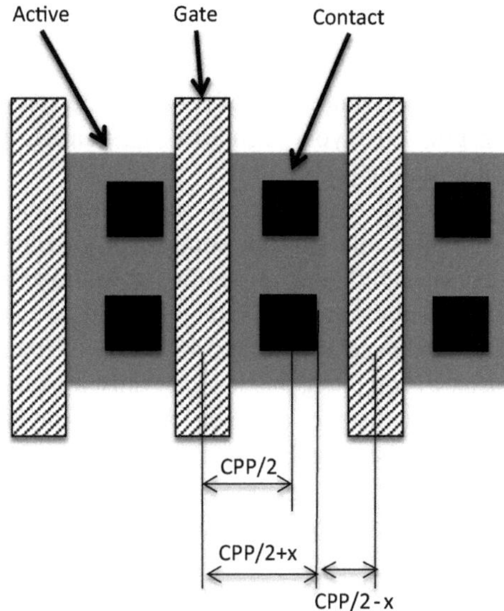

Fig. 3.15 Grid example showing poly to contact offsets

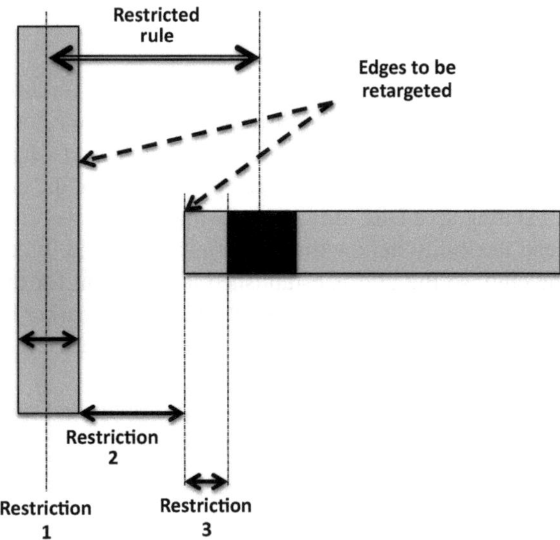

Fig. 3.16 Retargeting of edge positions in restrictive design based on 3 restrictive rules

All of the above grid definitions must be captured in the design rules and be supported by design entry tools. Because these grids are imaginary, CAD tools should provide options for turning on/off the grids on a per level basis, both at manual design entry to make the approach user-friendly, and for the automated checking tools, at verification time.

3.2.1 Grid Checking Versus Pitch Checking

A primary requirement of restricted design is the concept of checking for features "on grid." If layout regularity is essential, a grid check has to make sure that all features align on it.

A second requirement is an efficient and accurate way to check the pitch, i.e., the regular intervals. In contrast to grid, pitch is a relative measure of a feature versus its neighbors, rather than absolute grid spacing assigned to the whole layout. Pitch is a measure of optical influence, to control how features within a given interaction distance affect the printability of their neighbors. For example, in checking if a gate is on pitch, the designer must look at other gates within the specified optical interaction distance and make sure the gate is some equal increment of a pitch away. If the nearest neighbor gate is beyond that distance, then the gate does not need to be on pitch.

Another requirement soon to become critical is the ability of pattern matching – to disallow predetermined constructs placed in different layout environments as poorly manufacturable.

3.2.1.1 Grid Checking

Grid is hard-coded either for an entire chip or for a region of imaginary construct. The distance intervals don't have to be the same for X and Y. Grids for contacts, gates, and metal lines may relate to each other: if a feature is "on grid" at contact, it will align itself to the grid of the poly, such that the contacts land properly.

Setting up a DRC grid begins with defining its origin. Physical verification tools need flexibility in how the grid origin is defined. It can be defined by the extent of the chip, so the grid is established at one point for the chip and applied from that point across the whole layout, or for separate marker layers independent for the different regions. The grid in one region may not need to line up with the grid in another region.

Physical verification tools developed on design rules with manufacturing inputs must identify an array of "features of interest" to the fab and then measure them against the grid, such as:

- Polygon characteristic points (e.g., centerpoints or endpoints on one edge of a polygon must be aligned to grid)
- Centerline edge endpoints (e.g., points on the centerline of a polygon must be aligned to grid)

As a result, process engineers would be able to compare those characteristic layout points to silicon.

Another requirement are the "offsets", in which primary grid is established, but some features (e.g., edges of ends of lines) should not be aligned to the primary grid and must instead be aligned to the offset of that grid. Offset alignments occur

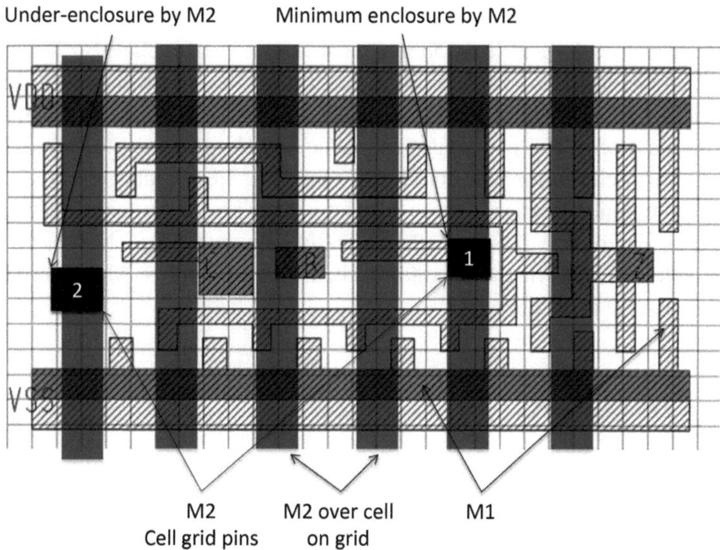

Fig. 3.17 Cell pin rectangles on Metal1 (1 and 2) must fully enclose Metal2 routing aligned to the Metal2 grid. Property values on error marker should help designers to determine how much positive or negative enclosure exists, so designers can determine how best to fix the violation (after [9])

because the primary grid relates to where the vias land. Metal lines that connect these vias need to line up to the vias on the primary grid. However, they need to enclose the vias on the line ends extending past these vias. In addition to specifying the offset to the grid for the alignment, physical verification must check whether the feature needs to be aligned to the offset on the outer side or on the inner side of the wire.

Examples of grid checks include placement rules for:

- Contact center points; must be on the X-Y grid (misalignment on one or both axes is an error)
- Gate vertices; must be on the X grid
- Gate facing active edges that enclose contacts; must be offset to the contact grid or must align with the contact grid not enclosing a contact
- M1 line ends; must extend beyond the contact by a specified offset to the contact grid.

All these new rules must become a part of DRC deck.

Determining how to fix grid errors during checking debug is difficult, since the grid is invisible and can be defined in many ways. The designer may not know how the grid is set up and how to find its nodes.

One example of grid checking is the condition that all cell pin rectangles must fully enclose at least one of the Metal2 routing grids (Fig. 3.17). Metal1 creates landing pads for the pins but because Metal2 routing is gridded, its wires will only be allowed on certain grid routing channels at the next layer. The landing pads for these pins must enclose at least one of those routing channels (grid "square"), to enable vias from Metal2 to Metal1 pin. Since this cell hasn't been placed in a design yet, it doesn't matter which routing channel it's going to be in, but it must fully enclose at least one. The challenge is that the Metal2 grid is not accessible when looking at this cell, as there is no Metal2 within it. New equation-based physical verification techniques enable these restrictions to be implemented because grid coverage (or enclosure) can be mathematically calculated. Equation-based checks can also calculate how much the edges of the pin need to move when there is a violation, to improve error debug.

3.2.1.2 Pitch Checking

Pitch is defined by adding the space between two features to the width of one of the features.

$$\text{Pitch} = \text{feature space} + \text{feature width} = \text{x} \tag{3.2}$$

Pitch is relative for the feature in respect to its neighbors. It differs from grid-based constraint: beyond a certain distance, pitch can be ignored. For example, the pitch may be correlated to the optical radius (influence distance of the reflected light during the optical process). By allowing features beyond the influence distance to be "off grid" in relation to the features within the influence distance, it permits for a less-regimented layout for the designer without affecting printability. For optical radius at ten times the pitch, features within ten pitches of feature of interest are expected to influence its print quality. Features beyond that radius are assumed not to affect the feature of interest. Because everything within the ten-pitch distance is expected to have impact, the neighboring features need to be placed on pitches from the given feature so that regularity and resolution are established in the light/dark contrast of the stepper.

Features considered as grouped i.e., within an optical radius of each other, are pitch-restricted only in relation to each other. Features in separate groups (i.e., outside of the optical radius) do not need to be constrained by common pitch.

Sample checks of features on relative pitch (Fig. 3.18) include:

- Metal routing wires within an optical radius
- All gates within the same active area
- All contacts within a specified interaction region.

Pitch has to be setup by CBC methodology. Relative pitch is not readily visible to the designer. If a pitch error occurs, the designer must know what the relative pitch is, and determine how a feature must be moved. There are no pitch markers on the layout, because it is mathematically calculated "on the fly" relative to a feature's neighbors, which makes pitch very difficult to visualize. When pitch errors occur, at the current level of software support, designers first need to physically measure features, then calculate the pitch, and then try to locate the pitch failure.

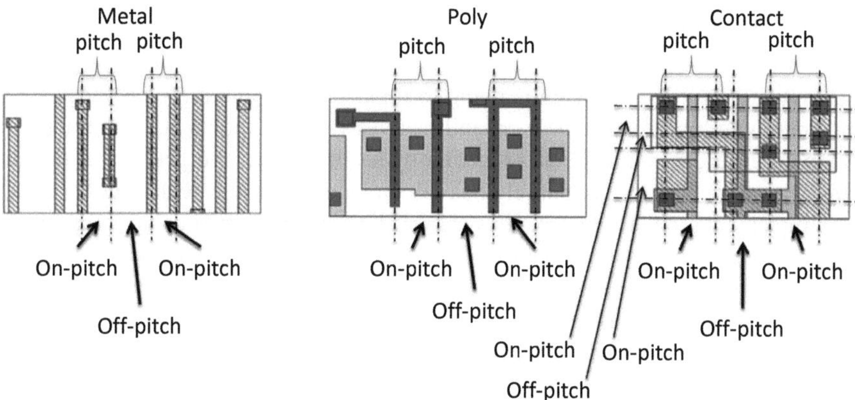

Fig. 3.18 Sample pitch checks for metal, poly, and contacts

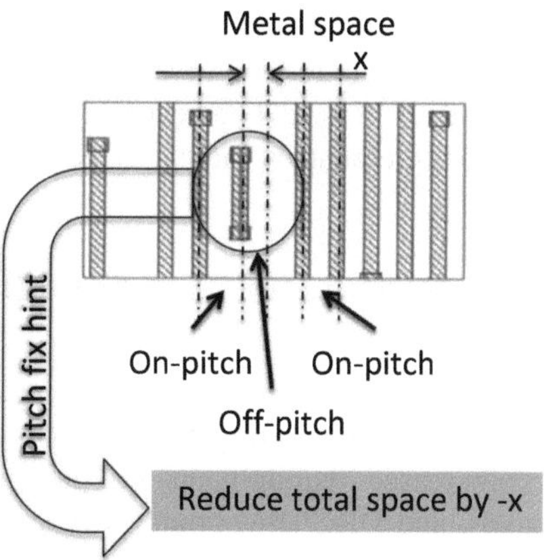

Fig. 3.19 Guidelines for pitch correction to help automated pitch checking

A DfM friendly PV tool not only needs to calculate pitch and perform a variety of pitch checks, but it also must be able to provide visual feedback to fix the errors, such as:

- Incorrect centerline to centerline distance of two features (physical measuring by designer)
- The percentage of the calculated pitch represented by the first measurement differnt from a multiple of 100 %
- If different from N × 100 %, this calculation is the percentage that the feature is "off pitch" (to sort and prioritize pitch errors)
- Calculated distance for the feature to be moved to be on pitch.

Equation-based physical verification should allow not only mathematical calculation of pitch, but show a designer how to calculate how much the centerline of the wire must be moved to fix the violation (Fig. 3.19).

3.2.2 *Summary*

In summary, grid-based layout is going to impact design and physical verification methodologies and require new capabilities in EDA/CAD tools at 22 nm node or below to ensure continued cost reduction by scaling. Pitch and grid are critical to effective restrictive design, but rely heavily on mathematical calculations for accurate implementation. Therefore, physical verification tools not only have to perform the design rule checks associated with pitch and grid, but also supply the designer with the information needed to debug and correct any errors associated with these checks. Prior generations of DRC did not have this facility (Table 3.3).

This approach is intended to reduce layout complexity in regard to the number of random rules and improve manufacturability. However, it quickly builds up new rules for layout regularity, which were not part of the system before. Trade-offs in design methodology, as well as an increased abstraction of a design rule in mathematical sense, from simple one-dimensional measurements to multi-dimensional mathematical relationships, would impact traditional capabilities of verification tools. Equation-based checks and advanced grid definition make restrictive design implementable today. Advances in error debugging, including mathematical property calculations and nearest grid point markers, have to enable the designer to meet new requirements without sacrificing time-to-market goals.

3.3 Routing for 28 nm and Below

The next level up in the layout hierarchy from the template – or RDR – based cell definition is the routing methodology.

Increased requirements for design logical functions and databases, multiple goals, and DRC/DfM complexity are creating significant routing challenges at or below 28 nm that require new approach to streamline:

- Global routing estimations
- Iterations of physical signoff and engineering change orders (ECO)
- Very large (one billion transistor) designs with multiple optimization objectives.

All these factors impede the quality, time-to-market, and cost targets. For 28 nm and below, designers must adopt new routing technologies that can solve for multiple design objectives within the scope of tool capacity, memory footprint, and runtime.

3.3.1 *Routing in Physical Design*

3.3.1.1 Restrictions, Resources, and Signoff

Design and DfM rules correct the layout for the parametric, systematic, and random manufacturing defects mostly related to the lithography process. The number of rules has doubled between the 80 and 28 nm nodes to ensure layout features known

Table 3.2 Examples of incremental design rules per-node required by foundries, which challenge signal routing (after [10])

Rule	130 nm	90 nm	65 nm	45 nm	32 nm	22 nm
Width based spacing	1-2	2-3	3-5	7	7+	7+
Min area	1 pitch	2 pitch	3 pitch	5 pitch	5 pitch	5 pitch
Cut number (Via)		1-2	4-5	5-6	5-6	5-6
Min step OPC		1	5	5	5	5+
Dense cell OPC	–	–	M1/M2	All layers	All layers (critical)	All layers (critical)
Pinch (OPC)	–	–	–	–	4	4+
Fat jog (OPC)	–	–	–	–	3	3+
Long (bar) via					Cut-to-cut	Cut-to-cut Cut-to-metal
Uni-direction rules	–	–	–	–	–	Yes
Discrete width spacing based on orientation	–	–	–	–	–	Yes
Double patterns	–	–	–	–	–	Yes

Table 3.3 Technology – dependent layout construction and verification

Technology		90	65	45	32	28	20
Construction	Template cells %	10	30	50	60	80	90
	Des-tech co-optimization	Memories			All circuits		
Verification	DRC	x	x	x	x	x	x
	Simulations		x	x	x	x	?
	RR (recommended rules)			x	x	x	?
	DRC Plus			x	x	x	?
	RDR (restricted design rules)				x	x	x

to affect yield are not introduced into the design (Table 3.2). Model-based DfM defects are more subtle yield limiters. While design is instructed to put layout regularity defined by RDR at the top of the architectural recommendations, for 28 nm, the foundries from their perspective need to continue to provide "recommended rules" (RR) in addition to the mandatory DRC rules. Their priority would reflect its relative impact on manufacturability, and add to RDR design rules [10].

Before committing to layout routing tools to perform a 'global routing' which estimates the available resources, it is important for these estimates to be accurate. This means more than counting the number of routing tracks across the chip that meet minimum spacing requirements. One must take into account resource requirements such as the effect of vias and stacked via arrays, blockages, staggered macros, design rule compliance, and signal integrity (SI) requirements like wire spreading, wire widening, and shielding. There may be only few RDR restrictions, but many recommendations supporting layout at this level.

(A) QA sequence

A key challenge at 28 nm is the decoupling of the routing and the layout signoff based on the quality check by verification engines. Typically, a router uses simplified DRC and DfM models for the trade-off between runtime and accuracy. Once the implementation is complete, the GDSII file is verified using signoff-quality DRC/DfM models and Standard Verification Rule Format (SVRF) rule decks. Historically, this approach worked well because the number of violations discovered at signoff was low. However, as technology rules get more restricted, there can be a growing number of DRC/DfM violations, and changes made to fix them can iteratively lead to new manufacturing violations.

Accordingly, there is a growing difference between the rules used during design of the routing and in the signoff. As a new process node matures, the foundry rule files in the SVRF language used by sign-off engines are constantly updated to address manufacturing issues. Consequently, the foundry signoff models and rule decks are intrinsically the most accurate and complete representation of actual manufacturing requirements. The rules used by the place and route system are simpler and may fall out of sync with the foundry rules. Further, at 28 nm and below, there are some rules that cannot be expressed in the simpler LEF language. As a result, the router will report the layout to be DRC/DfM clean, but signoff analysis may find a large number of issues.

DfM techniques, including metal fill/CMP, litho, and critical area analysis, are starting to affect the traditional design metrics like timing, power, and signal integrity. Usually, there is no automated way to repair DRC/DfM violations, and the flow requires the transfer of huge ASCII files between the implementation and signoff environments, which slows down design process. The design-then-verify flow that has worked in the past is increasingly unmanageable and unpredictable.

(B) Performance

Twenty-eight nanometer routers should support both gridded and non-gridded approach and use a universal connectivity model. It should also support sophisticated non-default rules (NDRs) and DfM requirements for advanced nodes including recommended rules, redundant vias, wire spreading and widening, and timing-aware metal and via fill. Because of the large physical size of many 28 nm ICs and SoCs, routing tools need to use multiple cores (CPUs), and physical memory very efficiently (Fig. 3.20).

The router should be able to reduce the runtime in spite of the increased number and complexity of DRC rules, by minimizing the number of operations during routing. The algorithms should use full DRC/DfM models for improved accuracy and minimum violations during post route optimization and signoff. The DRC should be based on polygon shapes rather than edge-to-edge checks to enable complex 28 nm rules effectively adhered to. The router should support both hard and recommended (soft), rules pre-prioritized by technology and manufacturing, and set up corresponding rule priorities for automatic routing repair.

A timing – and congestion-aware 3D global router is essential to determine the routing layer resources. It should use modeling technologies that ensure

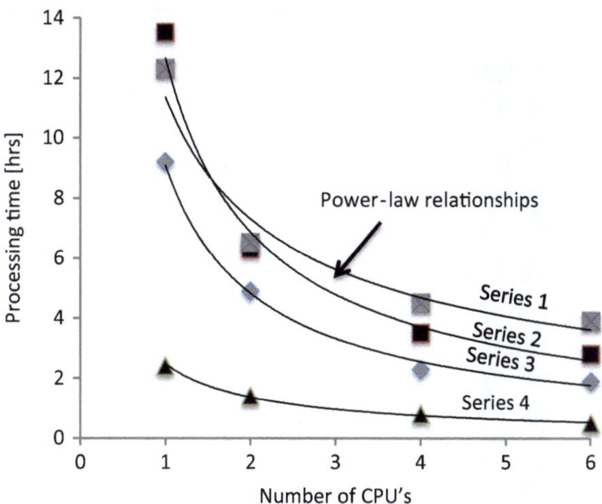

Fig. 3.20 Routing speedup with multi-CPU runs (after [10])

Table 3.4 Example of convergent routing flow

High capacity	Signoff timer	Variation analysis	Shape – based DRC engine	Global routing Track routing Route – based optimization Search and repair final routing Manufacturing optimization

that the resources required for vias or stacked via patterns are accounted for. The router should also be multi-corner, multi-mode (MCMM) aware and include signal integrity (SI) as an inherent cost function to achieve the best PoR and faster convergence. Routing engines (global, track, and final) driven by dynamic native SI analysis allow for faster design closure. Static signal integrity SI and proxy models, instead of native analysis, can be slower and less accurate. Newer routers have SI costing native to the routing kernel, which allows for dynamic, incremental, MCMM SI analysis. Using incremental on-the-fly extraction, polygon-based DRC analysis, and MCMM timing lets a router make quick decisions on issues such as increased spacing, wire spreading, and rerouting for critical nets, to achieve the goal functions (Table 3.4).

(C) Signoff

Physical signoff should be directly invoked from the place and route environment, to allow the router natively perform SVRF-based DRC and DFM analysis and allow designers to find LVS and DRC problems caused by mismatches between GDSII and abstract views.

Routers have an efficient and scalable data model to address growing design sizes at smaller nodes. The number of operations the router must perform at 28 nm is nearly four times more than at the 65 nm node. To maintain the routing runtime, clustering and filtering rules may be used. Rather than applying

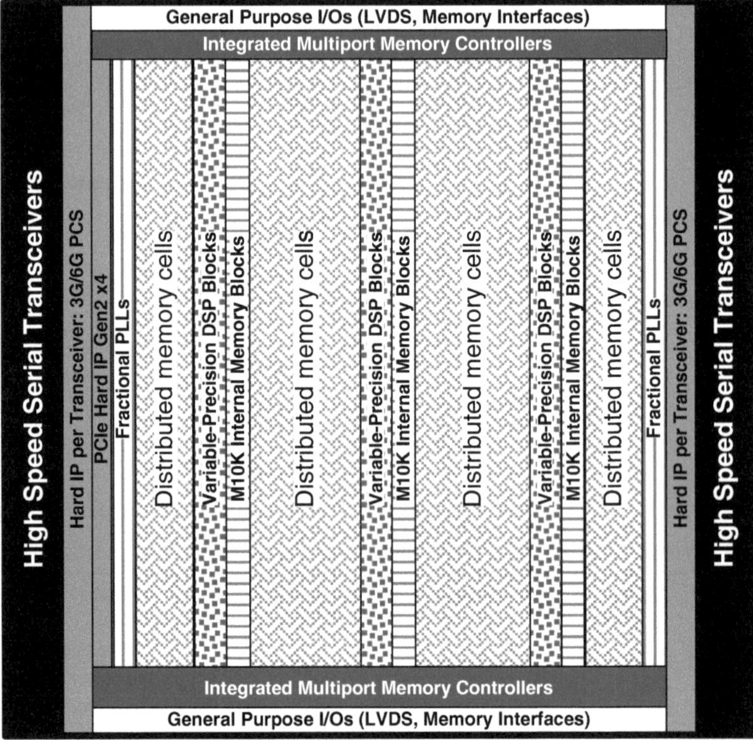

Fig. 3.21 Top-level device view of Aria V architecture (after [11])

each rule separately, a tool can detect rule commonalities and group them for processing.

With an efficient use of multiple CPUs, a significant speedup that can be achieved for different CPU configurations when the P&R architecture has an efficient data model and is built for maximum parallelism. Overall, routers for 28 nm must offer a flexible and powerful architecture to achieve optimal PoR across all design metrics in the shortest time.

3.3.1.2 DfM for Routing for Timing

Timing closure is of critical importance e.g. in high-speed FPGA designs. The routing congestion (local and global) and the ability to accurately time the logic to avoid violations caused by skews within the clock network, depend on the architecture and layout of the FPGA. DfM based on HDL coding shows how they apply to designs in 28 nm technology [11].

(A) Finding the shortest path
For an example abstract top-level view of FPGA architecture (Altera's Arria V, Fig. 3.21) laid out in a matrix of columns and rows, I/Os are framing the top and

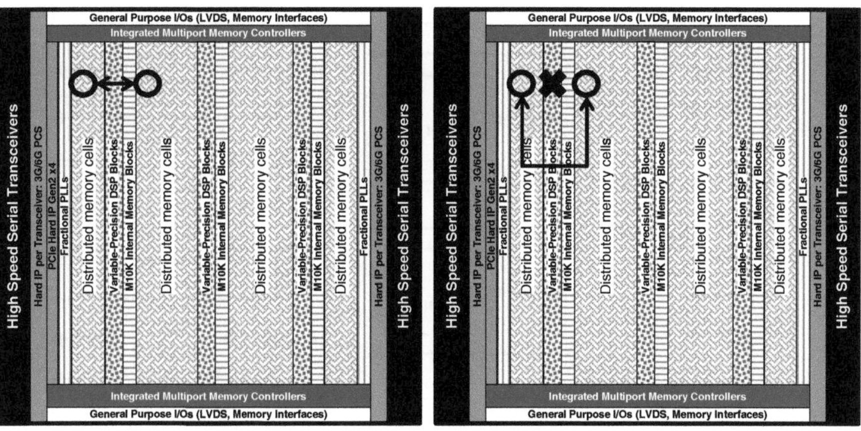

Fig. 3.22 Path switching due to congestion in an FPGA (after [11])

bottom of the device and serializer/deserializer (SERDES) transceivers are covering the left and right sides. Fractional phase-locked loops (PLLs) are colocated on the sides with the transceivers and in the center of the FPGA. Digital signal processors, adaptive logic modules (ALMs), and memory are distributed in regular columns throughout the FPGA. This column mapping of resources applies to the entire range of 28 nm products with the mix of resources (columns) dependent upon the FPGA features.

The layout of the FPGA is a regularly repeating matrix of resources, accessed via a mesh of interconnected fabrics, i.e., connected together through a series of horizontal and vertical pathways. In Fig. 3.22, the fitter tries to choose the shortest available path through the device but that path may be blocked because it is used by other elements in the design.

The fitter therefore acts like the GPS used in cars, trying to find the shortest path through the network by choosing between different paths based on available resources and constraints. The path selected will vary depending on how many paths it must choose from and the distance it must go.

(B) The highway is not always the fastest way (retiming the logic)

In 28 nm FPGAs, the amount of resources that can be placed within the mesh has grown dramatically. Devices are now approaching 1 M Les (Logic Elements) but the underlying interconnect mesh remains the same. This means that the cell delays have gotten shorter while the interconnect delays have gotten longer. Using the GPS analogy, the main highways (interconnects) are often congested because, as technology scales, there are so many more devices competing for access. Using local streets (local cell connections) can be much faster.

One way to avoid congestion is to collocate critical resources on critical pathways within the same cell. The shorter the distance the signal must travel, the less likely the fitter must use the longer interconnect lines that may be congested and thereby cause delays on the path. This is known as retiming the logic.

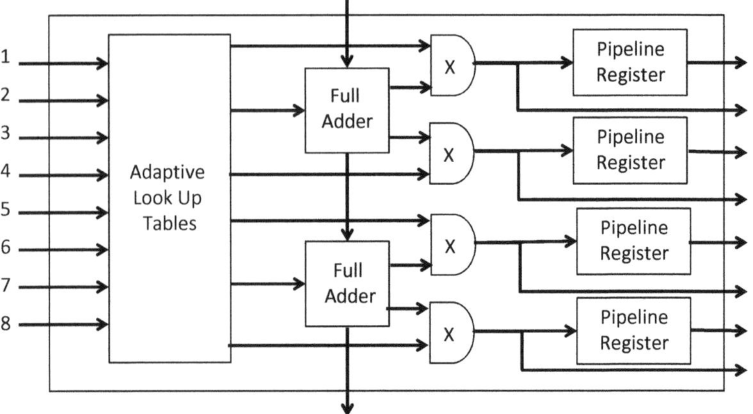

Fig. 3.23 An example of 28-nm adaptive look-up table (after [11])

The best way is to break up large chunks of logic and limit the number of dependencies in the HDL code. A good indicator of whether there will be timing issues, is the number of layers of logic between points in the timing paths. The more levels of logic, the more likely the resources to complete the function are spread out across the device driving heavier use of interconnect paths needed to reach them.

(C) Don't drive during rush hour (pipeline aggressively)
Using the GPS analogy again, the time to get from point A to point B in a congested area is dependent upon the time of day. During the end of a workday, a driver will encounter traffic. That same path is free of traffic at midnight if the driver can wait till then. In FPGA designs, the best way to manage congestion is to pipeline the design.

Pipelining a design adds latency in the system but avoids timing problems by changing the relationship between the clock and the data to ensure that they arrive at the destination at the same time. Pipelining can reduce the effect of longer interconnect delays not seen by the clock. Clocks travel on a separate network than data, so timing problems can occur if the data comes before or after the clock expects it (set-up and hold violations). By inserting pipeline registers into the path, the signal is retimed so that both the clock and the data appear to the next piece of logic at the same time. This is the preferred approach to handle long interconnect delays and an easy way to avoid timing violations.

Altera recognized the need for more pipelining in its larger FPGAs and has redesigned its LEs and adaptive look-up tables (ALUTs, Fig. 3.23) to include an additional two registers, for two reasons. First, the additional registers provide more resources so that normal logic functions are not impacted by the need for an additional pipeline register. Previously, an ALUT was not utilized when a pipeline register was allowed to retime a design. Second, this change allows for collocation of the pipeline register with the logic it is intended for, which is important because

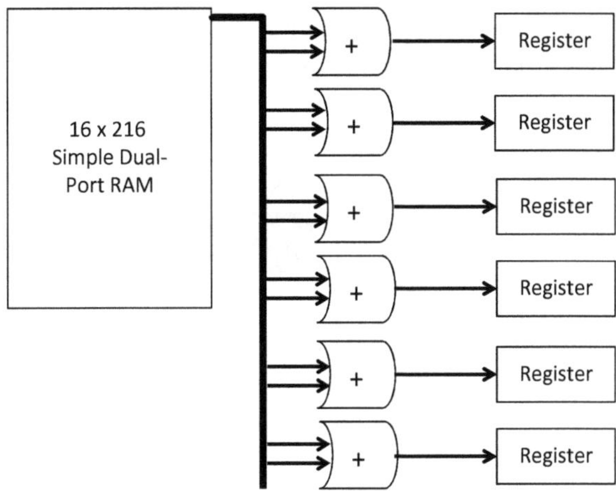

Fig. 3.24 16×216 simple dual-port RAM (after [11])

the pipeline registers remove long interconnect delays. Pipeline registers not collocated with logic would introduce yet another interconnect delay as the data must go to a register – which may be located at the other end of the device – and come back to the ALUT. The additional registers ensure that the pipeline register can be placed close to the logic, thus minimizing delays in the system.

3.3.1.3 Pipelining and Congestion

As pipelining example, a 16 × 256-bit dual-port RAM can be considered (Fig. 3.24). Due to high fanout of the RAM to multiple LEs, the speed on this path was restricted to 255 MHz. The analysis of the design, shows that the interconnect delays between the RAM and the adder account for 54 % of the total delay in the system.

The insertion of a single pipeline register before the logic (Fig. 3.25) eliminates much of the interconnect delay, allowing the design to run at 325 MHz. This difference of 70 MHz is a 27 % increase in performance [11].

(A) Carpool if possible (avoid high fan – out nodes)
   In logic circuits, often the same signal must arrive at multiple locations at the same time. Since the signal is launched from the same source, the fitter must find a path for all the copies of the signal to ensure that. This type of situation is called a high fan-out node. It may become difficult to ensure all the signals arrive at their destination at the same time because all copies cannot use the same routing. The best way to is to use manual or automated node replication in the tool such that 20 copies of the same signal appear on the same logic at the same time.
   The design is broken into two stages. The first stage is a 1-to-4 fan-out, and the second stage is a 1-to-5 fan-out. Adding the two stages together still provides the equivalent of a 1-to-20 fan-out of the signal, but with two separate and more manageable segments driving four or five paths through the FPGA rather than 20.

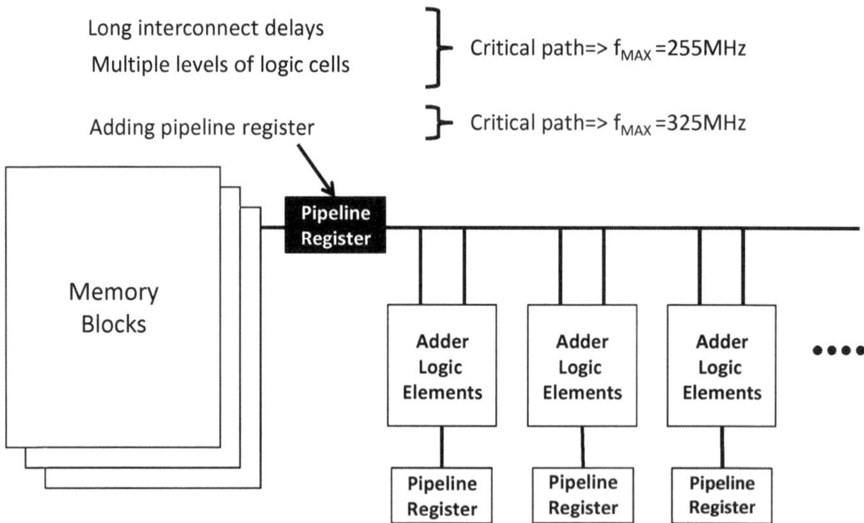

Fig. 3.25 Path analysis of simple dual port design (after [11])

(B) Reduce congestion
While it is important to retime the logic to avoid timing issues, one should optimize product architecture for congestion in the FPGA, caused by too many LEs competing for the same set of clock resources or interconnect lines. Because the fitter must trade off the needs of the different LEs, the more LEs are used, the fewer available options the fitter has for choosing optimal paths. It will be more difficult for designs 85 % full to achieve timing closure than for designs only 50 % full for congested designs up to 90 % full, improving timing may be done by partitioning and the management of clock networks [11].

(C) Think globally but act locally (use partitioning and hierarchical designs)
A hierarchical approach in FPGA design has many advantages. It allows for partitioning of the design into smaller blocks, for the use of incremental rapid recompile only on a portion. It helps the fitter focus only on the section of the design that may have changed, simplifying the routing while the rest of the design remains unchanged.

By focusing the fitter, timing closure becomes easier because the fitter does not need to consider all the paths in the design. When a user compiles an entire design, the fitter has to take into account all dependencies. This can lead to different routes every time the design is compiled and produce different timing results. Since the fitter does not know what was changed in the design, it will assume everything has changed and will use its algorithm to try to refit everything.

In addition, the overall design compile time will decrease when using rapid recompile because only the section that was changed is rerouted. A partition can be as small as a single ALUT and as big as the entire device.

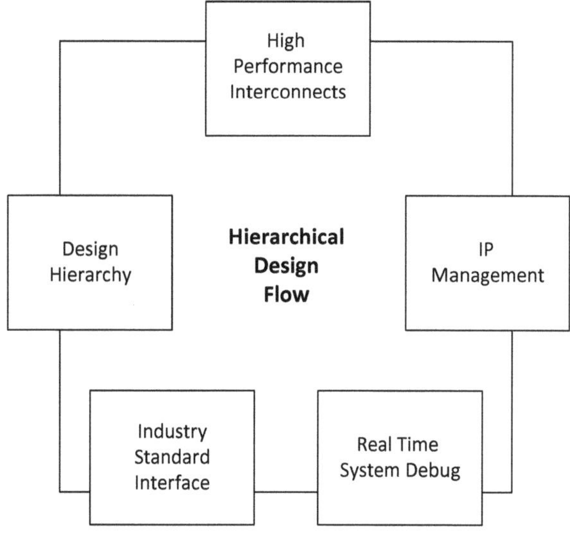

Fig. 3.26 Components of an integrated, hierarchical design/DfM flow

3.3.1.4 Traffic Management

A second structural method is based on hierarchical design to allow for the use of system test bench tools to quickly and easily connect different blocks of intellectual property (IPs) and integrate their debugging for virtual testing of the entire system using the ability to gauge performance and functionality of the design. Partitioning, i.e., breaking IPs into blocks for integration, enables hierarchical design flow, design reuse (IP can be preserved in blocks with the same routing), performance, and design to work at different locations on different parts of the product (Fig. 3.26).

(A) Making sure trains run on time (manage clock skews and clock networks)
For the different types of clocks e.g., FPGAs: GCLK, QCLK, PCLK, SCLK, ROWCLK, and VIOCLK, each clock from ROWCLK to GCLK covers a progressively wider area of the device. Unlike the interconnects, which were to avoid longer lanes (or highways) due to the amount of resources contending for access to them, the clocks are few in number. The larger clock lines such as GCLK allow quicker propagation, which limits clock skew and improves both set up and hold timings.

To get the clock signal from PLL to the internal resources as quickly as possible, means immediately transitioning from a local clock network such as a ROWCLK to a higher clock network such as QCLK or GCLK (clock promotion) which the fitter will attempt to do automatically. Where the clocks are restricted due to a software design constraint (SDC) file, the user must ensure that the clocks are promoted, within the regions they cover (Table 3.4) over the clock network (Fig. 3.27, Table 3.5).

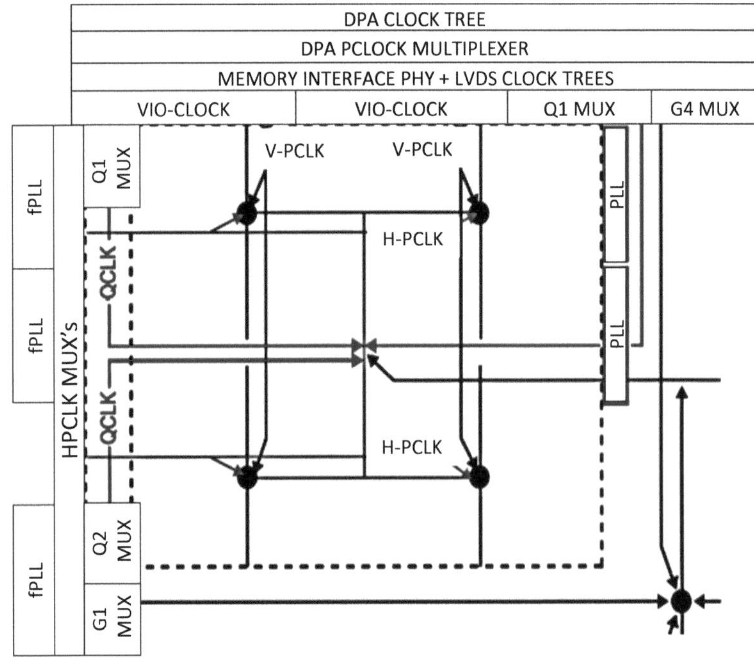

Fig. 3.27 Example graphical view of Clock Regions, PLLs, and clock multiplexers (after [11])

Table 3.5 Examples of different clocks and clock regions in 28 nm devices (after [11])

Network name	Clock region coverage/usage
GCLK	Device-wide network
QCLK	Quadrant
	Side-wide
	Spine-segment
PCLK	Spine-segment
	Vertical spine segment
	Horizontal spine segment
SCLK	Access 1/16 of core (4-spine segment device)
	Access 1/8 of core (2-spine segment device)
ROWCLK	Access core resource: LAB/M10K/DSP and HI0 register
VIOCLK	Access VI0 register

(B) Ask for directions (automated timing closure)
Even with the best planning and HDL code, closing the timing is a problem with large designs. Routes need to be identified to benefit from timing enhancements. To establish, which enhancement would help the best in any given situation, would be difficult.

Automated Timing Closure Analysis tool is designed to provide a set of recommendations what changes would most likely help with that circuit. It is often unclear where to start looking at timing failures, e.g.:

- Large clock skews
- Restricted optimization where the software was not allowed to retime or duplicate
- Unbalanced logic to benefit from retiming the path
- Region constraints where nodes on the path are locked to non-overlapping regions
- Partition constraints when a path crosses partition boundaries
- Too much logic for the given timing constraint
- Use of control signal paths on critical logic
- Reduced optimization because the fitter did not see this path as critical
- Interpath competition

The tool explains each problem and ranks recommendations so that the user can focus on those that have the biggest impact on the timing.

3.3.2 *Litho-aware PnR*

To avoid congested design configurations, PnR engines can be optimized for litho issues, such as via cells shape, routing uniformity and preferential direction, etc. [12]. This optimization can be made through lithographic simulation of Calibre LFD Kit, in order to have an early freeze of PnR settings. Because routing is automatically generated, rules cannot anticipate all the layout configurations with litho issues, and PnR hotspot detection and repair needs to be developed. The flow (e.g. Fast LFD Kit based on Calibre), contains an initial phase of hotspot detection and correction at the PnR step, and a second phase of litho simulation on corrected zones with correction rate verification for signoff.

Hotspot configuration is difficult to anticipate because it usually is a combination of problems in design density, clusters of minimum design rules, interacting layers, OPC engine limitations, 2D design configurations, unusual polygon shapes, etc. A design configuration would give rise to a hotspot not only because it is harder to print, but because the solution to improve its printability is not simple at OPC. The main issue with hotspots is that their correction is time consuming, especially for cases that need a specific correction. As a result, a hotspot is a design configuration with a lithographic issue such as difficult printability, small process window, etc., detected by post OPC simulation, requiring a non-negligible time of OPC script development.

Even if hotspots could be identified early by litho simulation, the best way is to avoid them at design level. OPC treatment for design configurations has become quite aggressive to improve patterning fidelity. Using DfM tools can help avoid hotspot configurations that require aggressive OPC modification possibly impacting key design parameters such as timing.

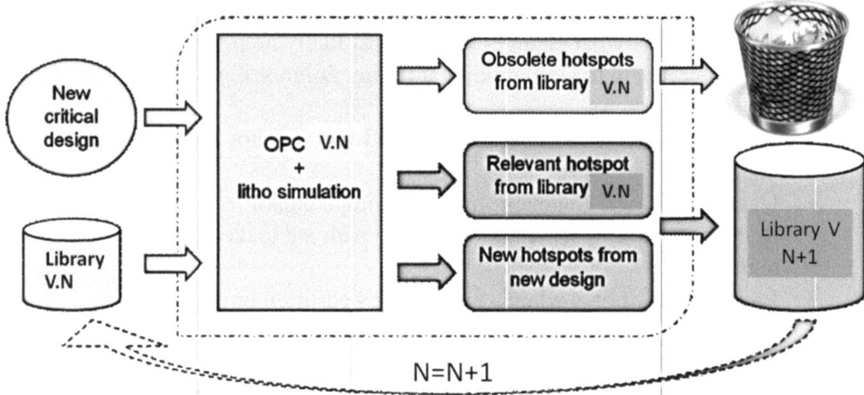

Fig. 3.28 The procedure of updating hotspot library (after [12])

To reduce production run time, the flow must focus on hotspot configurations that have already been examined, simulated with OPC tools, and flagged as critical. A hotspot library specific to a layer is required since libraries of different layers can be very different and have to be constructed at OPC level with optical model simulation. The goal of the flow is not to solve all of the litho difficulties of all designs, but to prevent the inclusion of critical and time consuming problems into layout. Improvements of PnR engines can remove some known litho-critical routing configurations, but new ones would arise. It is important to consider a hotspot configuration as a litho issue with a finite lifetime. As a result, libraries of hotspots need to be versioned and updated to maintain their relevance (Fig. 3.28). A flow may have a semi-automatic hotspot library construction phase to reduce the time needed for updates as much as possible.

3.3.2.1 Hotspot Coding

For every simulation issue, a layout environment surrounds the edges directly implicated in the hotspot. An intuitive way to describe hotspot configuration is to create a set of rules associated with the litho conflict. As DRC rules are widely used in PnR flow, coding DFM specific cases can appear simple and easy to deploy. However, proximity effects described by optical models of OPC engines are difficult to translate into design rules [13]. In fact, it is possible to find two configurations similar to each other but yielding different simulation results. For this reason, a hotspot would be represented by a cluster of rules which become very complex and may be not complete. This leads to a risk that the detection captures more configurations than needed. By definition, a rule will capture the exact environment described, which also means every other configuration that complies to this rule. If the configurations are similar to each other, it is safe to suppose that the simulations will be similar. However, there is no guarantee that every detected configuration will have a critical simulation value.

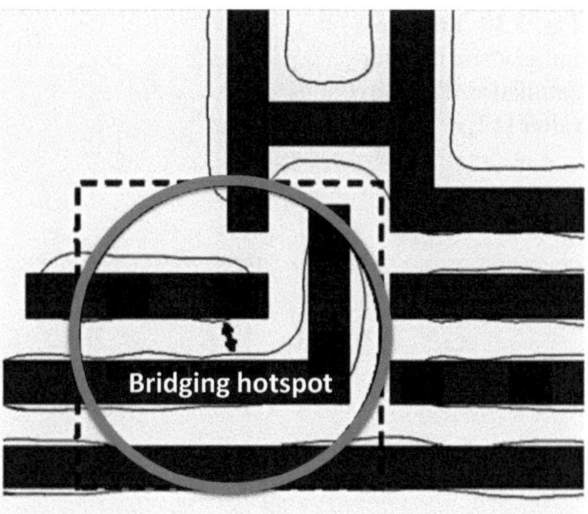

Fig. 3.29 Influence window of the neighboring pattern (dashed line) on the emerging hotspot (after [12])

Consequently, we can predict that a rule set will cover a certain percentage of realistic hotspots, plus a number of false defects.

One approach to avoid false defects is pattern matching. Hotspot description should cover every polygon within a certain window (e.g. dashed square in Fig. 3.29) surrounding the hotspot and making sure that polygons outside this window don't influence the simulation value. The window size is the only parameter that ensures that detection captures only configurations that will reproduce the original simulation problem, depending on the design, the optical model, and the OPC script. This approach is called exact pattern matching. The real shape of the influence window is very complex to evaluate so it is assumed to be a square for practical reasons. The circle illustrates the real influence window, and the dashed square represents the implemented window. Too large window size will result in pattern selection that will miss patterns with litho issues during pattern matching whereas a too small window will result in redundant pattern selection with majority of patterns not having litho issues.

Because similar configurations are expected to have similar simulation results, the introduction of fuzziness in pattern matching will allow a range of freedom for the edges of the hotspots. If an edge is allowed to move e.g. by ±/2 nm, simulating both placements will be sufficient to define if they are covering hotspots, and will ensure the purity of the detection (Fig. 3.30). One contour would represent the simulation of the initial situation; the other contour, the simulation of the shifted situation. If both simulations are problematic, then this and all the intermediate moves are kept for the fuzziness of this hotspot. If the fuzziness is correctly framed, it can help to spread pattern-matching coverage, and then to correct more hotspots.

3.3.2.2 Hotspot Libraries

A verification flow should reference hotspot library as often as necessary, based on input design database, for every layer. OPC combined with Litho verification would sort defects by categories and criticality (Fig. 3.31).

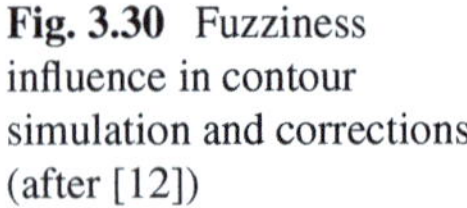

Fig. 3.30 Fuzziness influence in contour simulation and corrections (after [12])

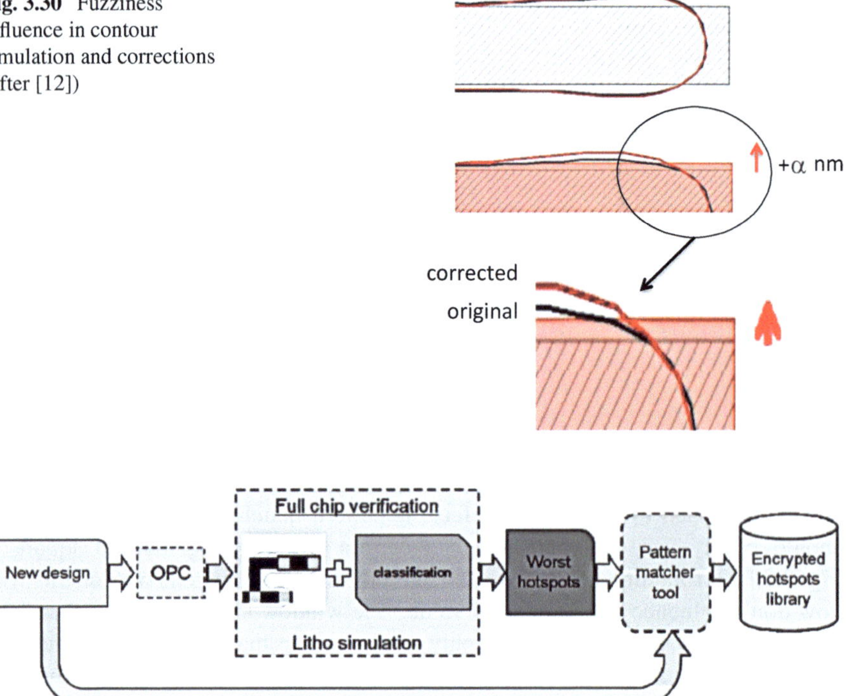

Fig. 3.31 Creation of hotspot library ([12], reprinted with permission)

All the information needed to identify how a library, such as the input design database and the OPC version used, is encrypted within the flow. A unique ID is also associated to each hotspot; the category (i.e. necking or bridging), and the ability to manage duplicate patterns. During litho verification, the same pattern can generate two different defects separated by less than the window size. In this case, a pattern-matching search is applied to test if the two hotspots have the same occurrences and only one of them will be kept in the library, to make sure that all the hotspots are unique.

All the hotspots coded for one layer will be independent of each other and for detection, the pattern-matching engine will search every hotspot independently to report every occurrence. As a result, one hotspot could be found 1,000 times while another not found at all.

Fast LFD Kit flow may use a two-stage approach. It takes into account that PnR tools are not able, for the moment, to make smart correction in order to avoid reappearance of a known hotspot or to avoid routing a new configuration that is as critical as before. It also understands that reroute is not able to converge into a new solution for some detected hotspots, because the design is simply too constrained.

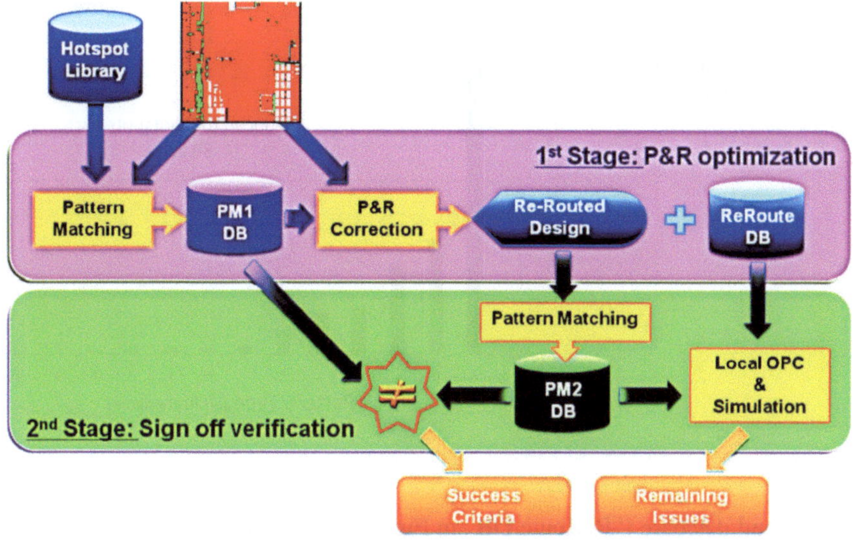

Fig. 3.32 Litho aware PnR LFD flow using a DfM kit ([12], reprinted with permission)

Another loop is then needed to frame the efficiency of the flow and to ensure the quality of the correction. The first stage takes place at PnR, to perform the detection and correction of hotspots from libraries (see top rectangle in Fig. 3.32). The second stage is done at sign-off, in order to check efficiency of the stage 1 correction by doing a litho simulation with the LFD framework (see bottom rectangle in Fig. 3.32). The advantage of this approach is that it combines quick turn-around-time of pattern matching with accuracy of integrating OPC and simulation.

(A) Phase 1: PnR optimization

For both phases, hotspot libraries are the same, but the way they are used is different. In the first phase, they are used to correct each hotspot occurrence. The result of pattern matching is output in a first database ("PM1 DB" in Fig. 3.30) and transferred from stage 1 to stage 2. The PnR tools perform their corrections based on the window coordinates of PMI DB. Since hotspots span several polygons, i.e. lines and spaces for metal levels, it would be hard for PnR engine to know where exactly the error is located. Consequently, once occurrences are found, an additional step is performed to create a smaller marker around the center of the window of interest. This center is where the hotspot was originally found, and depends on the defect category (necking, bridging or via coverage). For example, if it is a necking issue, the center will be located on the metal line, while for a bridging issue, the center will be in the space between two lines. After the correction, the tool will output the modified zones in a

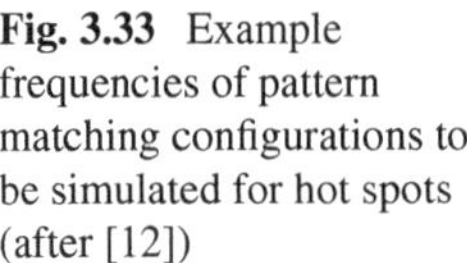

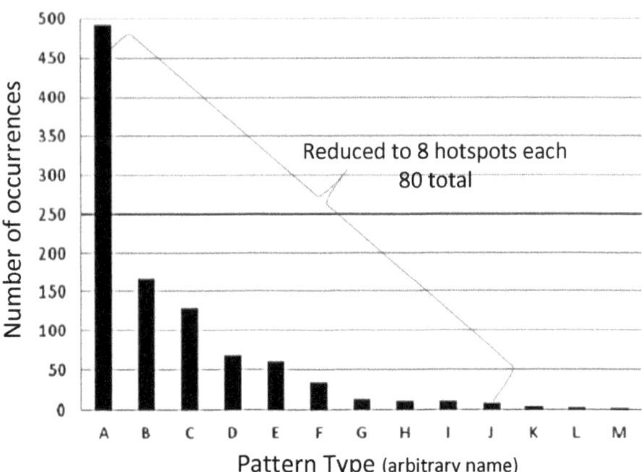

Fig. 3.33 Example frequencies of pattern matching configurations to be simulated for hot spots (after [12])

second database ("ReRoute DB" in Fig. 3.32), which will also be transferred from stage 1 to stage 2. This database of rerouted zones corresponds to the PM1 DB if each occurrence has been corrected, but one may expect that, for large design databases, some of them cannot be corrected because of density. The post correction design is then available ("Re-Routed Design" in Fig. 3.32), and will be transferred into the verification sign off.

(B) Phase 2: Verification sign-off

In the second phase, i.e., at sign off, pattern matching is redone on the corrected design, for two reasons. The first one is to clear the database of remaining hotspots, i.e. hotspots that have not been corrected ("PM2 DB" in Fig. 3.32). The comparison between PM1 DB and PM2 DB would allow evaluating the correction efficiency called "success criteria". The second reason is to apply simulation to remaining hotspots (PM2 DB) and on rerouted zones to keep control of corrected design. This way, one can check the criticality of remaining defects and catch new types of errors.

Even with such filtering, one can get thousand of areas to simulate. The different zones highlighted for simulation usually correspond to much fewer patterns with large number of occurrences (Fig. 3.33). Pattern matching should classify these zones to unique patterns to help reduce the OPC and simulation runtime. Because of the way hotspots are generated, the hotspot window can be used as the classification criterion. In (Fig. 3.33) hotspots A to M are found with different occurrences for the dispersion of the matching pattern. The total number of occurrences is over a thousand, by summing up all the columns of the histogram. Hotspot occurrences of patterns A to J can be reduced to eight hotspots each (eight combinations of rotation and symmetry) for the total of 80 hotspots (10×8). Pattern K can be reduced into five hotspots, pattern L into four hotspots, pattern M into two hotspots, whereas the remaining patterns (not shown for clarity) into one hotspot each. As a result, the 1,013 occurrences could be reduced to only 99 areas to simulate, corresponding to a 10.2× reduction in area.

Using the combination of pattern match filtering and classification, one has to do the OPC only on a very small portion of the design. This makes the sign-off possible on hardware available to designers.

In order to report all the errors, one also needs a post-treatment, in which all the hotspots are linked to their originating pattern, and then copied back to all the duplicates of the hotspots as found by the classification. This is done by comparison of coordinates and then rotating / mirroring according to the orientation of the hotspot.

The other advantage of this solution is that one can control the runtime, independently of the design area, bound by 8 × pattern type × runtime per pattern. Since we are using a dense simulation engine and all the patterns have the same area, the runtime per pattern is not dependent on the pattern type, which removes variability in the total runtime estimation.

(C) Hotspot and timing corrections

To evaluate the run time, place and route and signoff stages have to be separated. In the first stage, only pattern matching is performed, with a very short runtime. In the second stage, pattern matching is followed by simulation, which adds a non-negligible runtime, despite the approach described previously. Tests for several layers on different 28 nm designs showed that the LFD kit runtime is within 0.1 % or less of the overall DRC runtime.

Having the place and route tool correct the hotspots ensures that design constraints are respected [5]. The correction rate and the runtime on several designs showed correction rates above 99 %. The MDP effort depended on the number of remaining defects.

The timing and parasitic analysis before and after the corrections quantifying the impact of the manufacturing changes on the design performance (Fig. 3.34) shows the RC impact to be well-behaved, with smallest original values showing the highest percentage change. The changes of 80 % in resistance correspond to a reasonably low few Ohm at most. The changes in timing are centered on 0, with a spread of no more than 25 ps. The positive slack values stay positive, and the negative values are not affected (Fig. 3.34d) which was an indication that the design performance was not degraded by the litho fixes.

3.3.3 Conclusion

FPGAs exceeding 1 M logic elements (LEs), which can make PnR and timing closure a difficult process. Coding practices for the routing, such as retiming logic, pipelining where practical, and avoiding high fan-out nodes can make the problem much easier. Partitioning the design and good clock management at the start will go a long way towards avoiding issues once the design is complete. These suggestions are not all-inclusive and may not solve all timing issues where advanced tools should help the user.

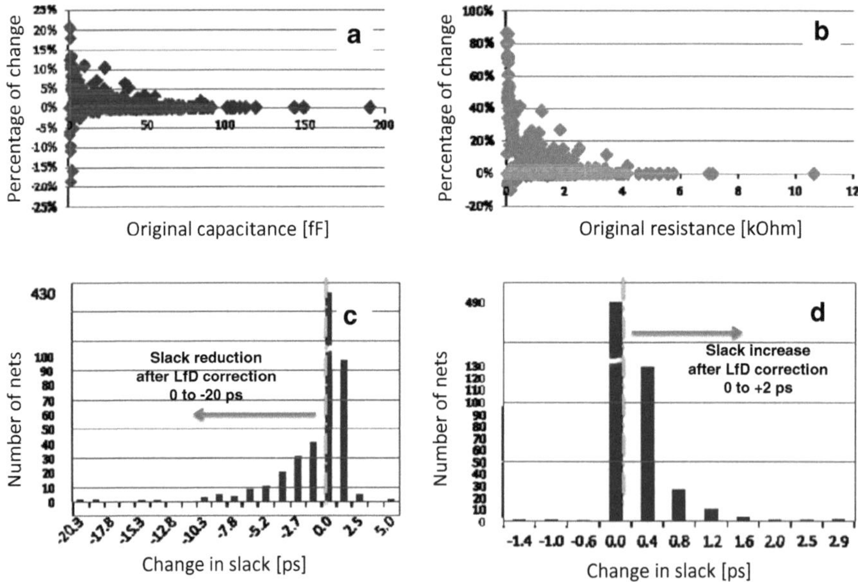

Fig. 3.34 Timing and parasitic changes after LfD correction (**a**), (**b**) percentage changes in C and R, (**c**), (**d**) number of nets which modified slacks ([12], reprinted with permission)

The next phase is to ensure that the PnR layout has no printability issues. Here, hot spot libraries are one key solution to ensure minimal impact on timing closure.

3.4 RDR Setup

Why is DRC inadequate for design signoffs? It uses a large set of rules to determine permitted designs. It is slow, labor intensive, ad-hoc, inaccurate, and restrictive. Layout engineering envisages replacement of DRC and printability simulation by a signal processing and machine learning-based approach for 22 nm technology nodes and beyond [14] by taking an AI approach to the problem.

Since the early days of VLSI, layouts have been drawn according to a set of design rules, which attempt to abstract what will or will not manufacture. These rules used to specify absolute minimal dimensions for circuit structures and absolute minimal distances to avoid shorts or defects. However, the current rules became relative to the layout environment and, therefore, complex. As a result, their enforcement can become extremely difficult. By nature, design rules are strict and excessively conservative, yet result in manufacturing problems. The rules are, in a sense, artificial and an arbitrary abstraction for the purpose of offering a simple drawing

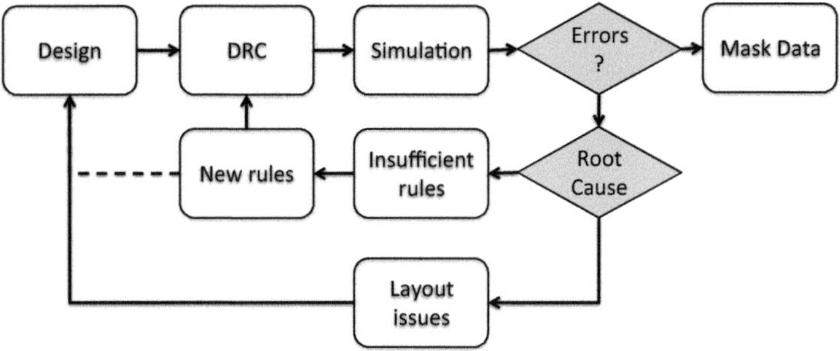

Fig. 3.35 Existing VLSI design cycle (after [13])

scheme and flow (Fig. 3.35). By removing this intermediate abstraction, better exploitation of new technology nodes should be achievable [15].

Over the past 10–15 years, practically every foundry in every CMOS technology node introduced SRAM memory cells, with violations of design rules negotiated and agreed with the manufacturing facility and containing so-called technology waivers. Their layouts are often drawn already in physical mask dimensions and no further post-processing is needed, with gains of 20–30 % area not uncommon. It is not unlikely that qualitatively similar results for other layout building blocks, including Standard Cells and titles of memory generators, are achievable. In Chap. 3.1.1.3, we called it technology design co-optimization, here, we define the rules to verify it.

Design Rule Check (DRC) pronounces patterns to be printable that do not actually print while those valuable from a design perspective which actually would print correctly, can be excluded. The number of rules makes dependency tracking hard. Even substantial extension of the rule set, does not result in a satisfactory yield and the design must be revisited multiple times before reaching production stage [1].

One can propose migration to a model-based approach using signal processing and machine learning-capabilities. The process would produce fast, accurate, autonomous printability prediction for optical lithography, built upon previous attempts to restrict the number of potentially allowed patterns and structures [1, 14].

In Artificial Intelligence (AI) parlance, DRC is a rule-based expert system deployed to aid a design decision process. Rule sets are traditionally derived from humans studying learning samples, not particularly adept at capturing multi-dimensional dependencies. In pattern classification, feature space metrics have largely replaced rule-based learning systems.

DRC can often be excessively conservative (not "letting things through") while also erroneous ("letting things through" that in the end do not print). Taking the lesson of AI – that statistical models usually beat expert systems, one can develop a model-based approach called P2C (Printability Predicting Classifier).

A proof-of-concept demonstrator can predict optical rule check ORC problems without the need to fall back on costly first-principle simulation. For one sample test site, the demonstrator can obtain an Equal Error Rate of a few percent.

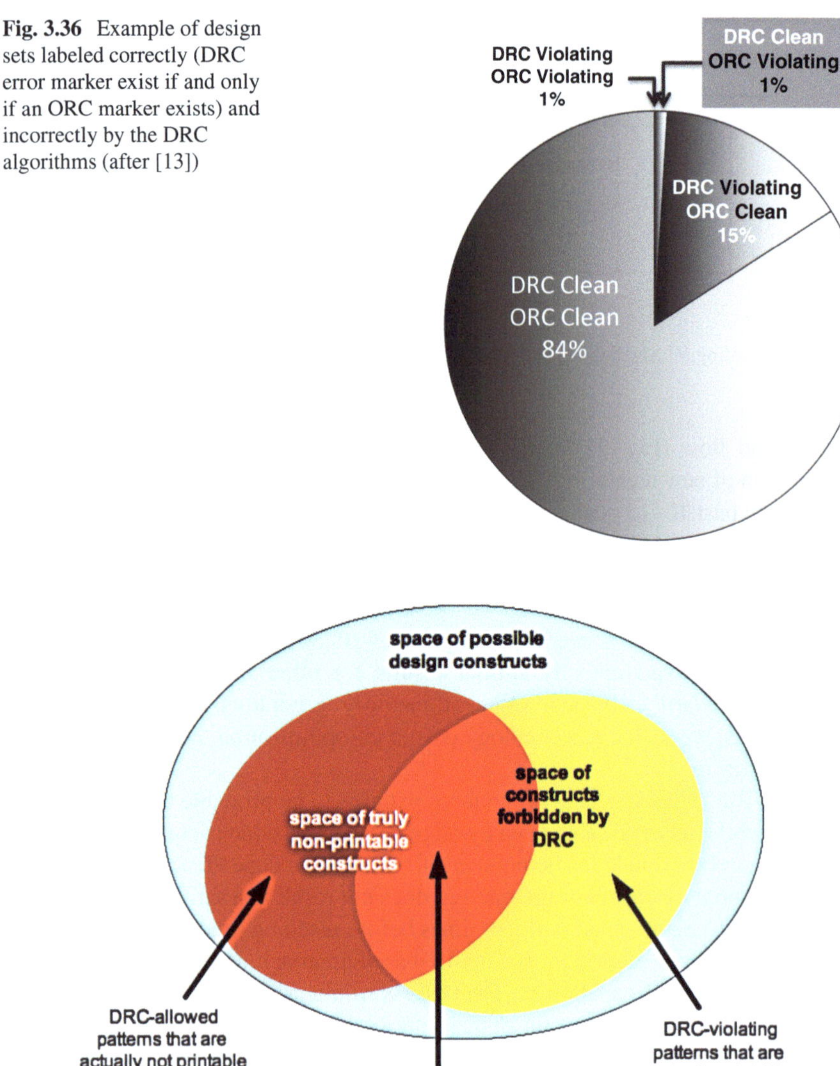

Fig. 3.36 Example of design sets labeled correctly (DRC error marker exist if and only if an ORC marker exists) and incorrectly by the DRC algorithms (after [13])

Fig. 3.37 Relationship between the global design space, the subspace of DRC-violating patterns, and the subspace non-printable patterns identified by the P2C ([13], reprinted with permission)

3.4.1 Transition of Design Verification Approach

(A) Design rule check (DRC) – the old paradigm
DRC is the first step in the modern VLSI design process (Fig. 3.35). It uses a large set of rules in order to eliminate from the layout design patterns that would be non-printable under almost any conditions. The rules may become

excessively complex and conservative so as to avoid yield problems in manufacturing. Although hard cases make bad law, rules are linked to printing problems encountered before rather than any underlying physical phenomena. They are also non-scalable to new technology nodes. There is uncertainty as to which old rules should hold, and new ones must be added through labor-intensive error analysis.

For sub – 28 nm designs, relying on DRC alone would be a conservative and erroneous approach. The requirements resulting from the optical properties of the printing process drive rapid escalation in number and complexity of the rules [12]. Creating rules requires a collaborative effort at the early stage of each technology node.

From the viewpoint of DRC accuracy, there are three main types of structures (Fig. 3.36):

1. Accurate: DRC error marker exist if and only if an ORC marker exists.
2. False Positive: the structure bears a DRC marker although ORC labeled it problem-free.
3. False Negative: the structure bears no DRC marker although ORC labeled it problematic.

As seen, DRC may commit both false positive and false negative errors. Manufacturing procedures are tuned to remain on the 'safe side', preventing false negatives (missed detections) at the cost of disallowing problem-free structures (false positive).

There are several solutions to improve alignment between DRC and ORC. One is to reduce drastically the number of patterns so as to restrict the design space to those shapes (e.g., using RDR rules) and constructs guaranteed to be printable (Fig. 3.39). A similar approach, which postulates radical layout regularization beyond 32 nm technology node, is referred to as prescriptive layout design [1].

Another option is to define DRC Plus rules through identification of classes of design patterns that are likely to cause printability errors. Those patterns are forbidden and removed from the design space. This is conceptually similar to the prescriptive design principle, but the application of the DRC Plus may still eliminate perfectly printable patterns from the design space, and may still allow non-printable patterns.

(B) A new rule-free VLSI design paradigm

In the language of AI, DRC as a rule-based expert system deployed to aid a design decision process [13]. It is a list of "if A then B" statements. A is a particular set of measurements of the design, and B is a decision regarding the printability of a given layout fragment. Traditional rulesets are derived from experimental observations of learning samples of those that do and do not pass. In this sense, it is a system describable in terms of machine learning theory.

While the space of A can be extremely diverse, B is basically equivalent to a classifier that decides between two classes: printable or not. Although this dichotomization could be executed by a rule-based system, such systems have

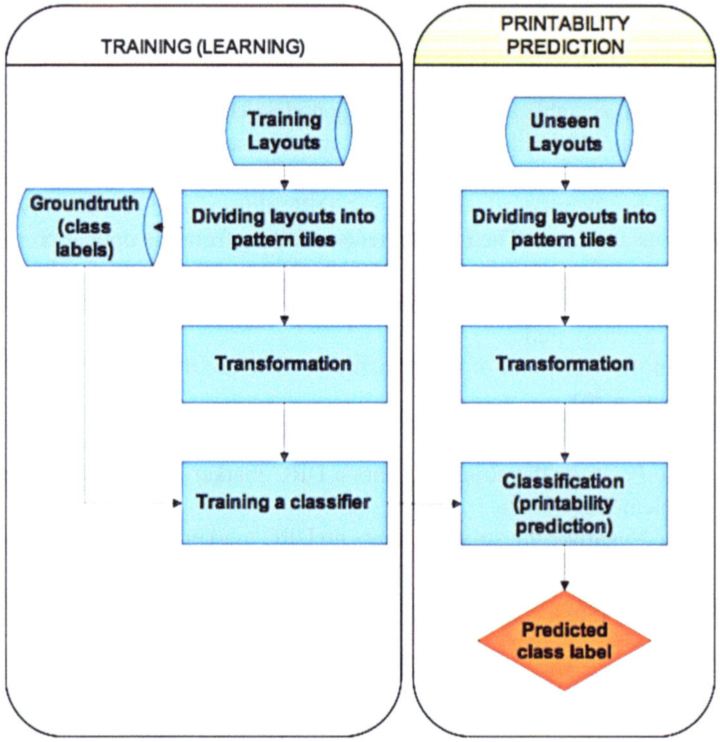

Fig. 3.38 Training and target deployment of the printability classifier (P2C) ([13], reprinted with permission)

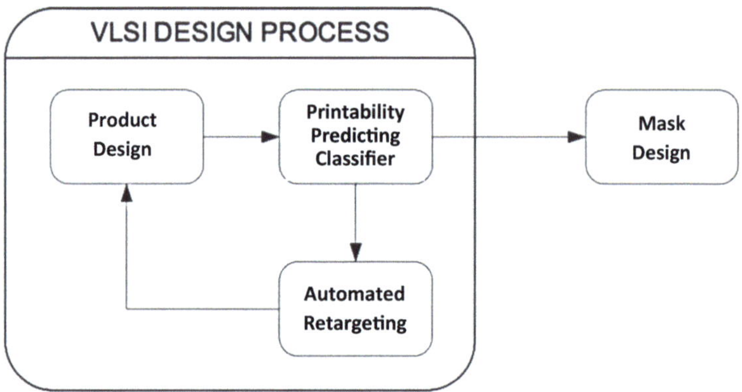

Fig. 3.39 Proposed VLSI design cycle with P2C and retargeting (after [13])

had limited success in pattern recognition [16]. In practice, rule-based learning systems in pattern classification have been largely replaced by feature space metrics. It seems intuitive to similarly replace a rule-based, traditional DRC approach by a learning, model-based method, such as P2C, whose learning and testing stages are depicted in Fig. 3.38.

(C) Learning stage

The underlying assumption of Printability – Predicting Classifier P2C is that shape printability can be treated as a stochastic manifestation of a series of deterministic physical phenomena during the lithographic process. As such, printability can be thought of as approximately correct learnable or Probably Approximately Correct (PAC)-learnable [17] and a classifier can be trained using observed samples, which originate from actual designs or from enumerated pattern sets. The training patterns are assigned ground-truth class labels based on full-chip printability simulations [18]. The objective is to minimize the empirical risk of misclassification of a given pattern, where the risk is a function of the error rate and error cost.

A representative set of synthetic VLSI patterns which cover the space of printable design shapes can be generated by tools for enumerated pattern generation. These tools can also generate patterns that purposefully violate DRC rules. We need to select those erroneously identified by DRC as non-printable, as well as the actual non-printable patterns erroneously passing DRC, simulating and examining the entire design space and the space of errors. At the end, one can be confident of having sufficient data to train statistical classification models and cut the complexity of the classification task.

The classification-based approach can predict printability faster, without recourse to full chip simulation. The major computational burden lies in the off-line training. The online classification of unseen patterns (printability prediction) requires computation of the sign of the distance from the decision hyper-surface (derived in training) to the point that represents the classified pattern in the feature space.

One way of incorporating the proposed model-based tool is as an offline printability check, where a readily available, pre-existing layout is checked for printability within the given technology node (Fig. 3.39). P2C is then used to scan the entire layout for printability defects, sending them, when appropriate, for manual or automatic retargeting. Online design can also be supported, where P2C is used as a fast tool for printability checking while design is underway. The tool can be made interactive, visually marking probable printability problems on the designer's desktop, for fast and accurate input of global layout design optimization. The interactiveness of the process may help designers looking at other factors, such as compactness and power.

(D) Printability-predicting classifier: sample demonstrator

As a proof of concept, one can build a simple P2C demonstrator, whose goal was to devise a system that can predict ORC problems using a trained classifier without the need to fall back on costly first-principle simulations. We focused on one type of ORC violation marker – the Line Width error. This occurs when the width of the wire printed on silicon falls below min CD value. Samples of the Line Width error from an enumerated synthetic layout can be used to extract a set of descriptive two-dimensional Discrete Cosine Transform features, and then train a sample P2C using the extracted features. The trained classifier would be applied to detect the line width errors in other structures, unseen during the training phase.

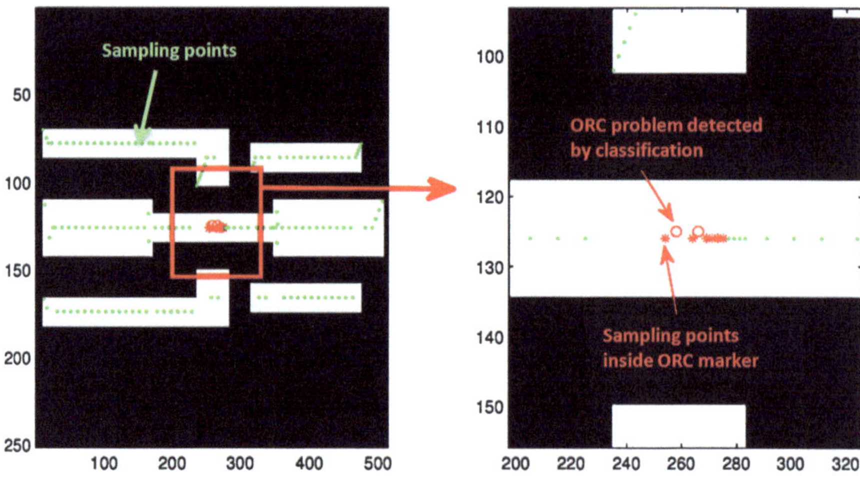

Fig. 3.40 A sample screenshot demonstrating the P2C concept ([13], reprinted with permission)

The ORC sampling points would be taken along the medial axis of the structure polygons (Fig. 3.40). The classifier goal was to detect the points with classification at each sampling point. ORC problem was marked by a circle. Accuracy was measured by comparing the number of the actual ORC problems (sampling points contained within an ORC marker) to the number of detected ORC problems. Using a sample test site, we obtained the Equal Error rate of ca. 4 %.

Based on these examples, one can conclude that DRC, a legacy of heuristic development in former technology nodes, is gradually becoming cumbersome and impractical. Expert systems applications, where long rule lists have been successfully replaced by more powerful and compact non-deterministic classification and prediction tools, should be looked into for alternative solutions.

3.4.2 *RDR Versus DRC*

To print a wide range of very tight pitches is no longer possible for subwavelength lithographic systems [19]. As discussed, restricted design rules (RDR) enforce a style of layout that is expected to be highly manufacturable by limiting the range of pitches. Below, we will consider tradeoffs between manufacturability and performance (measured as layout density, delay, power, etc.) comparing DRC and RDR rules, to minimize mask costs, maintain circuit performance, and enhance feature printability and reliability.

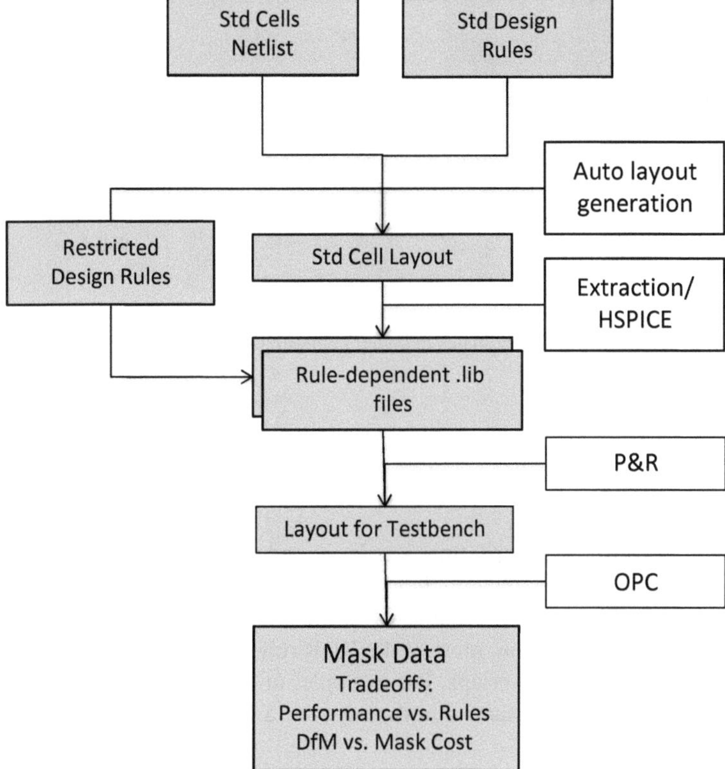

Fig. 3.41 ASIC design flow for RDR evaluation (after [19])

3.4.2.1 EPE Impact Evaluation

To evaluate the impact of restricted design rules, one may set up the design flow (Fig. 3.41) and then, starting from a set of default rules and a pruned standard cell netlist (BUF, INV, NAND, NOR, AND, OR, AOI, OAI), create GDS representations. After parasitic extraction, each cell is characterized for timing and power performance to generate a .lib file, to proceed to synthesis/place and route (P&R) [17].

Next, library generation is repeated with design rules altered through inclusion of a single candidate RDR, such as extra requirements for poly gate spacing, minimum line end extension, etc., to improve the printability and reliability with limited performance impact. Layouts and .lib files for different RDR sets, followed by synthesis/P&R, timing, power, and area reports are created after back-annotation for benchmark circuits, with circuit topology unchanged. The circuit is not re-synthesized with a new library but instead the gate-level netlist is mapped to a new .lib and followed with the back-end of the typical ASIC flow.

After circuits are placed and routed for each library, model-based OPC including line end correction, concave and convex corner correction instructions, changes

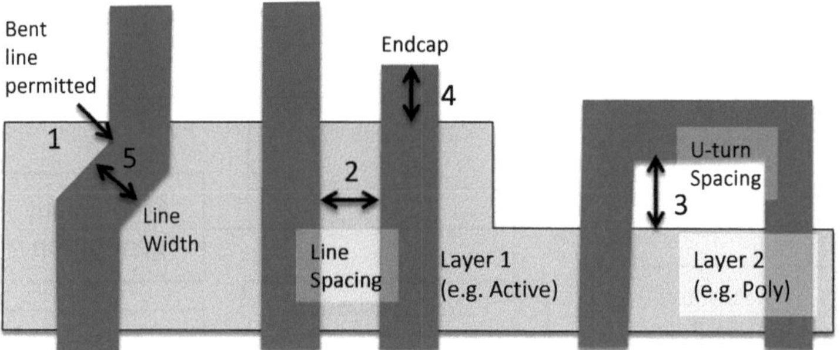

Fig. 3.42 Examples of RDR candidates in layout configurations (after [19])

circuit performance (delay, area, power) and manufacturability/printability/mask cost is measured on MEBES data volume versus histograms of resulting edge placement errors (EPE), etc.

A few rules critically impacting gate layer and transistor performance, such as spacing between features (the light field of a given feature is affected by the location of the neighbor features) may lead to CD variations that can result in loss of parametric yield. For that reason, most of the RDR rules deal with either intra-layer or inter-layer spacings and overlaps. For example, minimal gate overlap of diffusion (endcap) is critical as it ensures that the edges of a MOSFET maintain dimensions within the channel.

Starting from design rule set with all spacing rules at their minimum values and allowing all angled features, one can reduce it to restricted design rule sets e.g., by disabling bent gates, increasing minimum gate spacing, minimum spacing field gate to diffusion, minimum poly line end extension beyond diffusion (Fig. 3.42), etc. Once the form of the RDRs is decided, the range of values needs to be determined to drive printability improvements. The more conservative design rules improve the impact of larger spacings, but if the spacing becomes too large, it can become detrimental to manufacturability since lithography systems are not adept at printing intermediate pitch values [19].

One can use edge placement errors (EPE) to quantify how closely a printed feature reflects the corresponding designed feature. With EPE larger near the gate ends, the impact of CD on small-width gate variability is relatively larger since a smaller-than-nominal channel length leads to exponentially more subthreshold leakage [18], which points to line end extension rules as an important design restriction. A more restrictive minimum line-to-line spacing, the EPE distribution of a NAND2×2 (2-input NAND of size 2) without OPC shows a consistent left shift but after a certain value moves back to tighter distributions (Fig. 3.43). With off-axis illumination (annular or quadrupole), there could be several pitch ranges where the optical diffraction results in poor printed images (larger EPEs) for forbidden pitch ranges). The EPE (or CD) variation becomes smaller for isolated lines but the average value increases, due to the fact that the radius of influence of optical diffraction

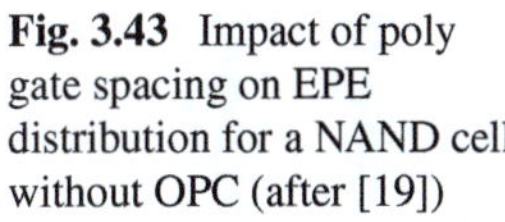

Fig. 3.43 Impact of poly gate spacing on EPE distribution for a NAND cell without OPC (after [19])

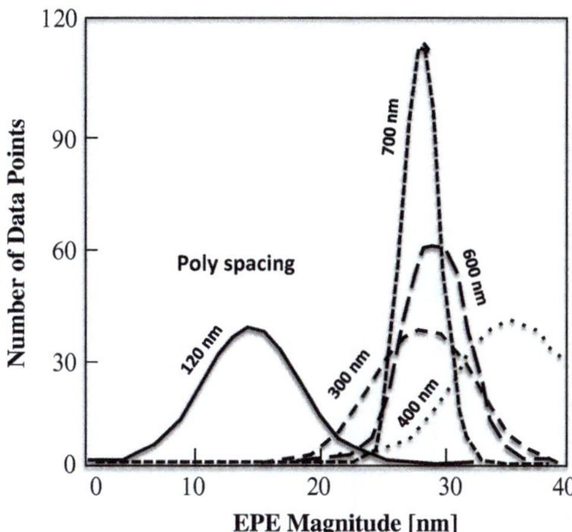

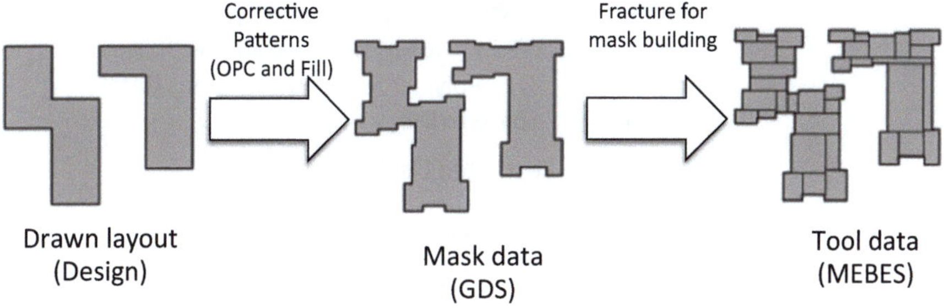

Fig. 3.44 Post OPC mask data preparation flow (after [19])

extends to ~0.6 μm and any pitch above that prints poorly [17]. One may define poly spacing at min L and other forbidden pitch ranges for RDRs.

3.4.2.2 DfM Cost Matrix Evaluation

While zero EPE is the goal for all polygons forming transistor gates, interplay of two edges (or EPEs) is needed to determine the CD, indicating the need to localize each EPE and match it with the EPE on the opposing side of the polygon. Considering that each transistor may have multiple CDs on the printed image (i.e. gate lengths would be non-uniform along the width dimension), an average CD for each transistor is integrated as the gate and active overlap area with the simulated printed image and dividing it by the measured gate width.

The mask writer (MEBES) data volume is often used as complexity measure of the resulting mask for the critical layer. An OPC or design-cost metric (GDSII files fractured into MEBES format, (Fig. 3.44), during mask data preparation) may became bottlenecks due to large figure counts from RETs.

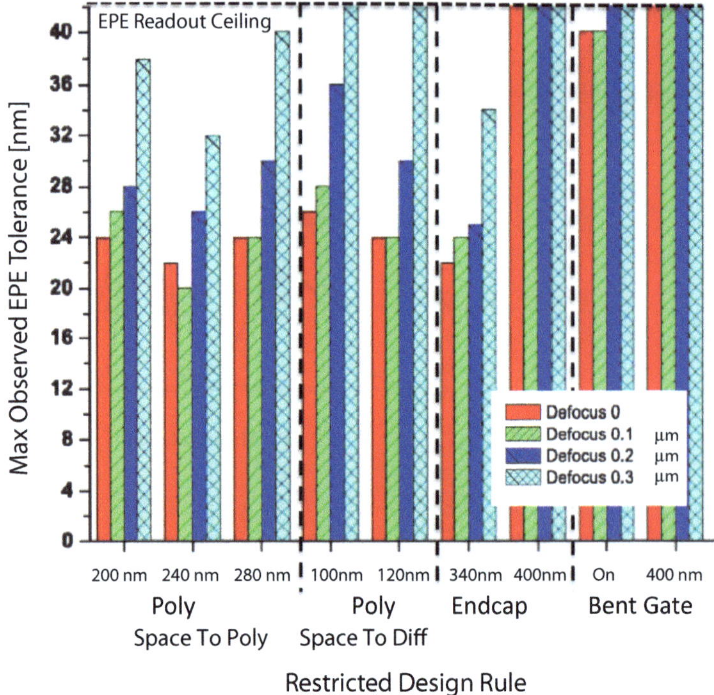

Fig. 3.45 Impact of defocus on EPE (CD variation) (after [19])

Defocus strongly affects the printability of fine resolution images because it reduces the process window for the range of exposure dose with acceptable image tolerance. When defocus exceeds a certain value, the printed features go out of the CD variation tolerance. Defocus values of 0, 0.1 µm, and 0.3 µm are examples of practical values to be tested in the experiments.

With constant defocus, as poly spacing increases, the maximum EPE tolerance (i.e. one observed for any gate in the specified design) initially either remains constant or decreases, demonstrating an impact on CD variation (Fig. 3.45). But then, EPE nearly doubles as defocus rises further [19]. Adhering to generic design rule set may lead to very large EPEs. Disallowing bent gates and other changes to the design rule set should improve worst-case EPE values. The best RDRs are the ones which increase the minimum poly overlap of active and relax the poly-to-poly spacing by 20 %. However, relaxation of design rules can actually worsen printability of some features.

To assess the impact of focus variation on CD, it is common to use aerial image models of optical effects. The intensity level for image simulation is fixed at the value which gives best shape for the isofocal spacing at best-focus, computed to about be 200 nm by defocus simulations of a simple test structure. For average CDs and their variation extracted from aerial image contours, the 200 nm poly-spacing

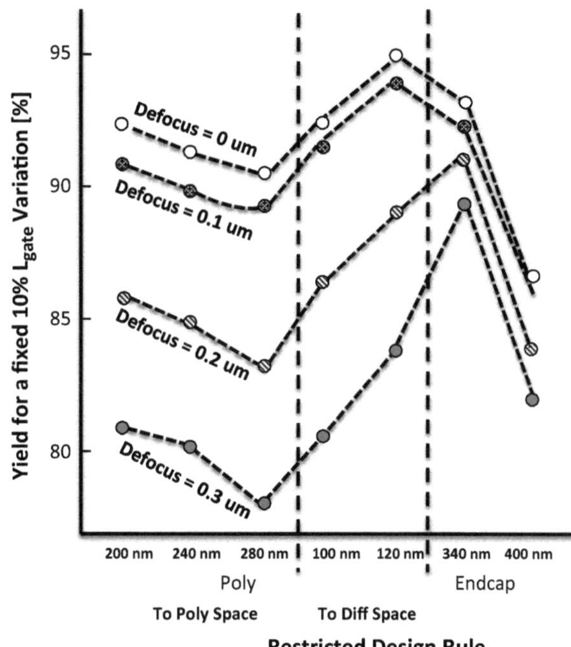

Fig. 3.46 Functional yield for a 10 % L_{gate} variation (after [19])

rule prints the best through-focus as it results in cell layouts with inter-device spacings closest to the isofocal spacing. An intelligent choice of the minimum poly spacing should be cognizant of the isofocal spacing to improve defocus characteristics.

3.4.2.3 Mitigating Defocus Impact with RDR's

The variability in physical gate length is allowed at 10 % by the ITRS Roadmap, which translates to an EPE of 5 % on each edge of the gate (it is possible for a printed gate to have larger EPE on both sides and still maintain a nominal L_{gate}). One can define functional yield as percentage of total gates that print with less than 5 % EPE for all fragments of the gate.

For any RDR set, the functional yield is sensitive to focus variation, because printability degrades for features out of focus (Fig. 3.46). However, the RDRs increasing poly line-end extensions reduce sensitivity of functional yield to defocus and larger process windows, thereby reducing manufacturing overhead/cost. At the same time, the bent gates with off-axis illumination produce a large number of substantial (>5 %) EPEs and are disallowed. Also, a standard design rule set identifies a high percentage of gates within the ITRS specification indicating it may be a valuable addition to the RDR rule set.

Isolated lines may be subject to more optical distortion than dense lines since lithography and RET recipes are not optimized for them. Although OPC corrects for the iso-dense bias, isolated lines with non-zero defocus print narrower or wider, depending on the lithography system. Scattering bars (SBs), modify the wavefront and reduce these distortions, but they add data volume for the mask writing equipment driving a 15–20 % increase in data file sizes.

3.4.3 RDR Categories

3.4.3.1 Single Pitch RDR

One option to restrict the layout is to tune the process to favor one particular pitch. In off-axis illumination (OAI), the angle of illumination to the mask is optimized so that one pitch can be printed perfectly, due to the diffraction of light. Although within a limited range of the illumination, the distortion caused by pitch differences may be compensated with other techniques such as SBs, designers still must work around the forbidden pitch range for better yield. A "single pitch, single orientation" rule (horizontal or vertical gate routes) is highly desirable even though it adds constraints in library design and P&R. A pitch larger than restricted by lithography may allow for a contact to be inserted between two poly lines. One can obtain a pseudo single pitch library with over 90 % of the gate pitches fixed at a single value, and the remaining few % among other values, due to limitations in the layout synthesis tools. Comparing the results with the reduced library (where OAI and OAI cell types are excluded, all RDRs are set at default, and scattering bar OPC set at defocus 0.1 μm), this RDR option shows good potential to reduce the 3σ L_{gate} variability and the data volume can be 25 % lower.

With all improvements provided by RDRs, there are performance penalties in timing, area, power, mask data volume, and parametric yield within a 10 % budget for all RDRs. The range of delay is small over all RDRs (5–10 % worst-case spread) while the area and power impact is larger (up to 20 % spread in both). The minimum poly_diffusion spacing is the most favorable rule for low data volume and high yield with acceptable performance. The line end extension rules exhibit very similar characteristics and robustness to process defocus. On the other hand, because bent gates may save area only at the expense of data volume and yield loss, they are prohibited in modern design rule sets.

3.4.3.2 DRC Plus: 2D DRC with Pattern Matching

Interactions between aerial image, photoresist, line-ends, resolution enhancement techniques (RET) and process parameters including k, NA, source wavelength λ, source shape and off-axis illumination, are all translated into DRC decks. The guarantee to circuit designers and layout engineers is well understood: follow these

DRC is and subsequent processing will manufacture the circuit to the drawn specification [19]. This abstraction is necessary to free designers to deal with other challenges in circuit design, such as area, capacitance, robustness, and delay. However, as devices continue to shrink and otherwise evolve from generation to generation, manufacturing challenges become more formidable and the clean abstraction provided by traditional DRC begins to drift in two ways.

First, while aggressive RET and OPC models are carefully tuned for the abstract 1D geometries of parallel lines and spaces, complex interactions of 2D geometries with various RET choices are difficult to measure, to model accurately, and to analyze. To simplify the task, one would impose rigid design rules which use highly regular structures for layout 2D interactions [21]. The drawback is the restricted design flexibility such that designers cannot optimize for other challenges of circuit operation (e.g. the parasitics).

For DRC to work well, the abstraction and the distinction between what is DRC clean or contains DRC errors to be fixed should be clear in the minds of layout engineers. With time, the design rule manual has become complex with cryptic rules to handle exceptional cases as they occur [22]. The exception handling procedure has met with mixed success. Although it typically resolves the issue at hand, the enforcement of a DRC rule may cause more complex 2D geometries to be generated. In effect, designers meet the letter of the law, without following or even understanding the spirit of the rule. This often leads to more exceptional cases added to the DRC manual, further increasing its complexity. One way to control the complexity is to specify a 2D geometry to be avoided, and minimize unwanted side effects.

The goal of DRC Plus is to provide a simple way of marking specific 2D geometries undesirable. Image-based pattern matching identifies such geometries [23] to be listed in the design manual. However, image comparison is not sufficient to separate good layout from bad layout. To apply a preferred DRC rule to the matched 2D pattern, one needs: (1) pattern definition identifying a problematic 2D configuration, and (2) a DRC rule identifying the desired solution only where the problematic configuration exists.

3.4.3.3 DRC Plus Features

One can identify simple layouts with sub-nominal, nominal, or relaxed line-end spaces (Fig. 3.47, Columns 2, 3, 4). The sequence of pattern match, followed by DRC rule verification on this pattern constitutes a DRC Plus rule. For the line-end in U-shape, the DRC Plus rule has a description very similar to a simple DRC rule, except it is annotated with a layout clip to describe the specific situation where it is applied, and a match tolerance.

The DRC Plus software should supplement existing tools (Fig. 3.48). For each DRC Plus rule in the deck, the pattern is matched against the target layout using the 2D Pattern Match Engine [23]. This produces a set of match locations represented as polygon markers linked to the DRC engine. The check result summary produced by the DRC engine would include DRC Plus results.

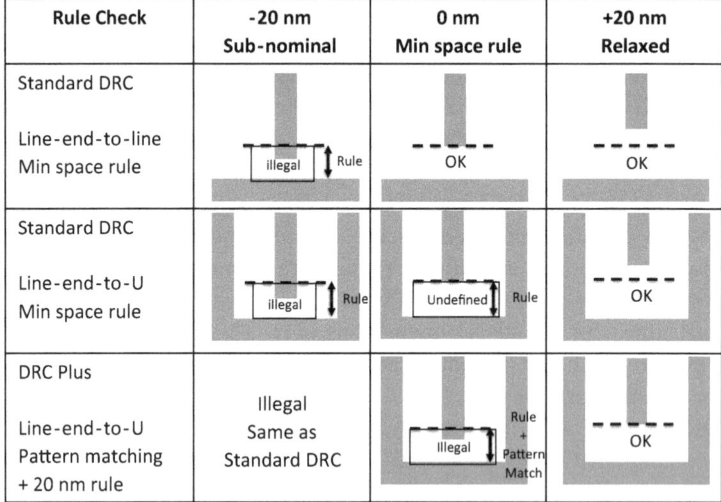

Fig. 3.47 DRC Plus versus standard DRC rule setup: line bridging prevented for small U-shape opening for min space rule (central cell) (after[19])

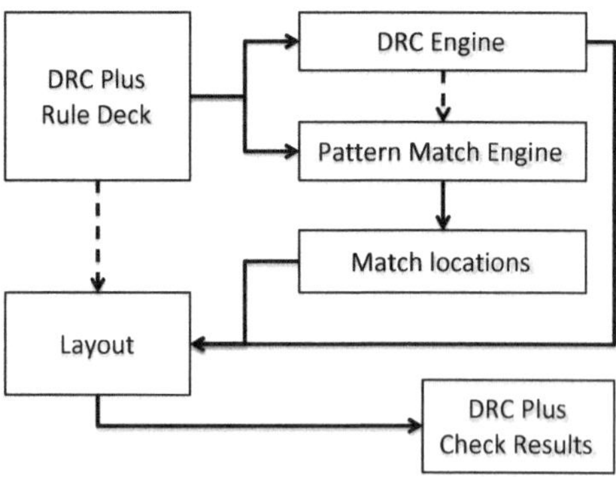

Fig. 3.48 DRC plus process flow (after [19])

Below, we compare DRC Plus to other DfM methods, such as DRC, preferred or yield design rules (YRC), regular or restrictive design rules (RDR), simulation based layout printability scoring methods, and simple 2D pattern matching (Fig. 3.49).

(A) DRC plus versus DRC
As a DfM technique, DRC Plus is most similar to advanced DRC. It operates directly on the drawn designs without simulation models, it results in the same pass/no-pass check result with an error marker denoting its location, and a simple description of the fix. With a high-performance 2D-pattern matching engine, its run-time is comparable with that of a standard DRC check. DRC Plus provides new functionality to the PV flow, by directly capturing and reporting marginal 2D situations, enforcing design in a simple, concise manner and improves the accuracy of design rule checks in predicting manufacturability.

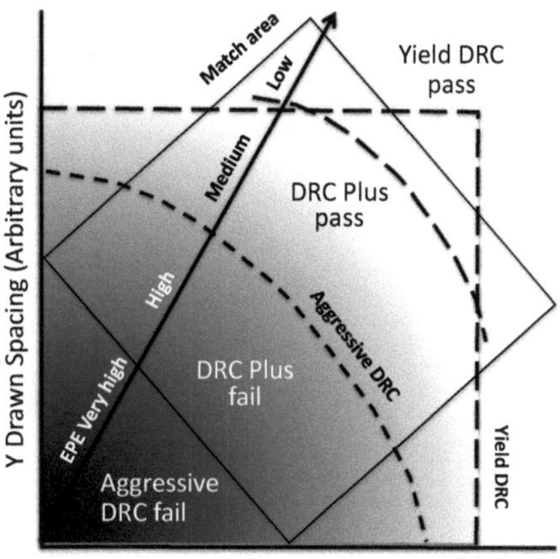

Fig. 3.49 EPE error ranges (very high to low) simulated in resist for various contact spacing rules: standard DRC, aggressive DRC, DRC plus (after [19])

This can be illustrated with a simulation in which the placement of a two contact minicell is systematically varied in a diagonal corner-to-corner configuration. Each two-contact pattern is passed through the standard mask generation flow, including target sizing, model-based OPC, and MRC. The resulting mask pattern is then simulated to determine final edge placement in resist (Fig. 3.49).

EPE in resist indicates that both contact holes, by symmetry, print smaller than they should, i.e. they are at risk of not forming a good electrical connection. To avoid this, one should apply minimum space DRC using a square metric (thick dashed line = Yield DRC). Everything to the top and right of that line meets the minimum space / Yield DRC, and consequently, should have no problems printing. However, this Yield DRC rule ignores white areas below and to the left of the line, which indicates that it is safe to push the contacts closer together when the contact pair is nearly vertical or nearly horizontal. If, instead, the Euclidean space DRC (labeled "aggressive DRC" in Fig. 3.49) were adopted, we could use this extra physical design space when the contacts are nearly horizontal or vertical, risking manufacturability problems when the contacts are placed diagonally. To remain conservative, the choice would favor field-oriented DRC, sacrificing design space. However, because in a vast majority of instances, minimum space contacts are placed vertically or horizontally, even a small amount of design space could impact the total area of design by several percent.

With DRC Plus, this choice is simplified: we can now use the minimum space aggressive CD, and then employ pattern matching to capture the undesirable corner to corner configuration to enforce larger minimum space, where Yield DRC has been replaced with a DRC Plus rule, a combination of a target pattern with two contacts placed just over minimum space, 45° angle with respect to each other, and related DRC rule with a larger minimum space.

The set of matching patterns form a region in design space, in this example, the diamond-shaped solid line Match Region. Its size is determined by the match tolerance of the DRC Plus rule, defined as the maximum percentage area

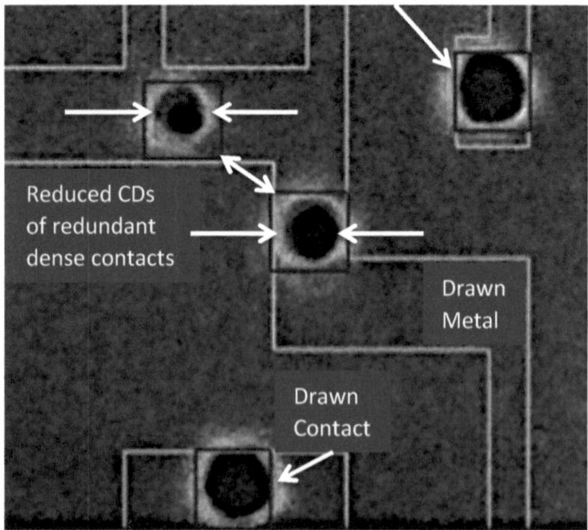

Fig. 3.50 SEM image of redundant diagonal contact pair at aggressive DRC for minimum space (after [19])

difference between the target and the considered pattern. As the match tolerance becomes larger, the Match Region becomes larger, the match becomes less deterministic, and more patterns are caught by the matching process.

Standard DRC may interact with DRC Plus for all designs that do not match the diagonal contact pattern. If they lie outside the match region, the aggressive DRC rule is applied. Inside the Match Region, the preferred DRC is applied. The combination of DRC and DRC Plus allows us to apply a simple DRC under most conditions, while capturing the manufacturability issues of a particular 2D configuration.

The behavior of DRC Plus can also be implemented by a parameterized DRC code. The pattern matching in DRC Plus is more visual and may be simpler to enforce specific instances where a yield-friendly set of DRC rules are needed. SEM image of an instance of the diagonal contact pattern at minimum space of aggressive DRC (Fig. 3.50) shows the small contact hole in resist, in relation to nearby contacts, at a corner of the process window. Ironically, these are redundant contacts, which would have been better left as a single contact.

In summary, DRC Plus is a refinement tool in the PV toolbox. Whether DRC Plus, pattern matching techniques or parameterized rules, would apply to fine-tune cutoff in specific instances, may depend on the RoI expectation.

(B) DRC plus versus recommended design rules

A technology rule deck may contain a DRC deck that is enforced, and a preferred rule deck with more or less aggressive design rules, which are recommended (RR) [24]. RR rules may be applied in non-critical design areas, but not enforced e.g. by an area id of a layout region. Pattern matching enables DRC Plus to specify situations when RR rules must be applied.

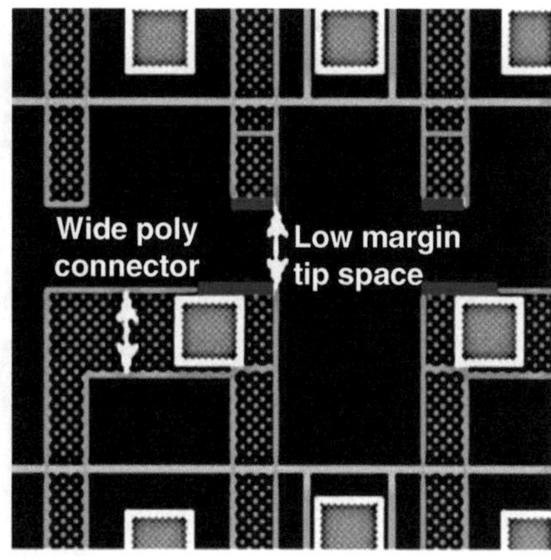

Fig. 3.51 Poly tip to corner spacing in single pitch-single orientation layout causing potential for pattern bridging (after [19])

(C) DRC plus versus regular design grid or restrictive design rules (RDR)
As technology shrinks continue, design rules imposing regular design grids for layout are needed to disallow problematic patterns, if at the expense design flexibility [21]. But even on a regular grid, difficult-to-manufacture 2D situations may occur (Fig. 3.51). The optical correction to line tips may compete with correction to the corners due to limited space on mask, which leads to greater (30–60 % more) line-end pullback than a normal tip-to-tip or tip-to-line configuration in single pitch, single orientation layout. In contrast, the more flexible DRC Plus would require RDR design rules only for known problematic patterns.

(D) DRC plus versus simulation-based layout printability and pattern matching
A variety of simulation engines help predict printability, in addition to relying on design rules [25, 26]. The complexity of computations and analysis often limits simulation to small-blocks of the design. It is also difficult to obtain an accurate model early enough. The simulation typically does not generate pass/no-pass criteria, nor does it offer suggestions how to fix the design. In contrast, DRC Plus with an efficient pattern matching engine can be applied to the large blocks, does not require complex models, and provides a pass/no-pass result with a simple description of the problem.

A single pattern match cannot determine manufacturability or provide a pass/no-pass criterion. As an example, if the image of tip and line only (Fig. 3.51) is used as a reference pattern, matching may not be able to identify the line-end space. In DRC Plus, the rule on pattern match regions is to apply match results directly to layout.

(E) Creating DRC plus rules
Similarly to standard DRC, DRC Plus operates on design geometries without understanding of OPC, mask constraints, or process. The components of a

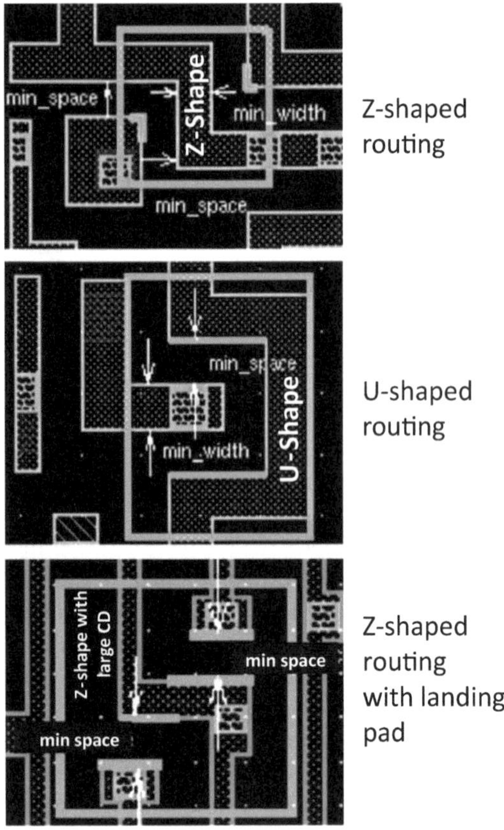

Fig. 3.52 Illustrations of DRC Plus rules for: Z – shaped routing, U – shaped routing, and Z – shaped routing with landing pad (after [19])

DRC Plus rule are a summary of these same manufacturability issues for the specified pattern (Fig. 3.52). DRC Plus is implemented based on analysis from other sources, including lithography simulators to check for layout geometries prone to hotspots in the order of increasing match tolerance. A tolerance threshold is identified beyond which either no hotspots are found or pattern affinity becomes irrelevant.

Causes of the hotspots are identified and new dedicated rules are checked to resolve the issue at hand, drawing on existing DfM recommendations. A design fix is then applied to pass the hot-spot related rule, and verified through lithography simulation that the hotspot has been resolved. The pattern, the matching tolerance threshold, and the preferred restriction are then captured in a DRC Plus rule and placed in the technology rule deck. The process is largely manual

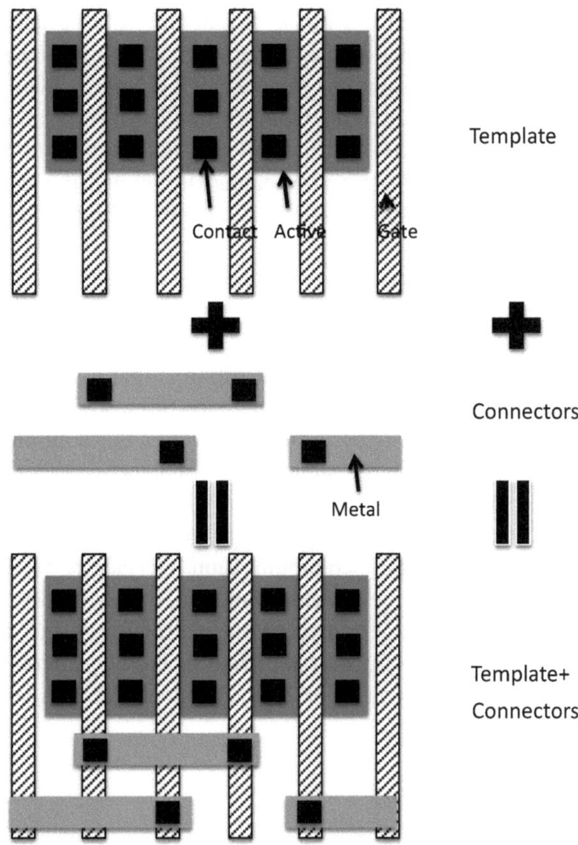

Fig. 3.53 Concept of layout template and connectors for active, poly, and contact layers (after [27])

with human judgment involved not unlike basic DRCs (Fig. 3.52). However, due to the pattern-matching requirement, DRC plus is less intuitive than DRC.

(F) Runtime of DRC plus rules
Though DRC Plus adds an extra pattern match step to every preferred DRC rule, for a highly efficient 2D pattern matching engine [19] the runtime does not generate much overhead compared to the standard DRC engine.

3.4.3.4 Pattern Count Reduction with Template and Connectors

Twenty nanometer technology introduces a significant departure from conventional technology development techniques. Double patterning with pitch splitting (DPT) is required below the 80 nm resolution limit. A design solution exploiting the concept of Templates and Connectors is used to define a standard cell library that can accelerate the introduction of 20 nm technologies [27] (Fig. 3.53).

Pattern count for a block of randomly placed cells depends on its size and is determined by the unique patterns in a 2-pitch interaction range. The number of unique patterns saturate within a block of approximately 40 × 40 μm, small enough to perform Source Mask Optimization (SMO) for all of them. In the case of RDR based solution, the pattern count does not saturate for block of 1 × 1 mm. The number of patterns from a non-regular design and the saturation size for such design styles is at least three orders of magnitude higher than that of RDR based design styles.

The library should be designed for a prescribed cell height e.g., of 640 nm to enable a scaling factor of 0.6 × compared to a 9-track library at 28 nm with horizontal/vertical decomposition style Double Patterning (DPT) at Metal-1 and Metal-2 layers. The drawback of this solution is that it requires minimal use of Metal-3 to route the flip-flops. Although both the 0.6 × scaling and the use of Metal-3 in some standard cells is not desirable, the fact that this solution enables a technology development path using the same fundamental technology as 28 nm provides cost benefit that scales in accordance to Moore's Law.

The design methodology using Template and Connectors has the potential to simplify DfM providing communication interface between design and manufacturing. By replacing the design rule paradigm, it may reduce layout pattern counts by two orders of magnitude compared to other regular design styles. It would also guarantee a correct by construction approach where all patterns appearing in the design are known upfront in a randomly placed block as big as few tens by few tens of μm.

3.4.3.5 Summary

DRC Plus is a helpful approach for capturing manufacturability issues in DRC clean 2D situations. It should be considered a transitional methodology between "classic" DRC on random layout and the rigorous implementation of template cells. Using a combination of pattern matching and DRCs, DRC Plus is easy to integrate into the existing PV infrastructure, as it reduces the coding effort for complex 2D situations. With fuzzy matching, it allows further investigation into close matching cases where printability may be marginal. The main challenge for DRC Plus is the creation and automation of the rules.

3.4.4 Variability Reduction for 28 nm

To assess the impact of process variation on circuit performance, product value, and return on investment, new metrics such as guardbanding, parametric yield at selling point, and inferred variation tolerance add to a comprehensive taxonomy of variations for 28 nm technologies and beyond. The variability impact trends can be summarized as follows [17]:

Table 3.6 EPE Min Corr methodology showing the number of cells with reduced OPC due to EPE - timing stack correction (after [17])

Test case	Source	Cell count
c432	ISCAS85	337
c5315	ISCAS85	2,093
c6288	ISCAS85	4,523
c7552	ISCAS85	2,775
alu128	Opencores	12,403
r4_sova	Industry	34,288

1. With technology scaling and fixed levels of process variability as mandated by ITRS, delay variation decreases, whether measured as the amount of guardbanding required to suppress it, or parametric yield loss.
2. For designs containing one dominant critical path, systematic WID variation does not affect yield, and design guardbanding is most sensitive to random D2D variation control.
3. Performance sensitivity to L_{eff} variation is reduced with technology scaling due to enhanced velocity saturation and a growing number of critical paths (NCP).
4. Under the same process variations, a larger NCP results in a smaller delay variation but a larger delay mean. Because of the shift in delay mean value, a larger selling point delay is expected.
5. For the same NCP, looser control of CD variability leads to a larger required design guardbanding accompanied by a larger mean delay. For ASIC designs, reducing NCP is the most effective way to achieve a smaller average delay.
6. The delay distribution shifts to higher means but is tighter as the number of critical paths increases. This effect saturates beyond approximately ten critical paths.

One way to reduce the impact of gate 3σ CD variation on chip performance while limiting mask costs and the complexity of OPC, is to leverage EPEs to correct OPC to meet timing specifications. An iterative linear programming can be used for slack budgeting for each gate mapped to allowable critical dimensions. EPEs generated from the CD budget and tags on gates would then indicate the appropriate level of OPC correction, leading to up to 20 % reduction in data volume and runtime reduction by up to 39 %, when the OPC =>timing slack flow is used (Table 3.6).

The systematic CD variations from OPC correction residues may cause performance degradation by alternating the top critical paths timing and ordering. For an automated flow for post-OPC timing, the analysis showed that over 20 % paths may not be reported in the "classic" timing analysis, no longer valid in nanometer-scale designs. Tagging gates on critical paths enables specific corrections to be applied selectively to these gates rather than attempting to reduce global CD variation. Integrating OPC into design flow would allow design-time optimizations aware of the manufacturing process.

For restricted design rules (RDRs) as a prevention to reduce lithography induced variation at the expense of design flexibility, the range of delay is quite small (5–10 % worst-case spread) while the area and power impact is larger (up to 20 % spread in both). The minimum poly_diffusion spacing appears to be the key rule to achieve low data volume and high yield or acceptable performance. There are good arguments to introduce small sets of RDRs to reduce cost of ownership, with limited impact on yield.

For upper level metals with only a second order impact on performance variation, global RDR deployment is not cost effective. To prevent yield loss due to complex 2D pattern features, a hybrid flow enhancing standard DRC with pattern matching for local RDR enforcement is preferred. This way, pattern dependent corner cases leading to (lithography) yield problems requiring significant effort in DRC enhancement, can be avoided at an early stage. Therefore, DRC Plus appears to be the solution of choice for metal layers in and below 28 nm designs.

The methodology based on minimizing edge placement error shows potential for reducing cost of OPC on multiple layers. It may be driven by the bounds of acceptable leakage power rather than by traditional delay uncertainty constraints and extended for field poly overlap with the contact to diffusion, metal1, metal2, etc. Post-OPC CD annotation should enable Post-OPC RC extraction and timing analysis for design-time optimizations with close-to-Si process feedback.

In summary, to reduce variability and detect yield detractor patterns, global RDRs on critical layers together with locally deployed hybrid RDRs for non-critical layers are a preferred method as they reduce design complexity. "DRC Plus" should include allowed design topologies for comprehensive quality and optimized Cost of Ownership.

3.5 Practical DfM Flow for Sub-28 nm Designs

3.5.1 Yield Impact

Yield enhancement with DfM techniques for 45, 32, 28 and 22 nm logic process may be expected to bring about 5 % improvement [28]. But DfM, although convincing in theory, is often considered difficult to implement into chip design. It is perceived as a risk to increasing the chip size to realize benefits, which are sometimes difficult to quantify and measure. The variety of solutions: DRC, DRC Plus, simulations, RR, RDR, etc., also add to the confusion and limit the scope of implementation for the individual design houses. Many EDA companies specializing in the DfM tool development have gone out of business and the application of DfM techniques on silicon for prior technologies has been limited. Recently, large IC makers started to combine a number of commercial and internal tools into a comprehensive solution that can be applied to the design library, IP and full chip in regular sequence or as retrofit to correct designs afterwards. The proposed tools include DfM kits, automated fixing, and DfM cleanup in post layout correction [34].

3.5.1.1 DfM Implementation by IC Product Lines

Most SoC (System on Chip) designs are composed from libraries (Standard, Memory compiler & leaf cells etc.) and IPs, which have to be released earlier. Therefore, defining DRC or RR rules as rule based DfM is important at the alpha stage (CBC), before DfM prevention (verification). Accordingly, process-aware CBC layout based on well-defined rules is validated by DfM kits. For the IP design, optimization for P&R tech file for DfM prevention and correlation with silicon results needs to be setup sequentially. If one lose this timeframe to integrate DfM with design tools before releasing them for the IP and Chip design, most IPs would be difficult to realign with DfM later. It means that some libraries and IPs would be difficult to place in chip design even if released to DfM corrections. Therefore, it is important that DfM prevention (CBC) and solution are done completely within the timeframe provided in the design flow [34].

During process ramping-up with test chip implementation, systematic, and parametric failures are often reported. At that time, feeding those results back to design is significant to complete DfM approach. Some DfM applications are required to dynamically prevent silicon failures while some others are to improve process margin.

Well-defined hard design rules (DRC), which include process behavior, are more helpful and impact design more than vague recommend rules (RR) [19]. Thus, defining a complete design rule deck with a lot of silicon results is important. But because of timeframe mismatch between rule definition and process development, model based DfM is often required to compensate weak points in rules and to efficiently deal with failures on silicon by assigning them back to design stage. As a result, most of 45 nm through 28 nm designs have been gradually optimized with DfM methodology (Fig. 3.54).

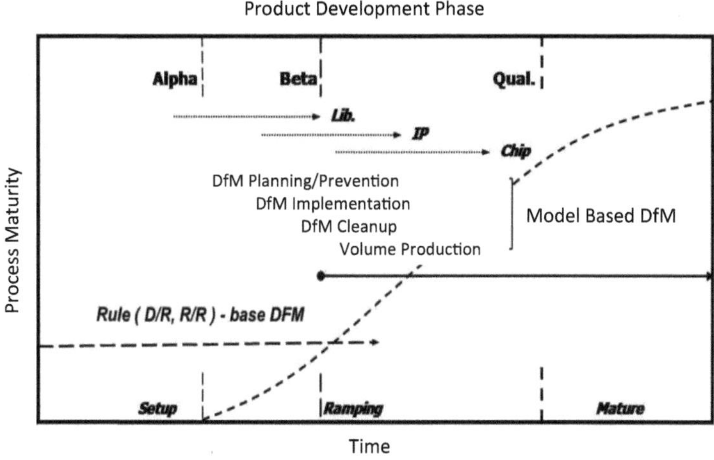

Fig. 3.54 General timeline of process development (after [34])

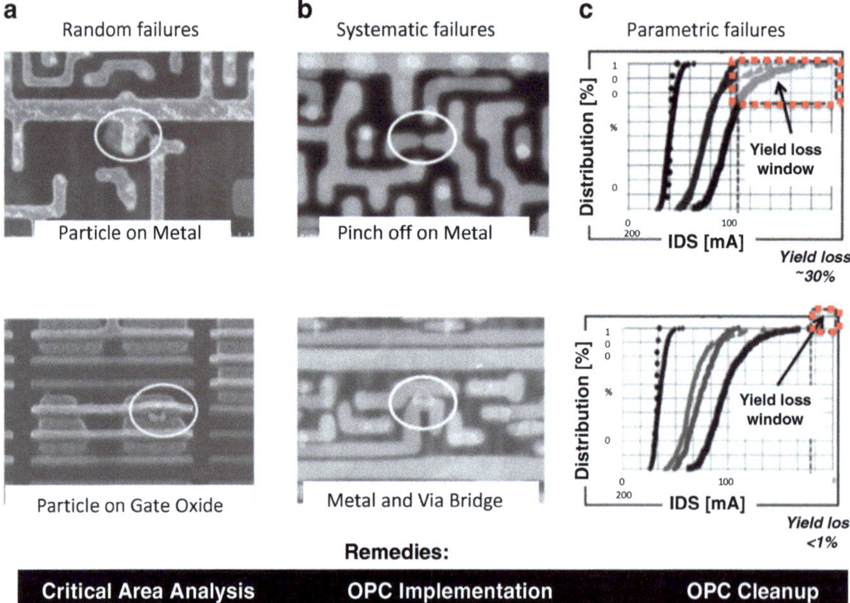

Fig. 3.55 Layout – related silicon failures and DfM remedies (after [34])

(A) Types of failures

General causes of yield loss are random, systematic, and parametric failures. Failures by particles (Fig. 3.55a) can be verified by PFA (Physical Failure Analysis) but there is no available solution to prevent them at design stage even if the root cause is found. However, knowing the particle sources, one can reduce yield loss through reducing critical area as DfM /CBC best practice.

Most systematic failures reduce process margin, which has layout or process dependency. In the past, most systematic failures showed hotspots on whole wafer due to the fact that most of them were caused by explicit layout errors e.g., related to OPC or mask fabrication. Current systematic failures are detected on wafers randomly due to low process margin (Fig. 3.55b). For key problems in classifying three types of failures [29], one can propose a ramp-up for optimization (Fig. 3.56).

As main part of DLY (Defect Limited Yield), random and systematic failures are numerous at initial process setup. Along with the process improvement and maturity, random failures, which can be controlled by process recipes & restriction of equipment usage, are reduced. Finding root cause of systematic failures is important to reduce process ramp-up. If we know the portion of systematic failures, root causes of parametric failures can be estimated from yield.

(B) Design to silicon through DfM

Relating all process information to designers is not required. DfM kits should include prevention, solution and cleamp to deploy into design domain. At present, most process information of DfM kits is linked to physical design as layout

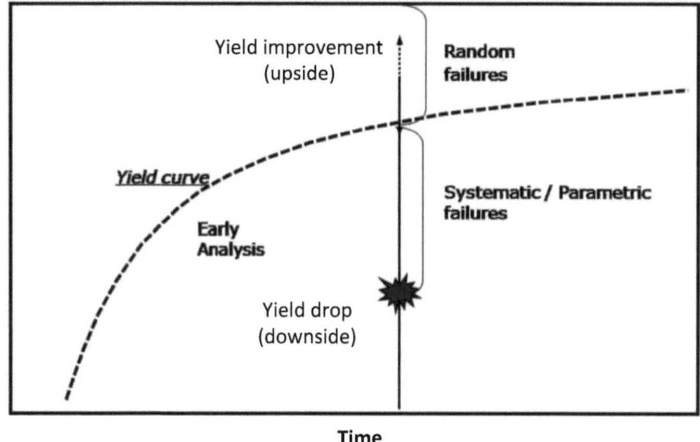

Fig. 3.56 Time – dependent yield ramp showing the impact of random and systematic failures (after [34])

can be modified by simulation before mask data is created. Henceforth, one can propose to develop electrical behavior oriented DfM kits (via implant and device simulation) for design targeting.

All process information about weak patterns, thickness variation and hotspots etc., should be added into Design Methodology through DfM kits [26], commercial and internal (Fig. 3.57, Table 2.2-1).

During process setup & ramp up, manufacturing weak points can be identified. Generally, model based DfM kits detect unknown hotspots and those results are needed for correlation between simulation and silicon. Especially, CAA kit can calculate doubled via and weighted critical area with distribution by particle size. Rule based DfM kits can be used as sign-off or verification, built with confirmed or known hotspots, which should be removed from silicon-based design. Naturally, both types of DfM kits improve process margin.

3.5.1.2 DfM Prevention Through CBC

To draw layout restrictively, one should efficiently use design rule commands or parameters embedded in parameterized cells. At metal levels, the yield issues are due to the routing. While both 45 and 32 nm technologies showed same trend of yield loss by particle (Fig. 3.58), CAA yield calculated by defect model based on manufacturing data, showed that yield loss by particles are less significant in BEOL (Back End of Line) than in FEOL (Front End of Line).

Less yield loss in BEOL could be caused by inadequate DfM defect prevention. Hence, optimization of P&R for the process-aware layout is becoming a main method to improve random or systematic failures.

Library validation (Standard, Leaf cells and I/O etc.) is another meaningful step in the design flow. However, one should not overlook the main factor about yield

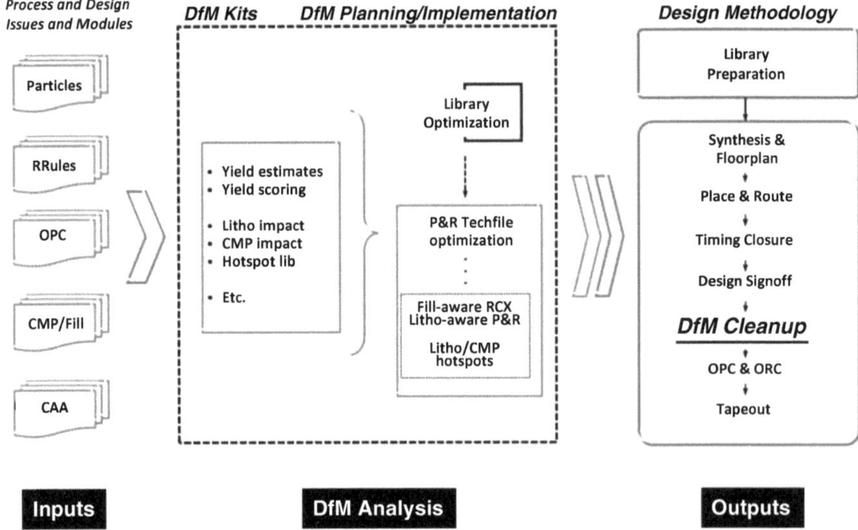

Fig. 3.57 Fab – sanctioned DfM flow from design to silicon (after [34])

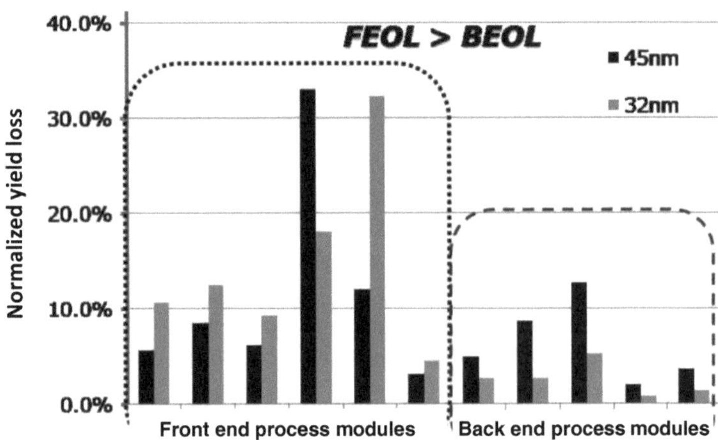

Fig. 3.58 Yield loss by process modules (after [34])

loss on the bit cells of memory. Thus, technology-design optimization for bit cell at the earlier design stage is required.

(A) Validations of custom design (standard cells, leaf cells & I/O, etc.)
Soft (non-mandatory) DfM rules cause implementation conflicts. It may be difficult to determine which rule implementation is sufficient without incurring unacceptable overhead. Another obstacle is the timeframe, which adds to the overall design cycle. For example, generation of double contacts in the standard cells is allowed by area within rule constraint, to reduce particle damage.

One can generate one Figure of Merit based on combining the benefits of various recommended rules in different categories. This calculation can be integrated into the scoring kit as commercial solution. An IP library would then be optimized using validation flow.

(B) Optimization of P&R tech file
Full chip level optimization involves enhancements of BEOL library, which must not result in full chip timing changes. Integrating DfM into P&R tools as "tech files" would allow to optimize metal layers without design iterations. Any DfM requirements that couldn't be achieved at this stage have to be later corrected with DfM cleanup.

(C) LUP (Litho unfriendly pattern)
Due to RET limitations for 45 nm and smaller CD's, there are significant patterning issues not detectable by DRC violations. Prevention of Litho-Unfriendly Pattern (LUP) in the P&R tech file is done by looking for specific geometries known to be difficult for RET and avoiding generation of these patterns during P&R phase. LUP should be defined carefully, to minimize P&R runtime. Also too many patterns cannot be labeled as LUP, as the layout work would become infeasible.

There are few types of LUP to avoid using P&R "tech files": T, H, U-shape. However, DfM prevention by CBC may still be insufficient to help designers. Optimized tech file should be applied from IP level to prevent all kinds of LUP.

Analysis of Middle-End of Line (MEOL) routing shows that most of yield loss by particles are due to metal short/open failures. To improve, metal widening and spreading has been applied to the routing to reduce critical areas as reported by the CAA kit (Fig. X), and to increase yield by 7.7 %. A decrease in critical areas for metal opens, and 0.7 % yield decrease was observed for critical area without iterations of the design flow on a test design (area over 10,000 um × 10,000 um) [34].

(D) Process hotspot repair
Although one can generate optimized tech files from P&R tools at the DfM prevention, by CBC, it is difficult to prevent all litho hospots completely. Instead, an additional PHR (Process Hotspot Repair) and Litho-aware P&R has been adopted during engineering phase to remove difficult to correct litho hot spots (Fig. 3.59). As the result, the average rate of fixing hotspots of 1× metal layers is over 95 %, if PHR was applied two times. It was also confirmed that there are no notable timing violations [30]. This solution also reduces burden for OPC.

There are many DfM kits to detect various hotspots. Typical model based DfM kits include Litho, CMP and CAA for unknown hotspot detection [31]. Rule based DfM kits such as Rule scoring, Pattern matching and LUP can also be used for known hotspots by comparing with pattern library. However, even with DfM as part of regular design flow and the sign-off, there are residual weak spots which should be addressed with layout optimization, without any material effect on the timing and within designer intent.

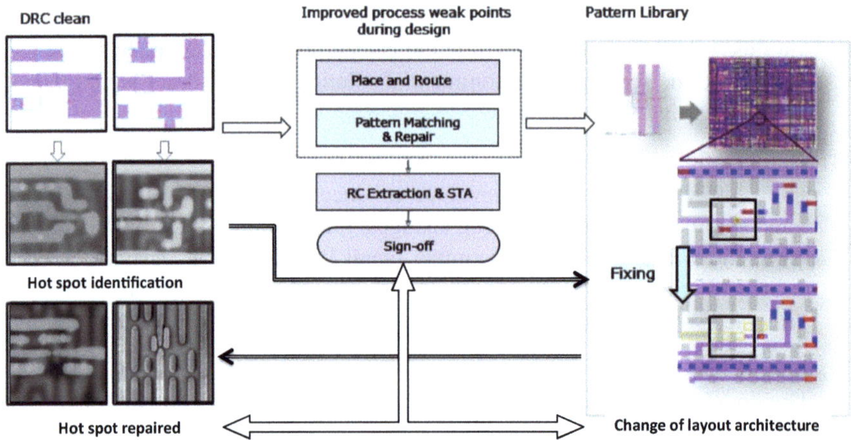

Fig. 3.59 Hotspot repair flow based on layout modifications (after [34])

(E) Fill-aware RC extraction

Another application of DfM is timing analysis. Advanced extraction of RC parameters using metal fill (or dummy) can provide more accurate timing model (Fig. 3.60a).

To calculate resistance and capacitance RC extraction flow should emulate fill information, from previous technology. However, emulation based extraction faces the challenge of technology scaling. Through timing simulation, the inaccuracy of emulated fill was over 5.0 % at 45 nm and over 6.0 % at 32 nm of test design capacitance [34].

Timing difference between real fill and emulation-based extraction (Fig. 3.60b). Confirmed meaningful difference showing that fill aware RC extraction can reflect realistic process behavior.

(F) Via position correction

In-line CD images in Nominal section of Fig. 3.61 show the effect of OPC related shifting of metal lines, due to asymmetric bias during OPC and corner rounding resulting from the limitation of the optics during patterning. This resulted in misalignment of vias generating manufacturing weak spots.

In cases where there is no room to apply symmetric bias, OPC aware design cleanup can nudge the placement of vias to better align intended features (Corrected section of Fig. 3.61). This is done without any design rule violations and no material impact on timing, due to the intelligent OPC-aware tool.

The overlap margin enhanced by Position Correction has been validated by SEM for 1× via layers on 32 nm product leading to yield improvement by 85 % defect reduction.

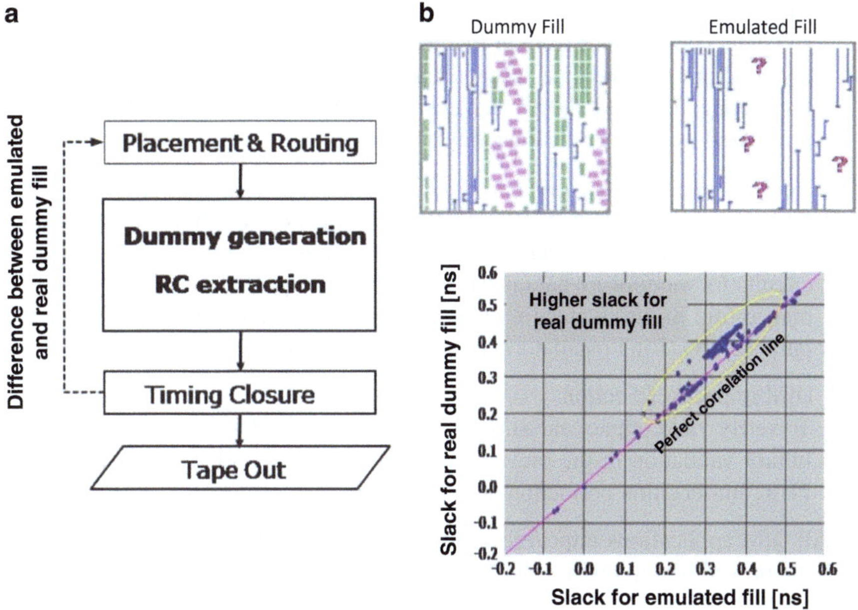

Fig. 3.60 Fill-aware RC extraction, showing small difference between fill and simulation, and impact on timing slack (after [34])

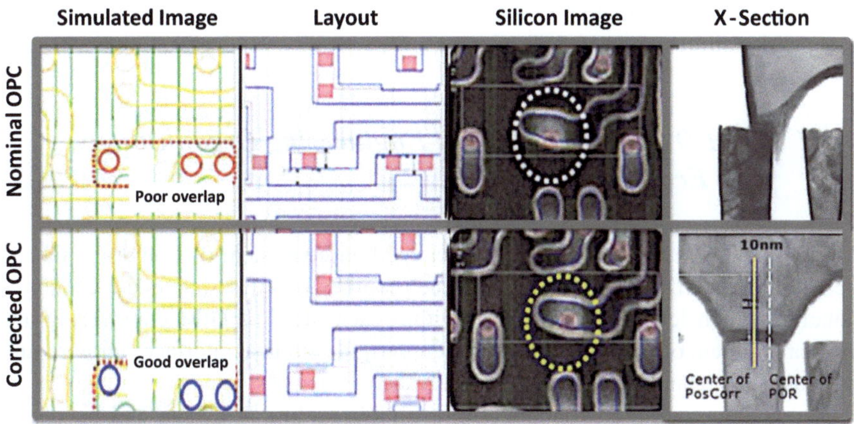

Fig. 3.61 Via overlap with nominal and corrected OPC method, with side view of overlap after correction (after [34])

3.5.1.3 DfM Diagnostic on Silicon

(A) Analysis of systematic failures

It is very difficult to understand causes of systematic failures in the logic area. Therefore, logic process yield ramp is done with SRAM test vehicles. The flow for analysis of systematic failures in logic area contains overlay analysis among candidate nets from scan fail diagnosis, candidate hotspots from DfM validation and defect coordinates from in-line inspection (a CD SEM image taken during manufacturing to indicate suspected defect locations). Hypotheses or historic knowledge for yield drop mechanisms can be translated to DfM hotspots, prioritized by the statistical overlay analysis.

(B) Analysis of parametric failures
Parametric data such as threshold voltage, frequency etc. should be tracked for multiple locations to understand OCV (On chip variation). Hence, unique IP circuits for monitoring parametric behavior are required, including AC performance (die to die, wafer to wafer & lot to lot variation/excursion). For this purpose, RO (Ring oscillators) have been widely adopted.

(C) DfM impact verification
To verify yield impact one should apply DfM methodology at chip level, with library validation in the prevention stage and checks related to BEOL. LUP, PHR, and Position correction developed with PFA results.

All DfM applications improved yield. Starting from LUP optimization of P&R, tech file proved best to restrict BEOL for process-aware design. Position correction was applied to both design data preparation stages, to misaligned via with metal which compensated insufficient OPC [34].

The second design with DfM has 30.3 % gain (ratio) of DLY compared to first design with no DfM. Design blocks showed more yield improvement proportional to die area indicating that DfM analysis is required to eliminate any systematic and parametric failures, even if TAT is increasing.

3.5.2 DfM Ownership Through Engineering Organizations: The Ecosystem

Moore's law of transistor cost scaling created an ecosystem of foundries, fabless design houses, IDMs, and suppliers. Each of these parties shared the economic benefits, which led to specialization within each layer of DfM engineering and vertical segment of the supply chain, wherein rigid technical interfaces allowed unidimensional technology trajectories to advance.

As we approach the economically justifiable scaling at 20 nm and beyond, a new set of challenges will require tighter integration along historically rigid interfaces between design architectures and process capabilities, process control needs and product sensitivities [35]. The sources of variability are magnified when aggressively scaling technology nodes are based on the same device architectures, processes, and layout design styles. Metal Gate/High-K (MGHK) stack at the 32/28 nm technology node will reduce variations due to random dopant fluctuations (RDF), but only for one or two generations. For the 15 nm technology, the only hope to limit RDF is to adopt novel device architectures such as FinFET or Fully Depleted SOI.

Aggressive scaling has unleashed the layout dependent variability, neglecting resolution limitations and the existence of stressors. Complex DfM flows attempt to model such layout dependent effects, but with poor accuracy. A pro-active DfM using pre-characterized circuit elements (IP templates) would try to mitigate systematic variations, which would have a prohibitive impact on 22 nm designs. Working around the lithographic limitations at 22 nm requires creation of a regular design fabric onto which one can efficiently map the selected templates using a limited number of printability-friendly layout patterns. The co-optimization of circuit, layout, and process is achieved by co-developing circuit functions, layout pattern library and lithography solutions. This solution replaces design rules for logic with a rigorously characterized, limited set of layout templates.

As discussed above, regular design fabrics successfully used for such co-optimization are based on gratings with deviations in forms of breaks (line-ends) in the infinitely long lines, enclosures (landing pads) around contacts and vias, or controlled wrong way routing (H-shapes), making sure that these deviations in the underlying fabric can be adequately printed using the same lithography setup, as the fixed pitch of lines.

To enable efficient circuit layout and limit the total number of patterns in the design requires a methodology for specifying locations at which selected deviations can be allowed. On the one hand, the use of regular design fabrics precludes the use of design rules. On the other hand, the usable design rules still need to be abstracted into the underlying fabric including the deviations. This shift from the conventional design rules requires a fundamental migration in the way layout designs are both conceptualized and implemented. Instead of defining minimum space and width rules for each layer, we now define the grids and pitches. In addition, the constraints for the deviations and interlayer dependencies must be specified, with a careful selection of the circuit topologies to map to the chosen set of deviations in the underlying fabric. It requires circuit and process engineers to collaborate closely to chose design pitches and grids to maximize design density and manufacturability.

Once the IP fabric is defined between process and design, the pattern constraints desired by its architecture are met in a production environment by using a combination of templates and connectors. Templates, i.e. fixed height cells, define specific pattern constructs always present in the design of the logic functions. A set of connectors join neighboring features in the template to create the logic function (Fig. 3.58). This enables designs with a limited set of layout constructs free from the technical and logistical problems of design rule based methodologies. It should also isolate library designer from the manufacturing challenges without limiting creativity in developing novel circuits and bring manufacturing experts close to selection of physical shapes to create specific circuit elements. The basic circuit-topologies are efficiently mapped to the patterns for process optimization supported by pattern specific rules. This approach can limit the total number of layout patterns compared to other regular design methodologies relying on DRC and RR.

The efficiency of DfM introduction depends on the structure of engineering organizations and the ownership definition. A comprehensive design technology infrastructure that encompasses all critical IC implementation areas to reduce design barriers and improve first-time silicon success is an Open Innovation Platform (OIP) [35]. It promotes innovation within semiconductor design ecosystem, design implementation and DfM branches, process technology and backend services. Key Open Innovation Platform elements are ecosystem interfaces and collaborative components. Defining DfM interaction flow is critical for shortening design time, minimizing time-to-volume, time-to-market and time-to-money. The Open Innovation Platform provides:

- The foundry segment's largest, most complete, silicon-proven IP and library portfolio,
- Advanced design methodology through reference flows, design for manufacturing and process design kits,
- Comprehensive design ecosystem alliance programs covering market-leading EDA, library, IP, and design services/suppliers.

3.5.2.1 Libraries and IP

Collaboration with ecosystem partners provides a comprehensive library and silicon-proven IP portfolio, checked against quality requirements and DfM standards to provide shorter design cycle times (Table 3.7). Portfolio includes IP macros/libraries across process technologies, including the 0.35-μm, 0.25-μm, 0.13-μm, 90 nm, 65 nm, 40 nm, and 28 nm nodes and 0.22-μm, 0.15-μm, 0.11-μm, 80 nm, and 55 nm half-nodes. IP ecosystem in 2010 included Soft IP through the alliance program. Soft IP has historically been process technology independent and not optimized for power, performance and area considerations. Given the ever-increasing need for first-time silicon success and early time-to-market, close technical collaboration between foundry and IP providers is imperative to optimize critical trade-offs (Tables 3.8 and 3.9).

(A) Standard cell and I/O libraries

The Design Infrastructure ecosystem has a goal to provide state-of-the-art standard cell techniques for process-optimized library portfolio. Designers choose from a large selection of standard cell libraries to meet their area, power, and speed requirements, or to provide the best tradeoff among the three benchmarking parameters. The libraries feature high density, multiple-threshold voltage, multiple Vdd back-bias control, low power control, and a proprietary routing methodology. Libraries also facilitate migration to half-nodes with limited additional design.

(B) Memory compilers

Foundries offer standard memory compilers and specialized memories for system-on-chip (SoC) embedded memory applications. The standard memory compilers include single- and dual-port SRAM, 1P/2P register files, and Via/Diff ROM compilers. The SRAM compilers use high-density bit cells for

Table 3.7 Technology dependent evolution of design infrastructure ecosystem companies (after [35])

		Technology node							
Cell type	Features	0.35 μm	0.25 μm	0.18 μm	0.13 μm	90 nm	65 nm	40 nm	28 nm
Standard cells	High speed			Dolphin integration	Arm	ARM,	ARM	ARM,	ARM
					Dolphin tech	Dolphin tech	Dolphin tech	Dolphin tech	Synopsys
					Synopsys	Synopsys	Synopsys	Synopsys	
						TSMC	TSMC	TSMC	TSMC
	High density	TSMC	ARM	ARM	ARM	ARM	ARM	ARM	ARM
				Dolphin integration	Dolphin integration.	Dolphin integration	Dolphin integration	Dolphin integration	Synopsys
				Synopsys	Dolphin tech	Dolphin tech	Dolphin tech	Dolphin tech	
					TSMC	TSMC	TSMC	TSMC	TSMC
	Ultra high density			ARM	ARM	ARM	ARM,	Dolphin tech	Dolphin tech
				Dolphin integration	Dolphin integration	Dolphin tech	Dolphin tech		
				Synopsys	Synopsys	Dolphin integration	Dolphin integration		
	Low power			Dolphin integration	ARM	Dolphin integration	Dolphin tech	Dolphin tech	Dolphin tech
					Dolphin integration	Dolphin tech	Synopsys	Synopsys	Synopsys
					TSMC	TSMC	TSMC	TSMC	TSMC
I/O library		TSMC	ARM, TSMC	ARM	ABI	ARM	ARM	ARM	Dolphin Tech
				Dolphin Tech	ARM, Cosmic	Dolphin tech	Dolphin tech	DOLPHIN TECH	TSMC
				Synopsys	Dolphin tech	TSMC	TSMC	TSMC	
				TSMC	TSMC				

(continued)

Table 3.7 (continued)

		Technology node							
Cell type	Features	0.35 μm	0.25 μm	0.18 μm	0.13 μm	90 nm	65 nm	40 nm	28 nm
Compilers	High density	TSMC	ARM Synopsys	ARM Dolphin integration Synopsys	ARM Dolphin integration Dolphin tech Synopsys	ARM Dolphin integration Dolphin tech Synopsys	ARM Dolphin integration Dolphin tech Synopsys	ARM Dolphin tech Synopsys TSMC	ARM Dolphin tech Synopsys TSMC
	High speed	TSMC		Synopsys	ARM Dolphin tech Synopsys	ARM Dolphin tech Synopsys	ARM Dolphin tech Synopsys TSMC	ARM Dolphin tech Synopsys TSMC	ARM Dolphin tech Synopsys TSMC
	Low power		Dolphin integration	Dolphin integration Synopsys	Dolphin integration Dolphin tech Synopsys	Dolphin integration Dolphin tech Synopsys TSMC	Dolphin integration Dolphin tech Synopsys TSMC	Dolphin tech Synopsys TSMC	Dolphin tech Synopsys TSMC

Table 3.8 IP Vendor and partner companies (after [35])

Type of IP	Supplier of partners
PLL	Analog Bits, ARM, Cosmic, Cadence, Mosys, S3, TCI, TSMC
DLL	Analog Bits, ARM, Dolphin Technology, IP goal, Prism, TCI
ADC	Cadence, Cosmic Circuits, Dolphin Integration, IP goal, S3 Synopsys
DAC	Cadence, Cosmic Circuits, Dolphin Integration, IP goal, S3, Synosys
Voltage regulator	Cosmic Circuits, Dolphin Integration, Synopsys, S3
ARM, ARM9, ARM9E, ARM10E, ARM11, Cortex families	ARM
4KE, 24K, 34K, 1,004K, 74K, 1,074k	MIPS
Ceva-X, Pakm, Teak, TeakLite	Ceva
ARC600, 700, ARC FPX, ARC XY	Synopsys
Diamond, Xtensa	Tensilica
USB2.0 & USB3.0	Synopsys, Snowbush, TSMC
DDR, DDR2, DDR3, Mobile DDR	Analog Bits, ARM, Dolphin Technology, Synopsys
PCI-e	Cadence, Snowbush, Synopsys, Mosys
SATA	Snowbush, Synopsys, Mosys
HDMI	Silicon Image, Silicon Library, Synopsys, Transwitch
10/100 Ethernet	Transwitch
HSTL, PECL, LVDS	Analog Bits, ARM, Dolphin Technology, Synopsys

Table 3.9 Examples of technology – dependent IP features (after [35])

	0.5 μm	0.35 μm	0.25 μm	0.18 μm	90 nm
Memory density	8K-512Kb	8K-4Mb	8K-4Mb	8K-4Mb	32 Mb
Standard bus width	8, 16	8, 16	8, 16, 32	8, 16, 32, 64	8, 16, 32, 64
Automotive macro	N/A	N/A	Available	Developing	Developing
ECC	N/A	N/A	N/A	Upon request	Upon request

maximum density on a smaller chip area. Both single-port and dual-port SRAM compliers generate instances of up to 16 Mbits. High current and ultra-low leakage bitcell help create single- and dual-port SRAM compilers for maximum speed or minimum leakage. In addition, redundant row or column features are available for fuse-based repair and yield improvement in volume production.

(C) Mixed-signal IP
Mixed-signal IP optimized for process cover a wide range of applications. Clock generating de-skewing PLLs provide competitive frequency range and jitter specifications. ADC and DAC feature a wide range of resolutions, sample rates, linearity, and area required for digital consumer and mobile applications as well as graphic processing and server chip sets.

(D) Embedded processor and DSP
Embedded processor partners provide complete IP support, with EDA, test, and software/hardware co-design systems, to shorten design cycles for first-time silicon success and ensure easy access to processor and DSP cores. For example, the single usage foundry program provides silicon-validated RISC processors for a low cost-of-adoption business model for licensable ARM cores at 90, 65, 40, and 28 nm process nodes.

(E) Electrical fuse
The Electrical Fuse (e-fuse) IP is one-time programmable (OTP) non-volatile memory (NVM) based on programming a fuse. Two e-fuse IP types – serial interface and random accessible, are used in chip ID, memory redundancy, feature selection, security keys, and parameter trimming. Starting at the 0.13-μm process node, high-density enhanced random accessible IP features an extremely compact bit cell size to facilitate redundant circuit design in high density embedded SRAM blocks. Highly reliable, low-cost e-fuses feature a low programming voltage and field programmability.

(F) High-speed interface IP
Process-optimized, high-speed serial interfaces provide top-end bandwidth performance. The portfolio includes IP for consumer connectivity, memory, computing, and storage interfaces, and wired or wireless connections, as well as high-speed backplane applications. These IP are silicon-validated across all targeted technology nodes.

(G) Embedded flash IP
Embedded Flash IP in non-volatile process technologies, with high-speed and low-power features, meet a wide range of applications. For example, IP with small sector sizes can replace EEPROM functions.

(H) Embedded DRAM IP
Embedded DRAM, a high-density memory IP for process technologies down to 40 nm, meets need for high-density, on-chip memory configurations, including both low-power and general-purpose derivatives, with memory density available in 1 Mb granularity, power saving options, such as sleep mode. The IP provides a pipeline and a flow-through SRAM interface with 32–256 bit data bus width for design flexibility. A built-in circuit extends data retention, reduces soft error rate (SER) and improves production yield.

3.5.2.2 IP Quality Management in Advanced Foundry Design

All IP libraries are rated for quality standards. The library and IP quality program is a structured assessment system to provide designers the confidence to use third-party IP (3PIP), increases first-time silicon success, and shortens yield ramp. The program consists of series of quality checks at five levels: Physical assessment, Pre-silicon Assessment, Typical Silicon Assessment, Split lot Silicon Assessment, and Volume Production. Assessment status reports are available on Design Portals.

Design flow infrastructure has been put together, based on Reference flow, foundry DfM, and process design kits (PDK's) across the ecosystem alliances.

(A) Reference flow

Reference Flow as a foundry design service standard is aimed at enhancing its DfM capabilities, resulting in a comprehensive portfolio encompassing multiple consecutive releases by the partner companies (Tables 3.10 and 3.11).

Foundry design methodology addresses critical design challenges driving advances e.g., for 28 nm design infrastructure. Reference Flow features various enhancements in 2.5 – dimensional and 3-dimensional integrated circuits (2.5D/3D ICs) using silicon interposer and design verification of through silicon via (TSV) (Table 3.12). The 28 nm model-based simulation, increased DfM speed and advanced Electronic System Level (ESL) design initiative enabling process technology PPA (power, performance, and area) are to be integrated into system-level design. Recent Reference Flow releases disclose 20 nm Transparent Double Patterning to build up 20 nm design capability within OIP.

The flow is a key collaborative component of OIP that accelerates time-to-market, improves return on design investment and reduces design infrastructure duplication. Extensive EDA partner collaboration is the key feature of Reference Flow, offering advanced design methodologies for risk reduction and easy adoption of foundry process.

The components include: ESL, SoC, Interconnect Fabric, Design Enablement, Low Power, Tuning, 3DIC TSV/ Silicon Interposer and SiP.

(B) Foundry design for manufacturing

The foundry Design for Manufacturing initiative to produce more good dice per wafer originated as a compilation of manufacturing data and development with ecosystem partners by applying manufacturing-related information to the design implementation stage. DfM Data Kits (DDKs) and tool-specific utilities enhance yields and accelerate time-to-volume. Foundry internal expertise and resources addresses design-specific DfM improvements prior to tape out.

Design for Manufacturing architecture consists of a DfM-Driven Desktop Approach, accessed through the DfM Toolkit from qualified third-party EDA tools and IP. The toolkit includes comprehensive data and utilities for design implementation including up-to-date advisories with descriptive guides, DfM utilities, and DDKs. Designers can implement DfM through qualified tools or through EDA partner DfM service.

Unified Design For Manufacturing (UDfM) architecture targets 40 nm and smaller process nodes and geometries. It was developed in collaboration with EDA vendors and other design infrastructure partners to provide a unified, encapsulated access to foundry data. It includes a DfM Design Kit (DDK) with an embedded software engine, an interoperable API, and process-related DfM data and models. UDfM inserts an exact copy of factory tool chain and process models into IC design tool chains, providing access to more foundry manufacturing data.

Table 3.10 Foundry design flow using mixed-signal process-design kit (PDK) (after [35])

	PDK					
Operation	Schematic entry ⇨	Simulation ⇨	Layout ⇨	Layout check ⇨	Post-sim ⇨	Release as IP Standard cell, I/O, or tape out
Supporting	Symbol and component	Simulation environment	Layout edit Tech file	DRC/LVS/ERC tech files	Layout parasitic	
Information files	Description file	SPICE model Netlist views	Parameterized cells Auto-router Tech file Layout utilities	LVS Netlist	Extraction (RC) SPICE model	

Table 3.11 IP Alliance and EDA partner companies (after [33])

IP alliance	EDA partners
ARM	ADTek
Arteris	Agilent Technologies
Chips&Media	Alchip
COSMIC Circuits	Apache
Delphia Technology	Arteris,
Ememory	ATopTech
FINALOG BITS	ATRENTA
Hitachi ULSI	Azuro
Imagination	Cadence
Intrinsic	Carbon
Kilopass	CIRANOVA
Cadence	CWS
MIPS	Dorado
Mixel	Dorado
MoSys SLi Silicon Library Inc.	EdXact
NANGATE	eSilicon
Novelics	EXTREME DA
Rambus	FASTRACK
Rapid Bridge	FORTE
RENESAS	GUC
Sidense	Helic
Silicon Image	ICC
Snowbush IP	IMEC
SOFICS	Integroud Software, Inc.
SONICS	Legend
Synopsys TCI	LORENTZ SOLUTION
Tensilica	MAGMA ANSYS
THETA	Mentor Graphic
TRANSWITCH	MunEDA
YMC	NANGATE
	Nanno
	Open-Silicon
	PGC
	SIGNITY
	SILVACO
	Solido
	Spring-Soft
	Synopsys

Table 3.12 Example TSV design verification flow

1	Design rule check on first device
2	Design rule check on second device
3	Design rule check associated with TSV of first/second device
4	Extract layout file of interface layer of first and second devices
5	Physical design rule check for interface layers
6	Connectivity check for interface layers

This "copy exact" method compensates for increasing manufacturing variances in advanced technologies, improves the efficiency of design solutions to fix the actual manufacturing hotspots (vs. simulated hotspots), and delivers a high level layout of accuracy. This DfM architecture has to be able to handle a very large dataset and design complexity.

(C) Process design kits
Design kits provide custom digital, analog, and mixed-signal/RF designers the head start based on symbols for each device linked to the device model and layout. The PDKs cover the entire design flow, from schematic entry, pre-simulation, layout, and layout check to post-layout simulation. PDKs cover CMOS, Si Ge, high voltage, and CMOS image sensors, encompassing technologies from 0.6 μm to 28 nm. The entire suite of technology and command files is posted online, for direct and e-mail technical support to increase designer productivity.

(D) Design ecosystem alliances
The IP Alliances take advantage of the industry's comprehensive catalog of silicon-verified, production-proven IP, and process-specific libraries to enable design reuse and integration. IP cores are validated in silicon through testchip prototyping service.

Expanding IP Alliance to incorporate a soft IP program will improve its readiness for advanced technology nodes and reduce time-to-market.

(E) EDA alliance
The EDA (Electronic Design Automation) Alliance, consisting of leading EDA companies, provides process technology files and PDKs. Alliance members work with design technology services to implement design methodology and Reference Flows. Through the Alliance, EDA companies gain access to technical insights to validate their tools and methodologies. The industry's most popular design and verification tools have updated tech files posted online.

(F) Value chain aggregators
The Value Chain Aggregator (VCA) partnership enables customers to benefit from foundry technology by leveraging partner expertise. VCA partners help extract full capabilities of technology and OIP ecosystem, and deliver them in the form of finished semiconductor products.

The VCA program extends foundry ability to serve the marketplace. VCA members as independent design service companies work to help system, ASIC companies, and emerging start-ups bring their innovations to production. VCAs integrate design enablement building blocks that are part of the Open Innovation Platform and provide specific services at each link in the IC value chain, including IP development, design backend, wafer manufacturing, assembly and test.

(G) DFM trivia
Successful architecting of DfM ecosystem depends on the common understanding of DfM facts among the players, summarized in a set of trivia, i.e., ten Semiconductor Manufacturing DfM Issues Every Designer Should Know [30].

1. Know your limits: DfM should help you understand the limits of your manufacturing partner and operate effectively within them.
2. Don't dump your tool flow: Find ways to make it manufacturing-aware.
3. Insist on accuracy: DfM-awareness helps only if the information is process calibrated.
4. Don't settle for error reports: Finding things like litho hot-spots is useful, but a method to correct the defect is better.
5. Know the terminology: Understand the difference between litho hot spots, critical area analysis (or yield hot-spots), and CMP variation.
6. Don't change careers: If a DfM approach requires you to become an expert in lithography or IC processing, run away. You shouldn't need to become an expert in semiconductor manufacturing to design manufacturable chips.
7. Examine pedigree: Look at a vendor's manufacturing competency. Vendors that don't have experience serving IC manufacturing are likely to stretch DfM rules.
8. Work closely with your fab: Unless design, manufacturing, and EDA are all coordinated, DfM efforts are likely to fail.
9. Read the fine print: Make sure you understand what you're getting. Are there models needed for a new tool? If so, who will build and verify them? What does post-sale training and support look like?
10. Don't take your eye off the ball: Always remember your goal – whether it's improved yield, lower power, or higher clock rate. And make sure your DfM tools help you achieve that goal. Don't be sidetracked by interesting analysis that won't allow you to directly move your design toward the goal.

(H) Design center alliance
The Design Center Alliance (DCA) is a network of qualified IC design centers that take design ideas from concept to finished product. DCA provides a wide range of IC implementation services to reduce design, manufacturing and schedule risks.

3.5.3 Summary

The ownership of DfM flow has been unclear for several technology generations. Here, the approach of Design-Foundry Ecosystem makes a good attempt to identify ownerships of the different DfM aspects. It is of increasing popularity due to the leveraging the diversified expertise and enabling the flow of information and products for improved yield at a lower cost.

The ecosystem should work to the advantage of both small and large players in their respective alliances.

3.6 New Process Effects Around 28 nm and Beyond

DfM challenge at 28 nm is not limited to dealing with ever – shrinking geometries and their printabilities. It has to provide solutions to new process effects impacting device physics.

The use of mechanical strain has become a significant booster for silicon CMOS technology over the recent years. Strain changes the semiconductor's band structure and modulates the conduction mass, intervalley and interband scattering properties of electrons and holes, to improve or degrade the characteristics for n- and p-type MOSFETs.

3.6.1 Implications of Mechanical Proximity Effects for Analog Design

Two significant small-geometry effects, well proximity and STI stress, need to be accounted for in analog circuit design [40]. For well proximity, device performance is impacted by layout features located near, but not being parts of the device. Bias points can shift by 20–30 %, causing potentially catastrophic failures and MOSFETs placed close to a well edge exhibit graded channels [41–46].

The pMOS conductivity is typically enhanced while nMOS conductivity is diminished by 15–20 % with STI stress [47]. Since analog designs often reduce footprint by sharing oxide definition (OD) regions, they incur parametric shifts due to STI stress. The well proximity may affect threshold voltage of MOSFETs more than 1 μm from the well edge. If not taken into account, current mirrors can be shifted out of saturation, leading to catastrophic failures.

Another small-geometry effect is poly-proximity. The polysilicon gate of a dummy MOSFET impacts the active MOSFET next to it. STI stress from active regions to the OD tiles used to maintain OD density for chemical mechanical polishing (CMP) impact neighboring MOSFETs. Because analog design depends on "matched" devices to attain very high precision, that impact, even if nominally small, may be non-negligible.

Matched devices, particularly current mirrors, are also impacted by systematic offsets due to well proximity. STI stress related to multifingered MOSFETs, and various layout configurations of a current mirror need to be evaluated to observe the effectiveness of STI stress offset mitigation techniques.

3.6.1.1 Well Proximity Effect

High energy implants form the deep retrograde well profiles are required for latch-up protection and suppression of lateral punch-through [44]. During the implant process, atoms can scatter laterally from the edge of the photoresist and become embedded in the silicon surface in the vicinity of the well edge [46] (Fig. 3.62). The surface concentration changes with lateral distance from the mask edge, typically over the range of 1 μm. This lateral non-uniformity in well doping known as the well proximity effect (WPE) causes MOSFET threshold voltage to vary.

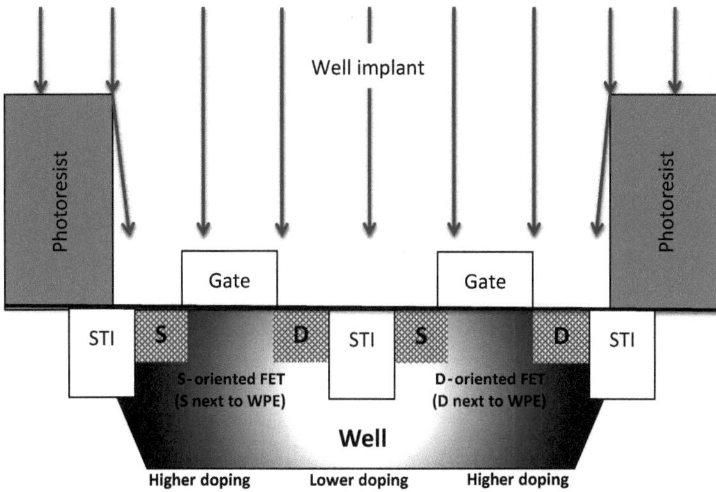

Fig. 3.62 Well proximity effect (after [40])

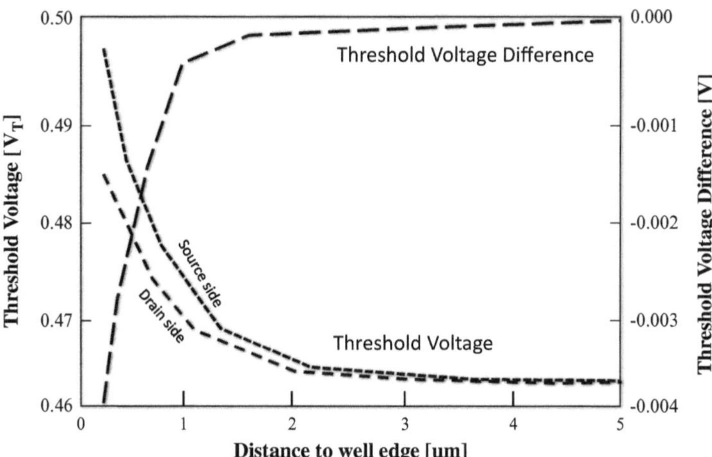

Fig. 3.63 V_T versus well-edge distance for 3.3V nMOS device (after [40])

The V_T may increase by as much as 50 mV as the device moves closer to the well edge also depending upon the S/D orientation, accounting for about 5 mV of offset (Fig. 3.63). WPE extends for at least a few tenths of a μm from the well-edge.

The saturation current offset between identical transistors in the array and the baseline device (Fig. 3.64) can have mismatches in the drain current as large as 30 % for $V_{gs} > V_T$, depending upon the proximity to a well edge. Although the V_T is consistently higher for devices with source oriented towards the well edge (S-oriented) in Fig. 3.64, the current for this device may actually be larger than the drain-oriented device in the near – V_T and above – V_T regions (Fig. 3.65).

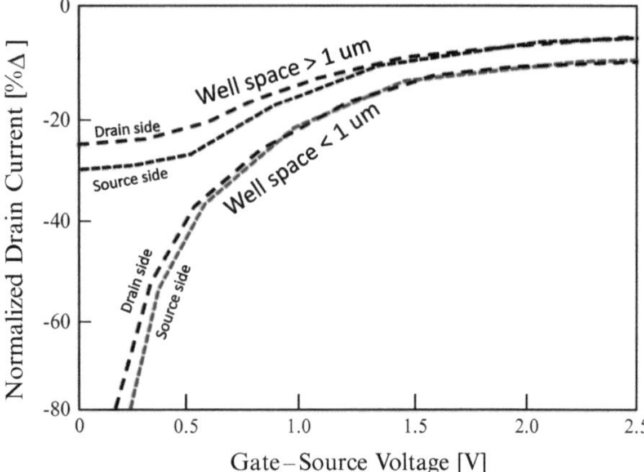

Fig. 3.64 The relative difference in drain current (I_d) versus gate voltage (V_g) for two well-edge spacings (after [40])

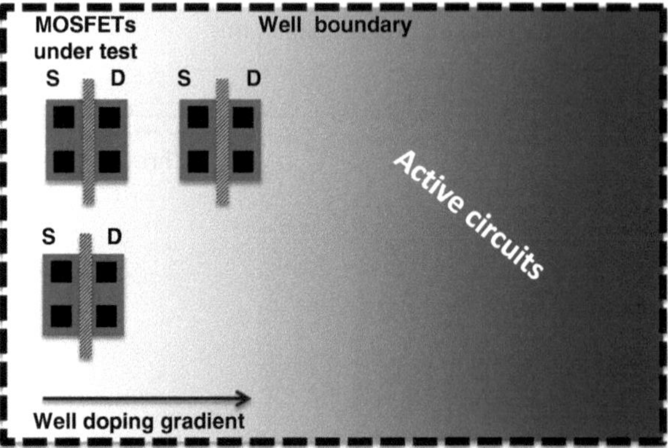

Fig. 3.65 A sample layout for a matched device near a well edge. Two devices oriented in the same direction S-oriented for matched devices to avoid S/D shadowing and pocket implant asymmetry (after [40])

WPE would introduce current offset into a current mirror where the well gradient is aligned with the direction of the channel length (Fig. 3.66) and multiple output transistors are distributed within a common well. The lateral current flows in devices oriented in the same direction, which is a standard recommended practice for matching. The reference transistor is far from well edge and has negligible well proximity induced offset.

For the current mirror, the output current versus V_{ds} at $V_{gs}=1$ V, $V_{bs}=0$ V in Fig. 3.67 follows the expected behavior for a device with negligible well proximity. This curve crosses 0 where the reference and output transistors are equal. For the WPE effects, the current offset is strongly dependent on the location of the output transistor. Even with more than 3 μm of distance between the well edge and the polysilicon gate, there is roughly 3–4 % offset in the current.

The S/D orientation dependency is created by the graded channel (Fig. 3.65). Normalizing the S-oriented device to the D-oriented device, both nMOS and pMOS I_d-V_g curves show the systematic I_d offset. If the channel dopant is graded from the source to the (left transistor in Fig. 3.65b), the highest dopant concentration sets the

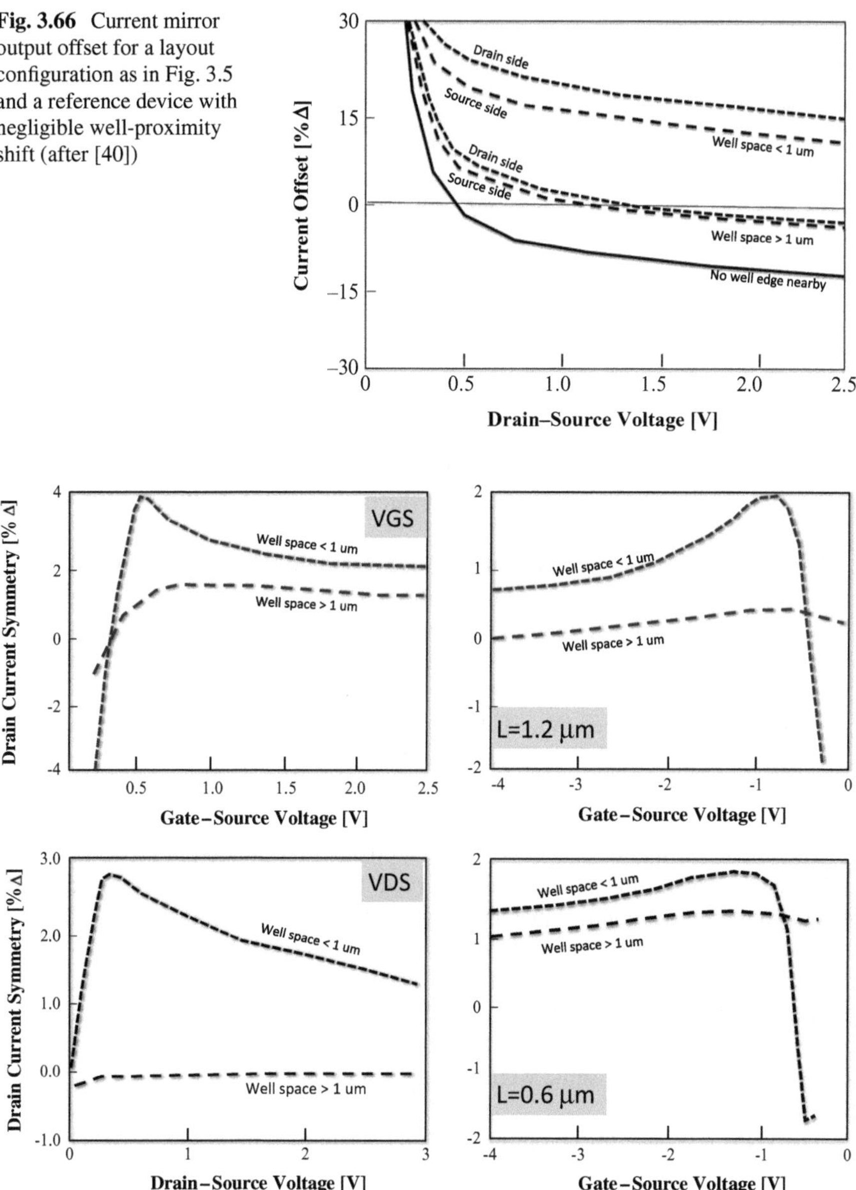

Fig. 3.66 Current mirror output offset for a layout configuration as in Fig. 3.5 and a reference device with negligible well-proximity shift (after [40])

Fig. 3.67 Source/drain asymmetries in drain current for different well spacings (after [40])

threshold voltage. At the V_{gs} where the heavy doped region begins to invert, the rest of the channel has already inverted. The resultant lateral electric field enhances the channel conductivity.

Scaling of the graded channel would result in bias conditions unchanged except for the gate length which would be reduced. As the channel length decreases, the source side of the left device in Fig. 3.65 remains in place, while the drain side is

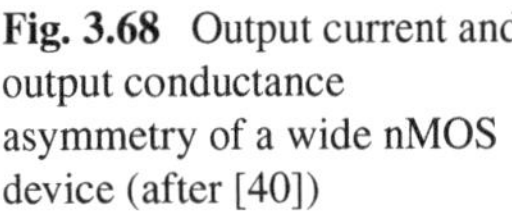

Fig. 3.68 Output current and output conductance asymmetry of a wide nMOS device (after [40])

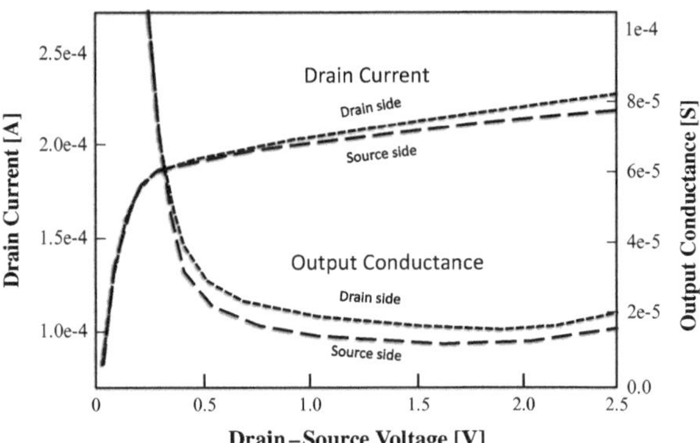

pulled to the left. The channel still has the same peak dopant concentration near the source but not as much in the graded channel. It has nearly the same V_T shift as the previous device but less g_m enhancement which corresponds to the cross-over point at a higher V_{gs}.

Digital designs, which almost always use minimum channel length, will likely only see an effective V_T increase but not the g_m enhancement. Analog designers, on the other hand, rarely use minimum gate lengths (2× to 8× longer than the minimum). Graded channel modeling is critical to analog designs with small well-edge distances, with a potential disconnect in characterization.

The linear region sweep ($V_{ds}=0.1$ V) would not exhibit the 9m enhancement that creates the cross-over effect in Fig. 3.67 since V_{ds} is low. Many devices are characterized by evaluating the V_T and g_m in the linear region only. Therefore, the cross-over effect and the graded channel could be overlooked.

The impact of the WPE-induced graded channel on the output current is minicling the behavior that should be observed for a graded channel device [50]. Although the output conductance has degraded for the D-oriented device, the shift appears to be only ~10 %, compared to S-orientation (Fig. 3.68).

Another layout configuration (Fig. 3.69) contains matched pair of MOSFETs oriented in the same direction and placed in a common well with symmetric space to well edge. A logical solution to the WPE it would contain devices oriented in the same direction, with the same well spacing values and the same S/D orientation to the well.

The WPE response for layout in Fig. 3.69 is summed up in Fig. 3.70. There are four combinations of the two source/drain orientations and the two devices (output and reference). The solid line represents the expected performance of the current mirror in the absence of WPE. WPE has introduced a large current mismatch, depending on the D/S placements. Even if both MOSFETs are oriented in the same direction with respect to the well-edge, the bias points are shifted because the dopant scattered from the well edge changes the output conductance.

SPICE models for well proximity consider the contribution of a well edge to be independent of orientation. This approach misses graded channel effect with implications for analog design, but including it would add significant complexity related to layout information.

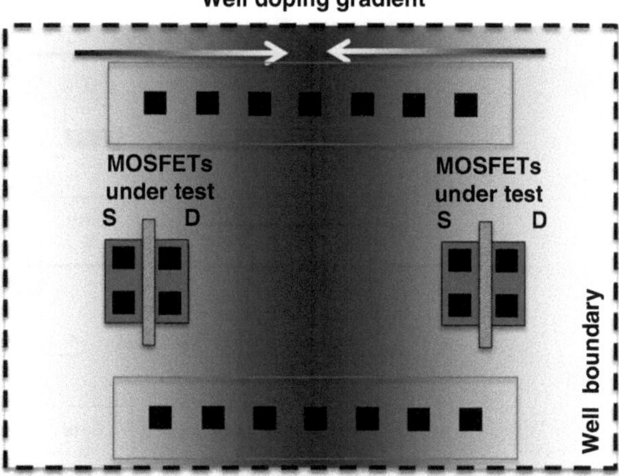

Fig. 3.69 MOSFET placement that exacerbates the source/drain orientation asymmetry due to the well proximity gradient, consistent with best practices but still susceptible to the well dopant gradient (after [40])

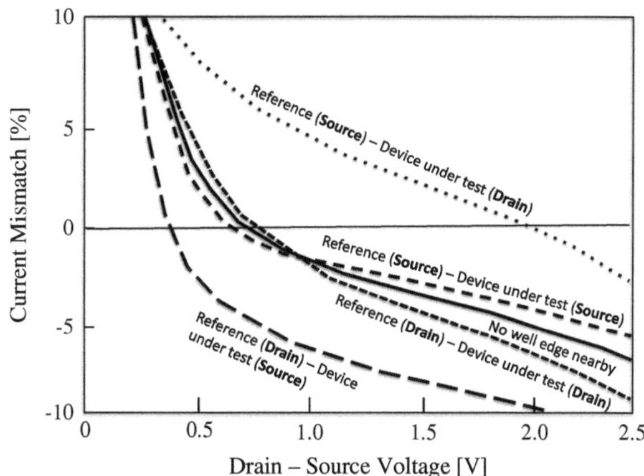

Fig. 3.70 Current mirror output current, normalized to the reference current, for various orientations of the S and D with respect to the well-edge (after [40])

3.6.1.2 STI Stress

The STI process leaves behind a silicon island that is in a non-uniform state of bi-axial stress [44, 46] shown to impact device performance, introducing I_{Dsat} and V_T offsets which must be included when modeling the transistor. The stress is non-uniform and dependent on the overall size of the active opening, meaning that MOSFET characteristics are a strong function of layout (Fig. 3.71). The residual stress and corresponding parametric shift can be described by two geometric parameters, S_a and S_b, representing the distance from the gate to the edge of the OD on either side of the device. MOSFET parameters such as V_T, peak g_m and I_{dsat} have been shown to vary linearly with the following function:

$$\text{Stress} = \frac{1}{S_a + \frac{L}{2}} + \frac{1}{S_b + \frac{L}{2}} \tag{3.3}$$

(where L is MOSFET gate length)

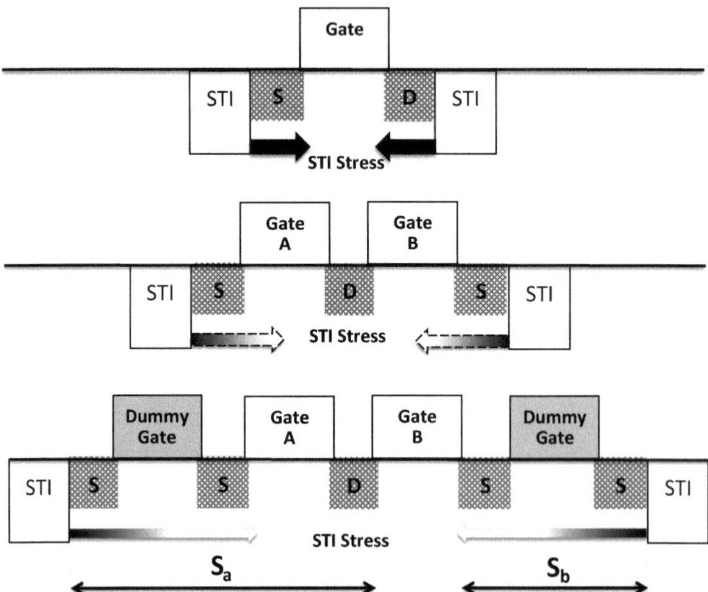

Fig. 3.71 STI channel stress, i.e., the combined effect of the size of the oxide definition (OD) region, the location characterized by the distances (S_a and S_b) of the gate edge to the OD edge and the size of the MOSFET (after [40])

incorporated into SPICE models. A device with large S_a is expected to be closest to a stress free device. There could be significant saturation current offset between the transistors in the array and the baseline device.

High V_{gs} causes an increase in pMOS and a corresponding decrease in the nMOS with decreasing S_a value because bi-axial compressive stress enhances hole mobility and degrades electron mobility. As the V_{gs} is lowered, the current offset in the nMOS devices increases significantly, especially for the devices with the smallest S_a. Changes in V_T due to STI stress are ascribed to stress enhanced/inhibited diffusion. The STI model included in SPICE models provide a reasonable prediction of the impact of STI stress, but does not include stress effects on effective channel length due to dopant redistribution. The g_m dependency on S_a, S_b flips direction as L decreases, indicating a non-constant $\Delta L = (L_{drawn} - L_{eff})$ competing with the p-type mobility enhancement. The model [48] would not capture this trend.

Layout configurations demonstrate the impact on output current and saturation voltage mismatch standard deviation and area for the non-cascaded current mirror biased with reference current.

There is substantial benefit to including dummy devices (Figs. 3.72, 3.73, and 3.74). A single dummy device appears to be very effective, while additional dummy devices offer marginal extra return. Without dummy buffers, the non-merged layout can shift the ratio to less that 1:3 overshooting planned design margin, and shifting the output device into the linear region.

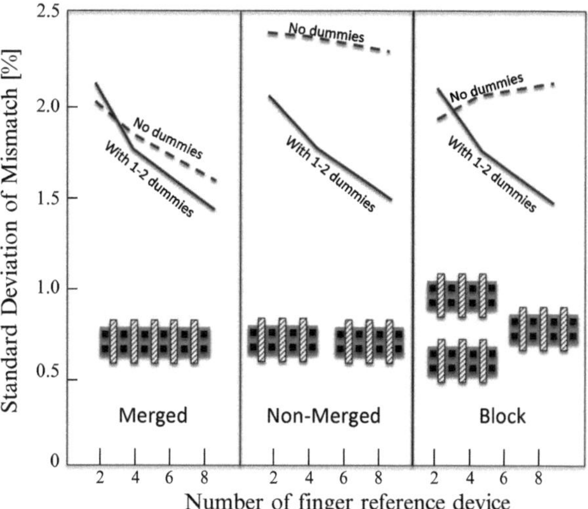

Fig. 3.72 Dependence of mismatch on the layout schemes (after [40])

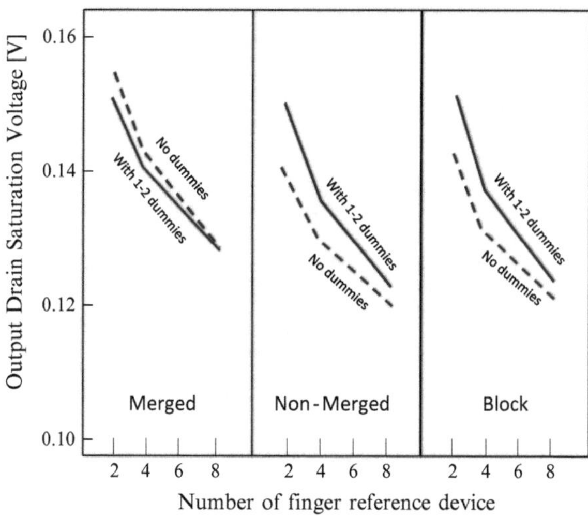

Fig. 3.73 Dependence of V_{dsat} on the layout schemes (after [40])

One difficulty in handling STI stress occurs in the CAD flow for analog designs. Multifinger layouts are common practice to reduce parasitic junction capacitances, save area, and improve the effectiveness of cross-coupling (common centroid) schemes. A single transistor divided into multiple fingers (Fig. 3.75) can be parameterized as a single instance MOSFET with lumped dependencies on the number of real or dummy devices. If the parametric shift due to stress is large, the bias point for each device is different and a single instance MOSFET cannot physically model the multifinger device. Merged, cross-coupled layouts cannot be parameterized since the connections are arbitrary. Each finger in the array has a different stress bias point. The alternative is to create a separate instance for each finger, but that would explode the size of the netlist.

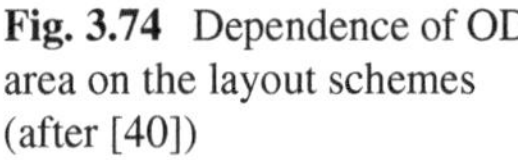

Fig. 3.74 Dependence of OD area on the layout schemes (after [40])

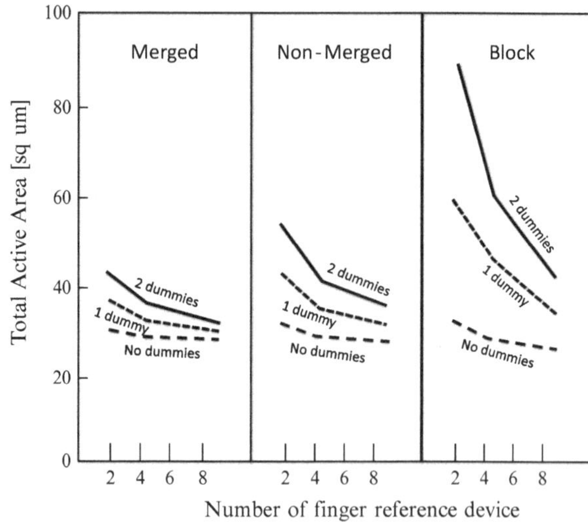

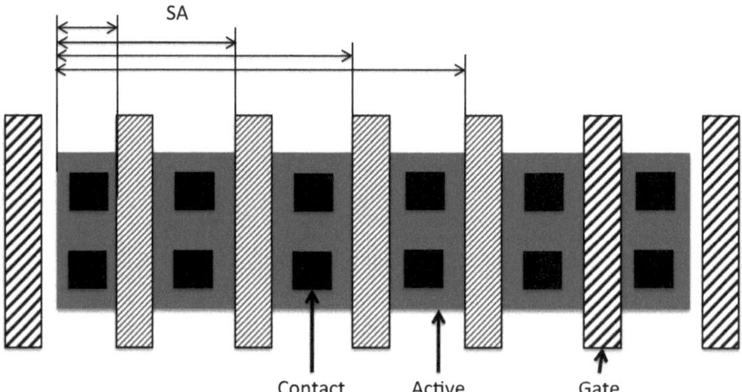

Fig. 3.75 Layout of nested transistor structure with dummy gates

An easy way around the WPE is to uniformly increase the active to well spacing. The STI induced offset can be controlled by creating identical blocks for current mirror ratios. But, these solutions consume area and analog design already lags in scaling with technology shrinks due to local variation requirements (i.e. matching).

Unfortunately, the problem is not conducive to CAD solutions. The proximity effects are not determined until the layout stage, but the impact needs to be considered at schematic entry and simulation. Backannotation of the layout attributes onto the schematic can assure consistency between layout and schematic, but understanding this interaction is tedious. Extraction of the S/D orientation from layout is problematic, since the MOSFET is symmetrical, and the S, D are arbitrary in layout.

Although models exist for proximity effects, the analog designer has no visibility into these within the CAD framework. No simulations can account for the source and magnitude despite being potential sources of catastrophic circuit failure.

3.6.1.3 Liner Boundary Proximity

STI-induced embedded SiGe (eSiGe) strain relaxation and dual-stress-liner (DSL) boundary proximity can cause channel mobility degradation for both n – and p-MOSFETs as they reduce channel strain along the <100> direction.

MOSFET performance has significantly increased by strained-Si engineering and the stress memorization technique (SMT) [48]. Among these, the DSL technique (both tensile and compressive liners on the same wafer to improve NMOS and PMOS performance simultaneously) has potential for large performance gains. For DSL, less attention has been paid to the boundary proximity effect, especially below the 45 nm node.

STI proximity was studied for n-MOSFETs with SMT and tensile stress etch stop liner (TESL) and p-MOSFETs with embedded (E) Si_x–Ge_{1-x}S/D (x=0.23–0.25) using energy-dispersive X-ray spectrometry EDS and compressive-stress etch stop liner (CESL).

Both the geometric phase analysis (GPA) and nanobeam diffraction (NBD) techniques can be used for accurate 2-D mapping of channel strain in MOSFETs. Strain scans were taken in the horizontal direction (or channel direction), approximately 10 nm below the top surface of the Si channel in the test structure (Fig. 3.80a), averaged over the 5-nm-wide area inside the channel.

The <110> strain distribution across the test structures, scanned using the NBD technique show the expanded lattices in ESiGe S/D applying compressive stress, resulting in compressive strain along the channel. Channel strain decreases as the transistor gets closer to active edge (or STI) resulting from eSiGe strain relaxation at the STI/SiGe interface. To quantify a decrease in channel strain, the average channel strain values for two different directions are obtained from both GPA and NBD techniques. Offsets (or strain loss) in the two techniques are comparable, indicating a significant amount of channel strain reduction, resulting from eSiGe strain relaxation at the active edge. The strain loss, can result in about 30 % reduction in channel mobility and a –15 % reduction in saturation velocity. While these numbers seem to be of arbitrary importance, one should note that, if embedded in the template cell architectures, they may actually impact the entire technology platform.

For transistors with different S_a values, as S_a decreases, I_{dsat} decreases approximately by 12 %, which agrees with the estimated velocity saturation reduction of 15 %.

Diamond ESiGe indicating ~66 % mobility degradation of the ISO test structure, relative to the MUL (multi-finger) test structure, results from STI-induced ESiGe strain relaxation in both the source and drain regions and roughly matches the estimated mobility degradation (lower by 50 %). Note that the estimated degradation of 30 % is doubled because the (isolated) test structure has minimum S_a values on both S/D sides.

Diamond ESiGe is a better performance booster than U-ESiGe because it provides better control of compressive strain in the channel. However, from an STI proximity effect perspective, U-ESiGe shows (U-shapted) less STI-induced ESiGe

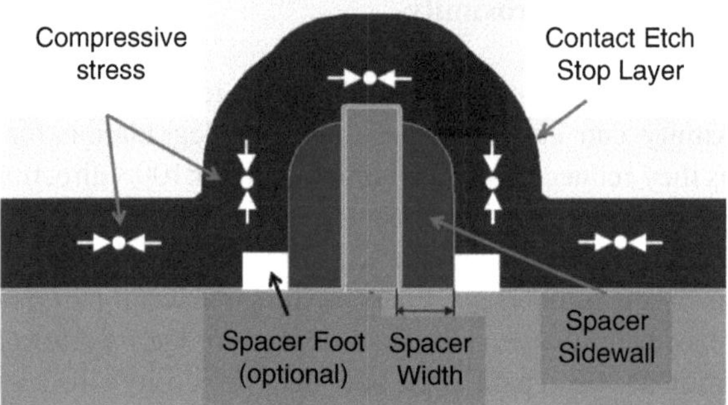

Fig. 3.76 MOSFET with tensile strained CESL and D – shaped spacers (after [52])

strain relaxation in U-ESiGe than diamond ESiGe (7 % less ID_{SAT} degradation in U-ESiGe, as compared with diamond ESiGe). Between the transistor performance and the STI proximity effect, depending on the shape of ESiGe, a device with diamond ESiGe can provide better transistor performance with enhanced p-channel mobility, but at the cost of more degradation due to the STI proximity effect compared with the U-ESiGe device.

DSL boundary proximity causes a decrease in compressive channel strain due to: (1) additional tensile strain from TESL in the proximity of DSL boundary, opposite to compressive channel strain, and (2) reduced contact area of CESL to substrate and/or amount of CESL.

For both n- and p-MOSFETs, a decrease in DSL gap causes the degradation of channel mobility. Although there is a more significant mobility reduction in p-MOSFETs (−25 % to −30 %) compared with n-MOSFETs (−12 % to −14 %), the amount of channel strain reduction is comparable between n- and p-MOSFETs. The negative/positive change in strain value means a decrease in tensile/compressive strain respectively for n/p-MOSFETs. This strain reduction can result in −12 % to −17 %/−5 % to −8 % of ID_{SAT} degradation in p/n-MOSFETs [48].

In conclusion, layout dependent channel mobility and strain characterized by strain measurement and electrical data show that (1) STI-induced ESiGe strain relaxation can result in a significant proximity effect in devices with diamond ESiGe and (2) DSL boundary proximity can significantly reduce channel strain, impact transistor performance variation and needs to be properly captured in SPICE models.

3.6.1.4 Stress Induced by Contact Etch Stop Layer in 22 nm Designs

One technique to introduce strain inside a transistor channel is the use of a silicon-nitride contact-etch-stop layer (CESL) with an intrinsic stress (Fig. 3.76). The nitride is deposited in a CMOS flow after the source/drain (S/D) and gate

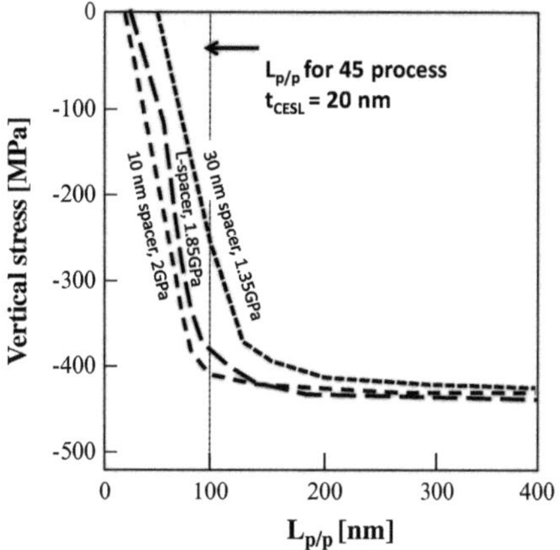

Fig. 3.77 Examples of vertical channel stress as a function of poly-to-poly length $L_{p/p}$ for various CESL thicknesses (constant spacer width $W_{Spacer}=40$ nm) and various spacer widths (constant $t_{CESL}=20$ nm) (after [52])

silicidation module, as a stopping layer for the contact etching between the first level of metal and the transistor's S/D and gate regions. CESL has a typical thickness of 20 nm and can contain up to 3 GPa of tensile or compressive stress (Fig. 3.77). The intrinsic stress inside the CESL translates into a stress on the MOS channel. A tensile stress improves the nMOS performance, while a compressive intrinsic stress is beneficial for the pMOS behavior.

Similar to the SiGe S/D-technology (compressive channel stress in pMOS transistors), CESL is a local stress agent. Both SiGe and CESL techniques have the advantage that the channel stress increases if the transistor length is scaled. But the CESL thickness, the transistor's gate length and height, the intrinsic stress in the CESL, the spacer thickness, and the active-area length and width complicate device modeling.

For a scaled nested-transistor structure, which is a chain of gates on one active area, separated by a distance of $L_{p/p}$, i.e., the poly-to-poly length building block, the amount of CESL between the consecutive poly lines is reduced. Stress simulations e.g. using finite-element simulator (Taurus Process of Synopsys) focus on short-channel transistors ($L_g<50$ nm), with the stress in the center of the channel, 1 nm below the silicon surface. The goal is to indicate how channel stress changes as a function of active-area length and to point out the essential differences between the parallel and vertical stress components induced by CESL in the channel. Typically, nested transistors have the silicon channel in a (110) direction on a (100) wafer and use polysilicon as gate material and silicon nitride as spacer material (Young's modulus values of 162 and 192 GPa, respectively). The silicon substrate is modeled as a mechanically anisotropic material.

(A) CESL stress physics
CESL intrinsic tensile stress is counteracted by stress in S/D, gate, and spacers, which is translated into a transistor channel stress. For tensile intrinsic stress, the CESL on the S/D regions pulls apart the poly regions, as well as the channel underneath. On the spacer sidewalls, the vertical shrinkage of the CESL leads to a compressive vertical stress in the gate material and in the

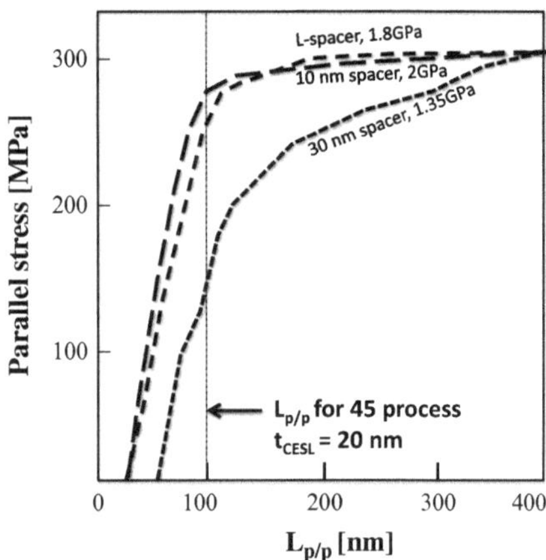

Fig. 3.78 Examples of parallel channel stress as a function of poly-to-poly length $L_{p/p}$ for nested structures with D-shaped and L-shaped spacer (after [52])

transistor channel. The CESL compresses the top part of the gate in the x-direction but has little or no effect on the channel stress, since for short channels, the amount of CESL on top of the poly lines is limited. In the z-direction, for wide transistors, the "plane strain" condition applies, leading to a small channel stress. Both the parallel and vertical stresses are generated by the interaction of the CESL with the gate topography. Stress techniques reduce the gate topography when the epitaxial layers are grown above the original S/D regions. As a consequence, the effectiveness of a CESL is expected to decrease.

The CESL-induced channel stress with +1 or +2 GPa of tensile intrinsic stress (positive stress values) results in a tensile parallel stress in the channel and a compressive vertical stress, z-direction, is negligible). Thicker CESL leads to higher channel stress, but it starts to saturate for CESL thicknesses above 40–50 nm. Increasing the intrinsic CESL stress to +2 GPa enhances the channel stress, changing the intrinsic CESL stress from tensile to compressive. This leads to compressive parallel channel stress and tensile vertical stress (Figs. 3.77 and 3.78).

Since a tensile parallel stress and a compressive vertical stress are favorable for the electron mobility, CESL with a tensile intrinsic stress enhances the nMOS performance. For pMOS, only the parallel stress induced by CESL is important, since hole conduction is insensitive to vertical stress. CESL with a compressive intrinsic stress provides compressive parallel stress in the channel and, therefore, improves the mobility in pMOS transistors.

Layout variations that leave the topography unchanged, are expected to have only minor influence on CESL induced stress. The CESL on the S/D regions is responsible for the parallel channel stress. For the vertical stress, the CESL material on the spacer sidewalls is crucial.

For large poly-to-poly spacings, the amount of CESL on the sidewalls remains constant if $L_{p/p}$ changes. As a consequence, the vertical channel stress is independent of $L_{p/p}$. However, if the poly-to-poly distance becomes small, the CESL on the spacer sidewalls starts touching. Further reduction of $L_{p/p}$ reduces the amount of CESL on the sidewalls and, therefore, reduces the vertical channel stress. To have a layout-insensitive vertical stress, $L_{p/p}$ needs to be larger than the "drop-off" poly-to-poly length $L_{p/p,0}$:

$$L_{p/p} > L_{p/p,0} = 2 \cdot t_{CESL} + 2 \cdot W_{Spacer}$$

where t_{CESL} is the thickness of the CESL, and W_{Spacer} is the spacer width.

When $L_{p/p}$ has become so small that there is no CESL left on the spacer sidewalls, the vertical channel stress is reduced to zero (Fig. 3.73, right). This occurs at the poly-to-poly length $L_{p/p,\sigma 0}$ when the spacers start touching:

$$L_{p/p,\sigma 0} = 2 \cdot W_{Spacer}$$

The intrinsic stress in the CESL has been adapted for each t_{CESL} to obtain a similar vertical channel stress at large $L_{p/p}$. The vertical stress remains relatively constant, as long as $L_{p/p}$ is larger than $L_{p/p,\sigma 0}$. Thicker CESL leads to a larger $L_{p/p,0}$ and is therefore more sensitive to layout variations. However, $L_{p/p,\sigma 0}$ is independent of the CESL thickness. For the 45-nm technology node [34], with a typical $L_{p/p}$ around 100 nm, even a 20-nm-thick CESL shows a reduction in the channel stress for a spacer width of 30 nm. Only a thin CESL with a high intrinsic stress can provide an $L_{p/p}$-insensitive channel stress.

$L_{p/p,0}$ is also a function of the spacer width W_{Spacer}. Reducing the spacer width makes the CESL-induced channel stress less dependent of $L_{p/p}$. Moreover, $L_{p/p,\sigma 0}$ is reduced for thinner spacers. For 20-nm-thick CESL, a very thin spacer is needed to have a layout-insensitive technology down to poly-to-poly distances of 100 nm. Thinner spacers using a lower intrinsic stress in the CESL still maintain the same vertical channel stress (Fig. 3.74b). This is more advantageous than scaling the CESL thickness, to keep a higher intrinsic CESL stress.

Layout sensitivity depends on the spacer architecture. Compared to the commonly used D-shaped spacers (Fig. 3.71), the simulated L-shaped spacer (Fig. 3.72) has a similar sidewall thickness (Fig. 3.73), but with a total foot width of 30 nm, it can be compared to nested transistors with 10- and 30-nm-wide D-shaped spacers. The spacer foot reduces the effectiveness of the CESL, such that a higher intrinsic stress is needed than in a 10-nm-wide D-shaped spacer. However, the L-shaped spacer is more effective in CESL stress transfer than the 30-nm-wide D-shaped spacer. Since the vertical channel stress is determined by CESL on the spacer sidewalls, layout dependent effects are expected to be valid as well for the L-shaped spacers (W_{Spacer} is the thickness of the spacer sidewall).

Strained CESL on the S/D areas is responsible for the parallel-stress generation in a transistor channel. Unlike for the vertical stress, where the channel stress remains constant as long as the CESL's on the opposite spacer sidewalls do not touch, scaling of $L_{p/p}$ reduces the amount of CESL on the S/D areas even for large $L_{p/p.}$ The reduction of CESL material on the S/D areas occurs in a similar way for every thickness t_{CESL}. For a fixed $L_{p/p}$, increasing the spacer thickness decreases the amount of CESL on the S/D regions, which leads to decreased channel stress. The condition for zero parallel stress is reached when there is no CESL material left on the S/D regions and is therefore given by (2), similar as for the case of the vertical stress.

The parallel channel stress starts to decrease for $L_{p/p}$<200 nm, down to about 50 % when $L_{p/p}$ decreases to 100 nm (Fig. 3.78). The channel stress is reduced by 20 % for a scaled nested transistor with an $L_{p/p}$ of 100 nm. Scaling the spacer width has the advantage that the same parallel channel stress can be obtained with a lower intrinsic stress of the CESL.

When an L-shaped spacer is used, the presence of the spacer foot reduces the parallel stress into the channel, such that a higher intrinsic CESL stress is required than for a 10-nm-wide D-shaped spacer. However, the stress transfer is more efficient than for a 30-nm-wide D-shaped spacer.

(B) Scalability of the strained CESL technology

The effectiveness of strained CESL reduces for shorter poly-to-poly lengths $L_{p/p}$. However, when advancing to the next CMOS technology nodes, several other transistor parameters will be scaled in conjunction. Typical nested transistors for the 45-, 32-, and 22-nm technology have spacer dimensions showing a significant impact on the effectiveness of CESL-induced stress, also depending on three spacer scaling schemes:

- Worst case, the spacer width W_{Spacer} is constant between the technology nodes,
- Nominal – a spacer that scales W_{Spacer} with the technology node,
- Best case, where the spacer width is zero.

In the coming technology nodes, the poly height will be scaled linearly with poly length, such that the gate's aspect ratio increases. It can be considered as worst case for the stress scalability, because CESL-induced stress decreases with poly height [52].

The bar diagram (Fig. 3.79) shows the absolute value of the channel stress in nested transistors. Maintaining a constant spacer width degrades both the vertical and parallel channel stresses. For constant spacers, decreasing the poly-to-poly length leads to $L_{p/p}<L_{p/p,0}$ for the 32-nm node (1) and, eventually, to $L_{p/p}<L_{p/p,\sigma_0}$ for the 22-nm node. Even for scaled spacers, $L_{p/p,0}$ does not scale to the same amount as $L_{p/p}$ when the CESL thickness is kept constant, leading to a slight decrease in vertical stress when scaling the technology node (Fig. 3.77). The parallel stress can be maintained (Fig. 3.78). For best-case spacers, the vertical stress decreases slightly when scaling the technology, while the parallel channel stress increases.

(C) Summary

Channel stress induced by strained CESL related to the poly-to-poly distance of nested transistors, has two main components: a parallel stress, in the direction of the electrical conduction, and a vertical stress, perpendicular to the plane of the wafer. The parallel stress is caused by CESL material on the S/D

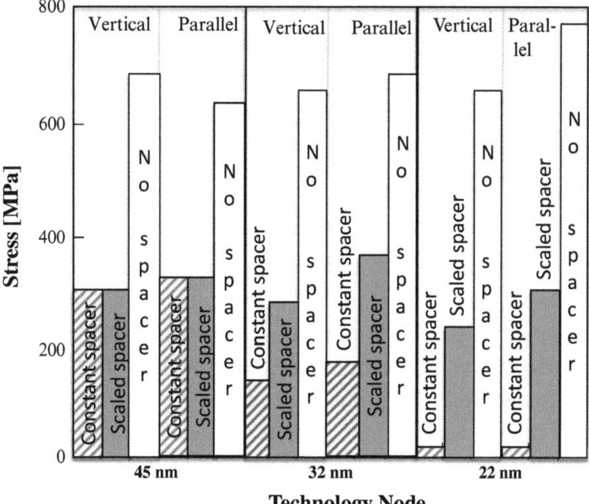

Fig. 3.79 Example values of Vertical and Parallel channel stresses in nested transistors of the 45-, 32-, and 22-nm technology nodes with D-shaped spacers, (CESL thickness 20 nm, intrinsic stress = +2 GPa)

regions, while vertical channel stress is mainly generated by CESL on the spacer sidewalls (as indicated by stress simulations of nested transistors with varying Lp/p, CESL thickness), and spacer width.

When scaling nested transistors from one technology node to the next, it is indispensable that the spacer widths are scaled proportionally to the gate length. If the spacer width is kept constant, the effectiveness of a strained CESL technology diminishes rapidly.

The layout sensitivity of the vertical stress depends both on the CESL thickness and spacer width. As long as the CESL on the spacer sidewalls from the neighboring transistors does not touch, the vertical stress is preserved. Therefore, thinner CESL and spacers make the CESL-induced vertical stress less sensitive to active-area length variations. As compensation for the thinner CESL, higher intrinsic stress is needed to maintain the channel stress, but thinner spacers can be combined with a lower intrinsic stress of the CESL.

Layout sensitivity of the parallel stress is a function of the spacer width but not of the CESL thickness. For 30-nm-wide spacers, the parallel stress is reduced by about 50 % in scaled nested transistors of the 45-nm node (Lp/p = 100 nm), even for thin CESL with high stress. This indicates that spacer scaling is a must to maintain an optimal parallel stress.

For L-shaped spacers, the thickness of the spacer on the gate sidewalls determines the layout sensitivity. These spacers have the advantage over thick D-shaped spacers that a lower CESL intrinsic stress is required to obtain a similar channel stress. The layout sensitivity is reduced for L-shaped spacers. Compared to thin D-shaped spacers with the same sidewall width, however, there is less stress transfer into the transistor channel.

When using a strained CESL-based technology both for nMOS and pMOS, such as a dual CESL technique, a very thin highly stressed CESL is not enough to be insensitive to layout variations. In particular, for pMOS transistors, aggressive spacer scaling is required to maintain the highest possible stress levels in the most dense layouts.

3.7 Conclusions

This chapter presents a snapshot of issues related to IC DfM for the 28 nm technology node and below. It is intended to focus on the DfM issues going forward, which are less precisely defined compared to "classic" DfM described in Chap. 2. The new DfM categories include pattern transfer DfM flow, CBC approach, mechanical stress, and more detailed disciplines such as OPC, RET, IP protection.

For the recent years, issues related to pattern transfer were dominated by few trends:

- The advanced equipment trend, based on developing the EUVL paterring, as the MfD based design transfer to wafer was trying to align with Moore's Law,
- The increased complexity trend, relying on the split pattern approach combined with advanced lithography techniques, as technology alignment to Moore's law through EUVL became questionable
- The restricted geometries approach, requiring simplification of the printable pattern, via design rule restrictions and CBC cell libraries, as the new lithography did not suffice to make progress aligned with Moore's law.

In this Chapter, we focused on the third trend, with the other two being widely discussed in the literature pertaining to manufacturing solutions. According to this trend, a designer should adhere to a set of design rules for preparing polygons for the initial layout that correspond to a desired circuit. The rules are formulated as two-dimensional criteria, related to overlay tolerance, critical dimension (CD), minimum and maximum spacing between polygon shapes, etc. They may be expressed in terms of tolerance bands around the desired design shapes. The designer will combine the requirements of the circuit logic with design rules, to arrive at an initial circuit layout, typically (two-dimensional) polygons. Design rules would include tolerances, constraints and other criteria related to performance and electrical characteristics, as well as manufacturability rules, for example, related to lithographic processes and overlay tolerances. An initial database is typically assigned the same polygon layout as provided by the circuit layout. The initial mask may be written out as a data set as input to the lithographers at the manufacturing site for further analysis and modification.

For the 20 nm technology nodes, even more critically than for the prior technologies, the circuit image on the photomask may not be reproduced precisely on the wafer, in part because of optical effects among transmitted and blocked energy. Here, similarly to the prior technologies, the process of modifying the initial design to form an actual mask layout would also include optical proximity correction (OPC) and resolution enhancement techniques (RET, incl. data preparation, "Data-Prep") in addition to all template – or rule based layout restrictions. OPC as the deliberate and proactive distortion of photomask patterns to compensate for systematic and stable errors may occasionally conflict with the design intent. It is still expected to consist of rule-based OPC done by determining the correctable imaging errors, especially for the layers with large CD's, calculating appropriate photomask compensations and applying the calculated corrections to the photomask layout. At the same time, template architecture would not save litho engineering from model-based OPC based on the capturing the imaging characteristics in a mathematical model that represents the complicated lithographic process. The two modeling steps: calculating the expected on-wafer circuit image to be projected by the mask pattern, and comparing the simulated image contour placement to the edge placement of the original pattern, iteratively adjusting the patterns until a suitable match of the simulated image to the desired on-wafer target pattern, within specified tolerances and other layout rules, is expected for both device and routing layers. The mask layout rules would also need to include mask house manufacturability requirements, which may need to be applied during the design of the circuit layout despite all the recent advances in RET methodology. It is expected that, for the 20 nm node, the on-wafer target pattern would be intended to reproduce the layout of polygons of the database layout which represents what the designer intends to be printed on the wafer.

IC DfM is no longer possible for the 28 nm nodes and below, without the insight into stress effects. Unlike OPC simulations, stresses are not intuitive and cannot be visualized or comprehended without a model. 2D scaling has to be subordinate to material interactions in 3D, across the temperature ranges.

This will be discussed in Chap. 4.

References

1. Liebmann, L., Pileggi, L., Hibbeler, J., Rovner, V., Jhaveri, T., Northrop, G.: Simplify to survive, prescriptive layouts ensure profitable scaling to 32 μm and beyond. In: Proceedings of SPIE, vol. 7275, p. 72750A (March 2009)
2. Liebmann, L., Baum, Z., Graur, I., Samuels, D.: DFM lessons learned from altPSM design. In: Proceedings of SPIE, 6925, 69250C (2008)
3. Perez, V., et al.: Convergent automated chip-level lithography checking and fixing at 45 nm. In: Proceedings of SPIE, 7275 (2009)
4. Hui, C., et al.: Hotspot detection and design recommendation using silicon-calibrated CMP model. In: Proceedings of SPIE, 7275 (2009)
5. Liebmann, L.: Layout impact of resolution enhancement techniques: impediment or opportunity? In: ISPD'03, Monterey (2003)
6. Webb, C.: Intel design for manufacturing and evolution of design rules. Proc. SPIE **6925**, 692503 (2008)
7. Pileggi, L., Strojwas, A.J.: Regular fabrics for nano-scaled CMOS technologies. In: Proceedings of ISSCC (2006)
8. Jhaveri, T., et al.: Maximization of layout printability/manufacturability by extreme layout regularity. Micro Nanolithogr. MEMS MOEMS **6**(03), 031011–033000 (2007)
9. Abercrombie, D., Elakkumanan, P., Liebmann, L.: Restrictive design rules and their impact on 22 nm design and physical verification, EDPS2009 (2009)

10. Volkov, A., Routing Technologies for 28 nm and beyond, Chip Design Magazine, Summer 2011
11. White Paper: Tips and techniques for 28 nm design optimization, Altera Corporation, November 2011
12. Beylier, C., Moyroud, C., Granger, F.B., Robert, F., Yesilada, E., Trouiller, Y., Marin, J.-C.: Fully integrated litho aware PnR design solution. Proc. SPIE **8327**, 83270A (2012)
13. Hurley, P., Kryszczull, K.: Replacing design rules in the VLSI Design Cycle, Proc. of SPIE **8327**, 83270B1–B6 (2012)
14. Dai, V., Capodicci, L., Yang, J., Rodriguez, N.: Developing DRC plus through 2D pattern extraction and clustering techniques. Proc. SPIE **7275**, 727517 (2009)
15. Russel, P.: Norvig, Artificial Intelligence. A Modern Approach. Prentice Hall, Englewood Cliffs (1995)
16. Duda, R.O., Hart, P.E., Stork, D.G.: Pattern Classification. Wiley, New York (2001)
17. Mitchell, T.M.: Machine Learning. Mc Graw Hill, New York (1997)
18. Cao, Y., Lu, Y.-W., Chen, L., Ye, J.: Optimized hardware and software for fast, full chip simulation. Proc. SPIE **5754**, 407–414 (2005)
19. Jang, J.: Manufacturability aware design. Ph.D. thesis, The University of Michigan (2007)
20. Taur, Y., Ning, T.H.: Fundamentals of Modern VLSI Devices. Cambridge University Press (1998)
21. Pileggi, L., Schmit, H., Strojwas, A.J., et al.: Exploring regular fabrics to optimize the performance-cost trade-off. In: Proceedings of the ACM/IEEE DAC (2003)
22. Webb, C.: Layout rule trends and affect upon CPU design, design and process integration for microectronic manufacturing IV. In: Wong, A.K.K., Singh, V.K. (eds) Proceedings of SPIE, 6156, 615602 (2006)
23. Eclair Pattem Matcher End User's Guide, Version 5.0, CommandCAD, Inc., March 2006
24. Yang, J., Cohen, E., Tabery, C., Rodriguez, N., Craig, M.: An up-stream design auto-fix flow for manufacturability enhancement. In: 43rd Design Automation Conference (2006)
25. Torres, J.A., Berglund, N.C.: Towards manufacturability closure process variations and layout design. In: IEEE EDPS Workshop (2005)
26. Torre, J.A.: Litho-friendly design: a necessary complement to RET. Microlithogr. World **15**(2), 10–13 (2006). 12-13A
27. Strojwas, A.: Cost effective scaling to 22 nm and below technology nodes. In: VLSI Technology, Systems and Applications, pp 1–2 (2011)
28. Hoppe, W., Roessler, T., Torres, J.A.: Beyond rule-based physical verification. In: Proceedings of SPIE, 6349(190) (2006)
29. Chiang, C., Kawa, J.: Three DfM challenges: random defects, thickness variation, and printability variation. In: IEEE Asia Pacific Conference on Circuits and Systems, 2006. APCCAS 2006, p.1099, 4–7 Dec 2006
30. Peter, K., März, R., Gröndahl, S.: Litho-friendly design (LFD) methodologies applied to library cells. Proc. SPIE **6349**(14), 63490E (2006)
31. Capodieci, L.: From optical proximity correction to lithography-driven physical design (1996–2006): 10 years of resolution enhancement technology and the roadmap enablers for the next decade, optical microlithography XIX. In: Flagello, D.G. (ed) Proceedings of SPIE, 6154, 615401 (2006)
32. Gianfagna, M., Liebmann, L., Pileggi, L., Hibbeler, J., Rovner, V., Jhaveri, T.: Greg Northrop, Private Communication
33. Jang, D. et al.: DFM optimization of standard cells considering random and systematic defect. In: ISOCC (2008)
34. Paek, S.W., Kang, J.H., Ha, N., Kim, B.M., Jang, D.H., Jeon, J., Kim, D.W., Chung, K.Y., Yu, S., Park, J.H., Bae s., Song, D., Noh, W., Kim, Y.D., Song, H.S., Choi, H.B., Kim, K.S., Choi, K.M., Choi, W.C., Jeon, J.W., Lee, J.W., Kim, K.S., Park, S.H., Chung, N.Y., Lee, K.D., Hong, Y.K., Kim, B.S.: Yield enhancement with DFM, Design for manufacturability through design-process integration VI. In: Mason, M.E., Sturtevant, J.L. (eds) Proceedings of SPIE, 8327, 832704 (2012)

35. TSMC Open Innovation Platform: www.tsmc.com.tw (2011)
36. Paek, S.W., Jang, D.H., Park, J.H., Ha, N., Kim, B.M., et al.: Enhanced layout optimization of sub-45 nm standard: memory cells. SPIE, 7725, 72751M (2009)
37. Kahng, A.B., Samadi, K.: CMP fill synthesis: a survey of recent studies. IEEE Trans. Comp. Aid. Design **27**(1), 3–19 (2008)
38. Simmons, M.C., Kang, J.H., Kim, Y., et al.: A state-of-the-art hotspot recognition system for full chip verification with lithographic simulation. SPIE, 7974, 79740M (2011)
39. Ha, N., Lee, J., Paek, S.W. et al.: In-design DFM CMP flow for block level simulation using 32 nm CMP model. SPIE, 7974, 79740W (2011)
40. Drennan, P.G., Kniffin, M.L., Locascio, D.R.: Implications of proximity Effects for Analog Design, IEEE CICC (2006)
41. Postnikow, S., Hector, S.: ITRS CD error budgets: proposed simulation study methodology. In: International Technology Roadmap for Semiconductors, May (2003)
42. Tabery, C., Page, L.: Use of design pattern layout for automated metrology recipe generation. In: Proceedings of SPIE: Metrology, Inspection and Process Control for Microlithography XIX, vol. 5752, pp. 1424–1434 (2005)
43. Hook, T., et al.: The dependence of channel length on channel width in narrow-channel CMOS devices for 0.25–0.13 μm technologies. IEEE Electron. Device. Lett. **21**(2), 85–87 (2000)
44. Su, K.W., et al.: A scalable model for STI mechanical stress effect on layout dependence of MOS electrical characterization. In: IEEE CICC, pp. 245–248 (2003)
45. Scott, G., et al.: NMOS drive current reduction caused by transistor layout and trench isolation induced stress. In: IEEE IEDM, pp. 827–830 (1999)
46. Bianchi, R.A., et al.: Accurate modeling of trench isolation induced mechanical stress effects on MOSFET electrical performance. In: IEEE IEDM, pp. 117–120 (2002)
47. Miyamoto, M., et al.: Impact of reducing STI-induced stress on layout dependence of MOSFET characteristics. IEEE Trans. Electron. Devices. **51**, 440–443 (2004)
48. Choi, Y.S., Lian, G., Olubuigde, O., Chung, J., Riley, D., Baldwin, G.: Layout variation Effects in Advanced MOSFECTs: STI-Induced Embedded Sige strain Relaxation and Dual Stress Liner Boundary Proxinuty Effect. IEEE Trans. Electron. Devices. **57**(11), 2886–2890 (2010)
49. Hook, T.H., et al.: Lateral ion implant straggle and mask proximity effect. IEEE Trans. Electron. Devices. **50**, 1946–1951 (2003)
50. Sheu, Y.M., et al.: Modeling well edge proximity effect on highly-scaled MOSFETs. In: IEEE CICC, pp. 831–834 (2005)
51. Kumar, D.V., et al.: Evaluation of the impact of layout on device and analog circuit performance with lateral asymmetric channel MOSFETs. IEEE Trans. Electron. Devices **52**, 1603–1609 (2005)
52. Enemon, G., Verheyen, P., De Keersgieter, A., Jurczak, M., de Meyer, K.: Scalability of stress induced by contact-etch-stop layers: a simulation study. IEEE Trans. Electron. Device **54**(6), 1446 (2007)

Chapter 4
New DfM Domain: Stress Effects

The mismatch of thermal properties among the IC component materials results in thermo-mechanical stress inside and around the devices [1–3]. It is tempting to divide the sources of such stress into the intentional and non-intentional ones or intrinsic and extrinsic. However, a better distinction would be whether we are able to take advantage of them in product implementation (intrinsic) or are they outside the device model space (extrinsic). When dividing them by the source of stress, one may identify the ones at die level, i.e., built into silicon (discussed in Chap. 3), and the ones at package level, i.e., between the chip and its package. A DfM methodology (Chip-Package Integration, CPI) for controlling stress, using design rules and material properties for both chip and package stack design, is required to span orders of magnitude of physical dimensions. It should not only comprehend the effects of mechanical stresses in electrical responses of the circuits, but also their reliability impact.

4.1 DfM of Chip-Package Integration

4.1.1 End of WYSIWYG in IC Design

The onset of 3D design architecture beyond the multi-layer structure at a single die level with planar technology, marks the end of the WYSIWYG (what you see is what you get) rules and checks. The out-of-plane issues, especially related to mechanical stress, are vital for successful continuation of technology scaling along the More-Moore paradigm. However, resolving these issues can no longer rely on visual assessment of design data, in as much as one can roughly predict for example for lithography, expecting generic line distortions (corner rounding and pullback). When thermal stress comes into play, visual bets are off.

Parametric variability due to mechanical stress across the IC volume has comparable impact on layout-driven variability as the lithography factors (which modulate device L

A. Balasinski, *Design for Manufacturability: From 1D to 4D for 90–22 nm Technology Nodes*,
DOI 10.1007/978-1-4614-1761-3_4,

and W) in 45 nm CMOS technology nodes and beyond. The combined use of low-k dielectrics in BEoL structures with harder interconnect materials in packaging (Cu pillars or Pb-free solder balls) is exacerbating CPI (Chip-Package Interaction) that cause delamination, cracking and/or fracturing of various materials. This limits yield and reliability of leading-edge products but cannot be readily imagined based on top-view 2D simulations enhanced by cuts in the cross-sectional pictures.

The challenges of managing stress/strain characteristics, including both impact on device mobility and material integrity, are concurrent with the Through Silicon Via (TSV) based 3D IC stacking technologies (here referred to as "TSS" for Through Si Stacking technology, [4, 5]. Stress related factors calling for a new DfM methodology make it necessary for design houses to manage mechanical stresses through design, rather than process 'knobs'. Process optimization for stress control in successful deployment of IC's relying on 3D interactions would assist fabless design teams to optimize design 'knobs' within the latitude allowed by the process technology (Table 4.1).

As discussed in Chap. 3, the controlled (intrinsic) stress at device level may be due to the Ge substitution of Si, to strain the lattice to boost carrier mobility, depositing stressed dielectric layers to induce compressive or tensile stress on Si, or on the interconnects to mitigate stress-induced voiding and/or electromigration. Device engineering has learned how to predict and take advantage of this stress without complex 3D simulations. By contrast, extrinsic stress at package level is more complicated as it involves a wider portfolio of more diverse materials. It may be caused by the CTE difference between metals (e.g., Cu or Al) and Si and/or glass through the thermal cycles associated with the BEoL, processing, or between the silicon die and package materials: underfills, solder balls, PCB substrate, plastic molding compound, etc. The stress would manifest itself through the thermal cycles associated with assembly and packaging processes (Fig. 4.1).

The incremental aspects of the TSS technologies exacerbate stress management challenges, as they may interact with and contribute to both the intrinsic and extrinsic stresses. For TSV filled with Cu, the CTE mismatch between Si and Cu and the typical BEoL thermal cycles make the TSV a significant source of stress, which calls for design rules regulating the placement of devices and metal layers in the proximity of the TSV.

TSS technologies relying on die-to-die bonding use microbumps (μbumps) deposited on the backside (die 1 in Fig. 4.1) to connect to the subsequent dice in a stack. The μ-bumps create stress in the same ways as the conventional Flip Chip (FC) bumps on the front side of a wafer. Stress from a μ-bump can interact with the devices of die 2, or, with thinning, with the devices on die 1. All layers in the stack are the potential sources of stress, exacerbated by their misfits in zero-stress temperatures and CTE values.

The μ-bumps to serve as a landing pad or to be used as backside redistribution and routing metals and dielectrics, are a source of stress impacting the overall stress distribution on the frontside. In addition to the direct stress effects of these incremental features (TSVs, μ-bumps, ...) there could also be stress interactions among the different features in Si and TSS technologies. TSS requires Si wafers to be thinned

Table 4.1 Properties of device, isolation, and interconnect layers used in the FEA model. Key features:
(1) The averaged strain energy density accumulated per thermal cycle decreased by a half when the Tg of underfill increases from 30 °C to 130 °C. For higher Tg, the underfill is more effective in suppressing solder fatigue failure.
(2) E had little impact on the fatigue behavior of solder joints under thermal cycling.
(3) Highest CTE yielded the shortest fatigue life of solder joints, the local CTE mismatch between underfill and shoulder bumps is important in determining the fatigue behavior of solder joints. The underfill with CTE closer to that of the solder joints produced a better solder fatigue lifetime

(A) Ranges of parameters for simulation of package stress

T_g	E	CTE	Underfill
30–130 °C	70–80 Pa	20–40 ppm/°C	Up to four types

(B) Material properties

Material	Young's modulus E (GPa)	Poisson's ratio	CTE (α) (ppm/°C)
IC materials			
Si	162.7	0.28	2.6
Cu	122	0.35	17
SiO_2	70	0.34	0.5
OSG (k~3.0)	17	0.3	8
MSQ (k~2.7)	10	0.3	10
ULK (k~2.4)	4	0.3	18
Solder materials			
Eutectic	75.84–0.152*T	0.35	24.5
(Pb) Lead–free	88.53–0.142*T	0.40	21.5
Underfill	6.23	0.40	40.6
	5.4 (T<Tg)		40 (T<Tg)
	~0(T>Tg)		100(T>Tg)
Organic substrate	Anisotropic elastic properties		16 (in plane)
			84 (out of plane)
			14 (in plane)
			64 (out of plane)
MSQ materials			
MSQ-A	2	0.35	10
MSQ-B	10	0.3	10
MSQ-C	17	0.3	10
MSQ-D	4	0.3	10
MSQ-E	4	0.3	4
MSQ-F	4	0.3	18
MSQ-G	4	0.3	26

to few ~10' s of μm. At this thickness, the redistribution of intrinsic stresses may take place, resulting in unintended stress relaxation at the surface, with possible impact on charge carrier mobility. The thinning process (removal of several 100 μm of substrate Si) may leave damage on the backside of the wafer, which could also result in stress redistribution. Handling very thin dice through the stacking and packaging processes involves bonding/debonding of a carrier material, which could induce

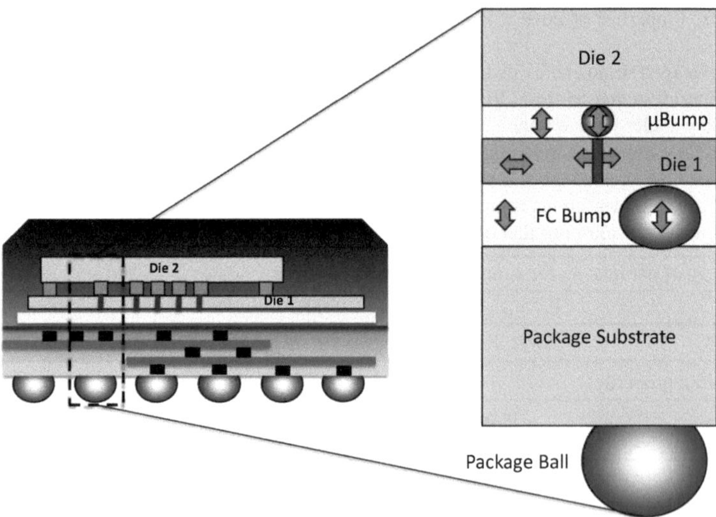

Fig. 4.1 Example of 3D IC for TSV based technology (after [2])

different sources of stress, especially for thinned Si wafers, which, in turn, will warp and bend much more than the wafers with standard thickness.

3D stacking involves assembly processing that gives raise to interactions between multiple dice and between die and package substrate. Thermal mismatch associated with the μ-bump attach or underfill cycles creates an opportunity for mechanical stress development for all units contained in the package.

4.1.2 Mechanical Stress in Through-Silicon Stacking

One can consider the stress effects as a new class of process-design co-dependence, driven by both process (temperatures, materials, …) and design (sizes, layout geometries) parameters, which cannot be modulated entirely by either process or design 'knobs'. Historically, for managing mutual process-design impact semiconductor industry used a rule-based (RB) or a model-based (MB) approach to restrict feature sizes, spaces, pitches, etc., in a verification-centric design methodology. The RB approach is simple, but results in large margins, as it tends to bucket a range of process-design interactions into categories that are constrained by one-size-fits-all assumption. The (MB) approach is more accurate, but more complicated. An example is SPICE modelling of electrical device characteristics, a simulation and optimization centric design methodology. It eliminates excess margins, but is intrinsically complex and hence it is used only when the expected margins required to use the simpler RB approach become prohibitive. This highlights the need for rule criteria definition for the physical mechanisms in and around the IC die: what stress levels and distributions can be permitted and which ones should be made illegal (Table 4.2).

Table 4.2 Effect of TSV configuration on residual stress (after [1]) which can be used to build rule decks: σ_{rm} - mean radial stress, $\sigma_{\theta m}$ - mean loop stress, σ_{ZM} - mean z- stress, σ_{zf} - circumferential stress

	Max stress change			
Parameter change	σ_{rm}	$\sigma_{\theta m}$	σ_{zm}	σ_{zf}
Via diameter: 5–15 μm	+41 %	+24 %	+80 %	−9 %
Cu fill ratio: full fill to ¼ fill	−24 %	−54 %	−23 %	−11 %
Cu volume ratio: 0.5–30 %	31 %	+27 %	+82 %	−25 %
TSV aspect ratio: 4 to 20	−3 %	+3 %	−1 %	+9 %
SiO_2 thickness: 0.15–0.45 μm	−25 %	−26 %	−35 %	−3 %

4.1.2.1 Rule – and Model Based Paradigms

(A) Rule based paradigm
It is the goal of DfM to control the stress effects via a set of design rules. For example, the Keep Out Rules would prohibit the placement of active devices within a certain given distance from the TSVs, based on perturbation, which may charge carrier mobility (earlier engineered carefully with spacer architecture). Similar rules could be defined for the placement of μ-bumps, FC bumps or any other TSS features. Primary concern is that the attractiveness of 3D TSS technology is related to the integration of multiple dice, obtained from different sources, built in different technologies, and assembled in different assembly houses. The issue becomes then, whose rules are applied and who is responsible for the rules that manage the interactions between several dice in the System – in – Package (SiP) stack. The rules that dictate the placement constraints for μ-bumps would be a function of technology and design (node, thickness, number of metal layers, die size & floorplan, design & layout, see e.g., Table 4.3) and the packaging process (materials, process parameters: temperatures, pressures, …) used to assemble the multi – IC SiP product. Ultimately, the liability is with the organization that integrates the SiP, which may not have access to the material, process details, and control "knobs". In practice, this approach may result in conservative margins and in iterative product-specific learning. Rule-based paradigms for products manufactured by a distributed supply chain results in technical, commercial, and business risks – especially at the front end of a technology development and for disruptive technologies, such as 3D TSS stacking.

(B) Model based paradigm
In the MB paradigm, one has to describe the stress characteristics of various materials and features at an end point of a given set of process steps. A simulation engine fueled by these models is required to allow designers to assess the net stress effect of a given design configuration. This infrastructure calls for collaboration across the entire supply chain. MB approach is generally scalable and it allows flexibility required for design space exploration to optimize 3D TSS designs, with different die size and type combinations, die-to-die alignments, layout and placement packaging choices, etc. It can be explored to optimize the TSS design, including all the mechanical stress interactions.

Table 4.3 Fracture toughness of silicon (after [1]). K_{IC} - stress intensity factor, G_{IC} - material toughness

Si Direction	K_{IC} (MPa · $m^{1/2}$)	G_{IC} (J/m^2)
<111>	0.83 to 0.95	4 to 5.3
<100>	0.91	4.9
<110>	0.94	5.2
Polycrystalline silicon	0.94	5.2

Rules to replace full model based approach with a rule – based one would be too complex or may even become self contradicting. Managing the flow of information between the process entity (foundry) and without exposing the proprietary details of a given process technology is handled by 'design kits' that e.g., include standard BSIM models for SPICE simulations. The range of responsibilities of the various entities in the supply chain required to support the model-based paradigm is a better fit to the design – technology – packaging interactions, than what is required to support the rule-based paradigm.

(C) FEA simulation setup

The challenge of managing mechanical stress is not new, and simulators based on Finite Element Analysis (FEA) exist in the microelectronics industry. Unfortunately, FEA simulators are typically challenged if they have to deal with a large span in dimensions or cross – correlations (e.g., from nm to mm) from one type of simulation to another. This is resolved by sub-modeling cycles, where the results of macro-scale simulations are progressively imported as boundary conditions for analyses of finer spatial granularity.

FEA simulators such as Ansys or Abaqus have been used for focusing on soft packaging materials with plastic or viscoplastic behavior, low glass transition temperatures, etc. The Si die is typically modeled as a monolithic volume (Fig. 4.2). A simulator, which models charge carrier mobility, is required to derive MOSFET properties. The front – end materials and process steps required to model on-Si effects would be of high hardness compared to back – end materials that show dominantly elastic behavior capable to support temperatures associated with BEoL processing (~400 °C). For stress analysis of device performance, the simulation needs to resolve features of the order of a transistor size (~10 nm). As the range of stress interactions in Si is typically of the order of ~μm's and package level features contributing to the stress of the order of 100 μm, the number of polygons that have to be analyzed becomes extremely large. Even with the sub-modeling methodology, FEA is incompatible with analyzing all of the stress effects within the Si design environment. The simulator should interface to design data in GDS format to capture the entire die design (layout) for selective implementation of compact modeling methodology.

TSS Design Environment requires a stress DfM flow to help with design process, not just rely on a specific, single stress simulation tool. To avoid yield (or reliability) impact by mechanical stresses, a new methodology that addresses all interactions associated with TSS needs to include [6]:

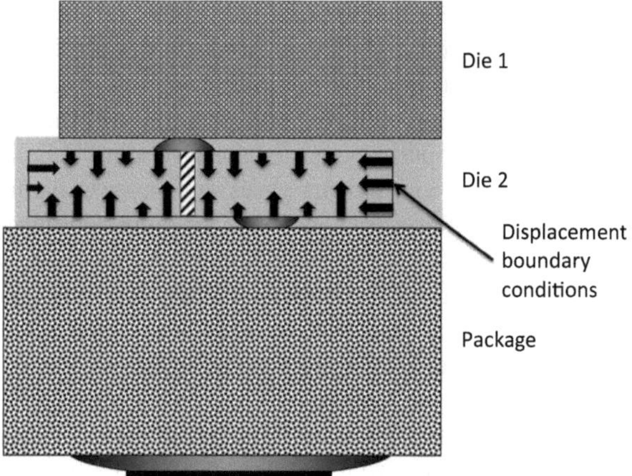

Fig. 4.2 Cross-sections of two dice in a package with stress levels (*arrows*) depending on displacement boundary conditions

- Pro-active simulation rather than reactive observations, i.e. a DfM sensitivity analysis during the design phase,
- Electrical modeling of material integrity as well as device performance (traditional CPI is not sufficient),
- nm to mm scale features to capture Si and TSS layout characteristics,
- All data about physics and material properties, i.e. the information cannot be single – point experimental since the existing experience base with disruptive technologies such as TSS is not sufficient,
- All data for elastic materials used in Si process technology as well as for the plastic materials used in the packaging,
- Formats and conventions used in Si design environments and their intersects with electrical design considerations,
- New degrees of freedom of TSS technology, compatible with die stacking using multiple orientation stacking technologies.

(D) Material properties and process characteristics
Simulation of mechanical stresses is driven by a set of material properties, mainly:

(a) Young's Modulus (E)
(b) Poisson Ratio (ν)
(c) Coefficient of Thermal Expansion (CTE)

which are the basic ones for each material in the IC and its package. Each of them has its temperature profile (including transitional temperatures) and also other dependences making the simulation process very complicated and the results not always intuitive. For sub-μm structures, material properties are likely to depend on the size. For polycrystalline materials, their microstructure has also to be considered. Parameters of softer materials (typically a non-linear function of temperature) need to be determined and stored in matrices for the use in the equations at several temperatures.

To make matters worse, accurate-stress modeling requires also a detailed process history. The stress (force) – strain (displacement) characteristics are a result of stress redistribution through the thermal cycles associated with the manufacturing of the die and assembling of the stack in a package. Some stress material properties e.g., for the insulating ILD layers, depend on the process chemistry (e.g., plasma gases and process parameters etc.). Since access to this type of process conditions is required, stress modeling has been a domain of the TCAD class of simulations not practically accessible to fabless entities. This is a barrier for implementation, since the foundries do not tend to release process data. One work around is to express material parameters through a 'Stress Free Temperature'. The effect of the cumulative process history is reduced to the aggregate "materials characteristics", at equivalent zero film stress temperature, a calculated parameter. TCAD simulation would be required to extract or derive it by fitting a model to measured data.

The fitting terms in the compact models for the Hot Spot Simulators need to be calibrated using FEA tools and material properties need to be measured. Some of these properties can be comparable to textbook values while other characteristics need to be calibrated for a specific film or structure. Measurement test vehicles are required for validation.

Metrology to be developed include the direct measurement of strain in a Si device and the corresponding correlation to the charge carrier mobility. Data generation needs to use global measurements (e.g., unpatterned wafer bow to obtain directly the mechanical strain) and local physical measurements (e.g., μ-Raman, TEM, etc. to measure local strain in silicon) in addition to new test structures for the characterization of electrical parameters (e.g., mobility gauged by strain).

These requirements are parallel with the requirements for foundries to provide stress DfM Design Kits (DDK) with information analogous to calibration data for lithography and Critical Area needed for process simulations. DDK kits are now readily supported and the simulations are mandatory for tapeouts for the 40 nm CMOS technology and beyond.

Incremental validation of simulation results versus measured material parameters will be accomplished by test chips that compound the various sources of mechanical stress and evaluate the net effect on device performance and/or material integrity, evolving as 3D TSS technologies proliferate, and that the stress related challenges become visible.

(E) Summary

Managing mechanical stress and its impact on the physical and electrical integrity of 3D-stacked semiconductor products are driving the methodology for TSV die and TSS stack design. A stress – impact mitigation DfM flow, based on a set of specialized EDA simulators and 'Design Kits' that include material characteristics will depend on industry guidance factors:

1. The uptake of the 3D (SiP) integration technologies, which would bring mechanical and thermal management challenges to the forefront, with a structured stress

management methodology as an essential part of mitigating the risks of the adoption of these new technologies. If 3D IC integration technologies are not adopted by the mainstream, then this capability may not be required.

2. Model-based rather than rule-based paradigm for stress related process-design interactions. The rule-based approach is simple and may be preferred by the industry at the expense of margins.
3. DfM methodology: enabling design teams to do trade-off analysis to mitigate implementation risks. The alternative would be a traditional TCAD-like methodology based on computational power not readily accessible to the fabless portion of the industry.
4. DfM support from the supply chain e.g., EDA companies developing simulators for the foundries to calibrate the models and provide Design Kits with the involvement of the academia. Pull from end user entities is recommended.

The barriers are similar as those faced by the DfM tools a few years ago. The need for a structured methodology to deal with mechanical stresses will increase over time if the complexity of the mechanical stress interactions will grow with the diversity of stacked IC products along with the scaling of the TSVs and related dimensions.

4.2 Strain Engineering in Devices

In Chap. 3, we discussed examples of stress effects on MOSFETs related to the new materials, processes, and device scaling. The other aspect of this issue is the generic MOSFET behavior under stress driven by the IC layout and technology. Therefore, we will continue this discussion here, focusing on the individual device level.

Strain engineering generates tensile and compressive stress states in NMOS and PMOS transistors. Their characteristics most affected by the stress are the low-field charge carrier mobility μ_0 and the zero-biased threshold voltage for a long-channel device (V_{T0}), as they depend on material properties, distribution of the dopants, and geometrical factors. Due to lattice changes by strain in the Si transistor channel, the band structure of Si and consequently the effective mass of charge carriers and scattering processes, affect V_{T0} also by the modification of the dopant distributions caused by the stress-induced retarding or enhancement of the dopant diffusivity [7].

Different compact models can address stress simulations in active devices. A model may have full-chip simulation capabilities to predict intra-channel stress components in each transistor. Stress sources such as epi Si-Ge, liners, tensile contacts, STI, are located inside a floating window (area of interaction) surrounding each gate (Fig. 4.3), characterized by different sizes of the windows depending on the intra-channel stress on the stressor size. It is important to know how would the stress generated by CESL, epi-Si, Ge, S/D and STI, saturate as the stressing agent increases (refer to Figs. 3.76, 3.77, 3.78, and 3.79 for MOSFET level stress).

For the calculation of stress components, the simulation tool generates the cut-lines, which cross the transistor channel and define a "2D" structure (Fig. 4.3). It then takes into account interactions of materials along the cut-line and traction with the

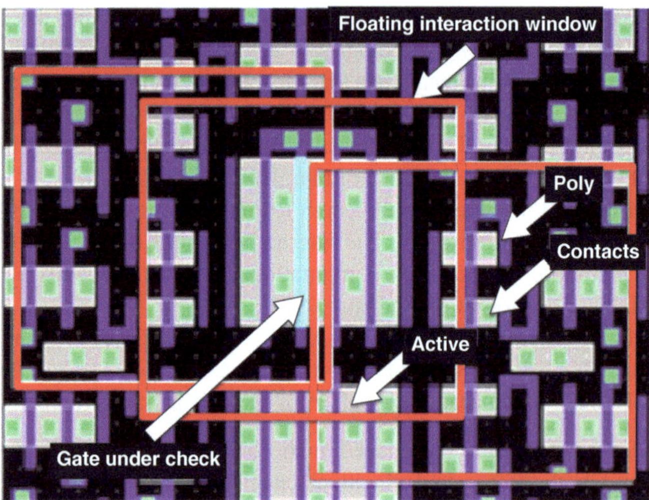

Fig. 4.3 All stress sources inside floating windows for transistor channel simulations (after [2])

underlying silicon substrate. Due to the different mechanical properties of the segments in the composite layer associated with each cut-line, the boundaries between segments experience displacements. The simulator generates transistor-to-transistor stress distribution using an analytical solution of several elasticity problems corresponding to different stress sources [8]. The layout-induced stress is caused mainly by a shrink or an expansion of the stressor volume. In order to describe these effects in the layout-scale, the compact model introduces initial "internal" strains in every segment occupied by a particular stressor. For epi-Si 1-x Ge_x source/drain, a relative difference in linear dimensions between the S/D space is occupied by the relaxed (zero stress) $Si_{1-x}Ge_x$ and the actual (slightly deformed) S/D space. In the case of STI, it is the relative difference in linear dimensions between the space that the STI oxide intents to occupy after cooling down from the deposition temperature, and the actual available space.

The compact model uses these initial strain values as calibration parameters determined at the calibration step from the best fit between measured and predicted transistor characteristics. The relaxation of the "internal" stressor strain ε_0 generates displacements of the segment boundaries. Due to this, a shear stress is generated in addition to a normal stress caused by interactions between neighboring segments in the cutline. A value and a sign of the resulted stress can be found from the force balance equations. The compact model reduces a system of such second order partial derivative equations to a system of linear equations for extraction of the edge displacements for all segments inside the cutline.

All three stress components: longitudinal, transversal and out-of-plane, are calculated in accordance with the Hooke's law. There is correspondence between two

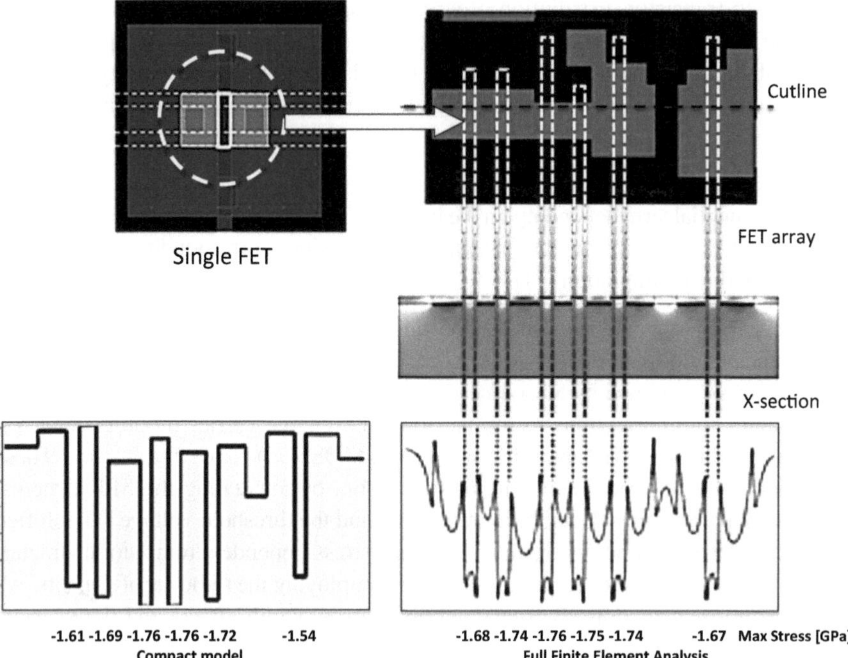

Fig. 4.4 Dividing FET channel into segments. *Dots* indicate the channels. A correspondence between longitudinal stress and the calculation methodology (after [2])

sets of the longitudinal stresses generated by the tensile-stressed liner calculated with COMSOL FEM tool and with the compact model (Fig. 4.4) [9]. For the edges of active areas, the model first calculates displacements of the channel edges and then the strain component inside the segment.

The intra-channel stress components calculated in accordance with the Hooke's law show different distributions of the transversal stress inside transistor channels of the die depending on whether package-induced stress does or does not contribute to the total stress.

The calculation of the transistor-to-transistor stress distribution generated by TSVs is also done based on force balance equations. The difference is the generation of the initial strain inside every segment where a semi-analytical solution for the thermal stresses in a TSV structure is used [10]. The stress distribution near the wafer surface around the TSV differs from the stress calculated in accordance with the classical solution of the 2D plane-strain Lame problem [11]. The near-surface stress distribution in the via environment is a linear combination of the stress distributions from the solution of the 2D plain-strain problem. It is a semi-analytical solution, in which both ends of the via are subjected to pressure of the axial stress in the via generated in the 2D problem. That solution obtained in integral form can be simplified for regions located in close proximity to the surface like transistor channels. In such an approach, the model calculates the strain distribution across the layout as a superposition of strains generated by all TSVs. Subsequently, the final transistor-to-transistor distribution of stress components is calculated allowing the initial strains to redistribute depending on the layout geometry. To quantify the impact of stress on mobility, the compact model uses the piezoresistivity [12] with its sensitivity coefficients. The stress effect on the threshold voltage of a MOSFET is modeled as a combination of:

- A linear term representing the total hydrostatic stress
- An exponential term depending on the hydrostatic stress at the moment of activation/anneal of dopants in the shallow junction extension, corresponding to the stress effect on dopant migration [11].

Different types of dopants in NMOS and PMOS are characterized by different dependencies of dopant diffusivity on local stress and stress gradients [13].

Once stress generated by all sources is calculated, the compact model converts the stress values into corrections to the μ_0 (low-field charge carrier mobility) and V_{T0} (zero-biased threshold voltage for long channel MOSFET) for each transistor. These corrections will be used by the circuit simulation, by annotating the SPICE netlist with instant parameters: the mobility multiplier and the threshold voltage (V_T) shifter. These calculated parameters are the most basic stress-dependent transistor characteristics, the same for different types of transistors employing the same set of dopants. All geometry impacts such as short-channel and narrow-width effects and the possible difference in the transistor types are taken into account at circuit simulation.

The variations in I_{dlin} for NMOS and PMOS transistors for a small portion of a layout (~4,000 devices) in a close proximity of TSVs can be due to epi Si_{1-x}, Ge_x, S/D, CESL and STI, but also to TSV and packaging. The stress generated by TSV and the package can change the total stress level by several time and also reverse the direction of its impact on I_{dlin} from positive to negative.

Accounting of TSV-induced stress generates a small number of peaks additionally to the original sorted distribution of I_{dlin} caused by layout-induced stress sources (Fig. 4.3, also discussed later, Fig. 4.43). These peaks describe changes in I_{dlm} for the devices located in the closest proximity to TSVs (order of magnitude of the TSV radius). Other devices demonstrate the unaffected values of I_{dlin} and K_T. If package-induced stress is added, the distributions are shifted for both NMOS and PMOS transistors in the direction of increased I_{dlin} and transistor-to-transistor variations are generated.

Sorted I_{dlin} for NMOS devices for the layout-induced stress show that the addition of TSVs results in almost the same I_{dlin} distribution, as in the case of layout-induced stress, but with slightly modified edges. This reflects I_{dlin} modifications for the devices next to TSVs. The package-induced stress further modifies the previous distribution generated by package-induced stress only.

The physics-based compact model links package-scale FEA and chip layout (GDSII, OASIS, etc.) and represents an extension of the analytical model for the assessment of the layout-induced stress [14]. The major difference between these two models is the way of introduction of the initial stressor strains, either as parameters to be extracted at calibration stage, or as the results of the FEA-based simulations at package-scale and die-scale steps, depending on material data (Table 4.1). The elastic modulus of materials (e. g. BEoL low-k/ULK and metal structures) can be determined applying nanoindentation and/or AFM-based techniques.

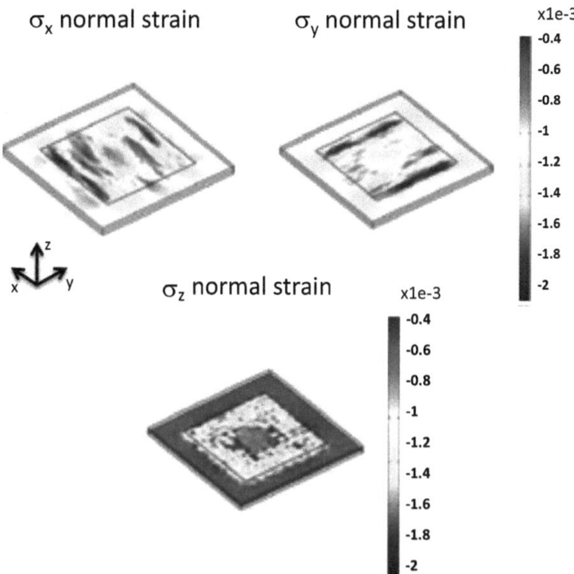

Fig. 4.5 Die Scale simulations of strain across device layer (after [2])

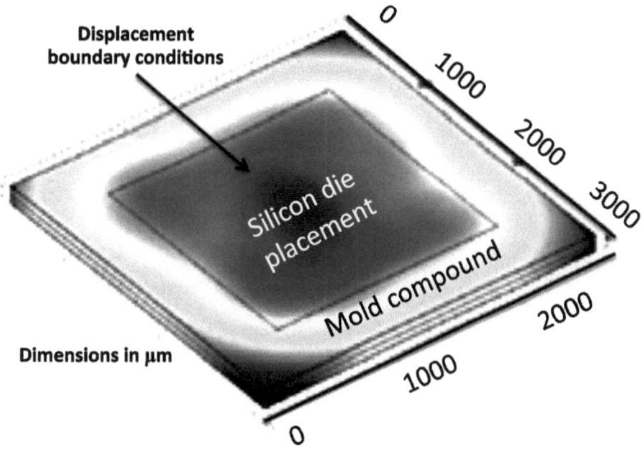

Fig. 4.6 Die scale simulation: displacement boundary conditions (after [2])

Simulation INPUT:

- GDSII – layout of the device in the vicinity of the TSV structures,
- Across-layout distribution of the package-induced strain components,
- Mechanical properties of relevant materials.

Simulation OUTPUT:

- Intra-channel stress components for all transistors in design over the die in package (Figs. 4.5, 4.6, and 4.7),
- Instant parameters: mobility and threshold voltage variation (MULU0 and DELVT0) for every device in the analyzed design for annotation of SPICE netlist for further circuit simulations.

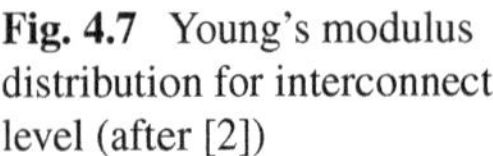

Fig. 4.7 Young's modulus distribution for interconnect level (after [2])

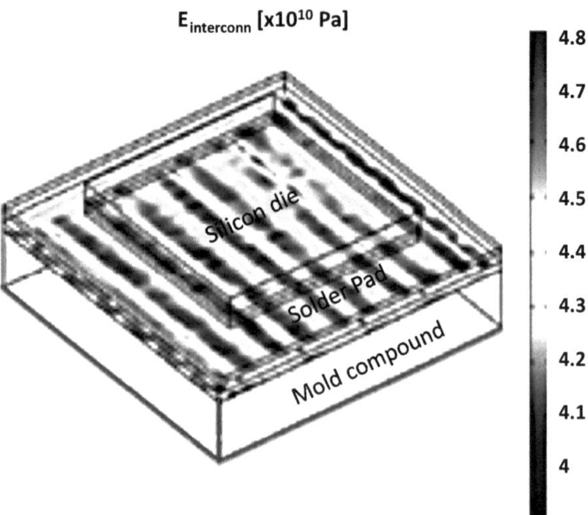

Fig. 4.8 Example of MOSFET test structure showing the parameters to be varied for stress studies

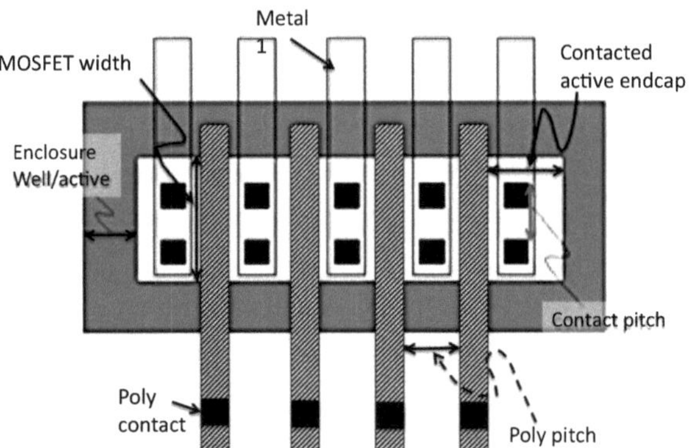

4.2.1 Calibration and Validation

The design verification flow should be capable to analyze any 3D die stack to determine the locations of out-of-spec excursions in the device characteristics caused by mechanical stress. The compact model should be calibrated for any specific technology node and materials such that the predicted characteristics should fit the measured one.

Major target of the calibration is optimization of model parameters for across-die distributions of transistor intra-channel stress components and conversion into stress-modified transistor electrical characteristics. A dedicated test-chip should contain a distribution of inter-channel stress values, with varying number of fingers, sizes and shapes of test structures, spaces between neighboring test structures, distances to the well edges (Fig. 4.8) etc., in several locations of the die at different layout densities. The compact model has about 30 parameters to be calibrated. Each gate should be measured independently (decoupled) of others. Calibration module should use the least-squares regression to determine the unique set of values that fits the measured data. For a reliable calibration of all model parameters, a set of V_T and I_{dlin} measured for about 100 gates should be developed.

For 3D IC, the total stress generated inside transistor channels can be divided into stress by layout-induced sources (including TSV), and stress generated by die thinning and stacking. Different parameters are associated with these two components. Since only several calibrated parameters are associated with the stress resulting from die thinning and stacking, one should perform the majority of characterizations on the thick die and only some measurements on the identical thinned and stacked die. To achieve the best results the calibration should consist of:

1. Device characterization of the unthinned die before stacking,
2. Characterization of the same device on the thin, stacked die.

Firstly, one needs to provide a capability for calibrating model parameters associated with the layout-induced stress sources and compare the calculated device characteristics with the measured ones. If I_{dlin} and V_T are the parameters of interests, then the measured device characteristics are interrelated with the calculated stress by the following equations [2] (Table 4.4):

$$
\begin{aligned}
I_{dlin}(\sigma) - I_{dlin}(0) &= \left(\frac{\partial I_{dlin}}{\partial MULU0}\right)_{\sigma=0} \left[R_{UO}\left(1 - F_{UO}(\sigma)\right) - 1\right] \\
&\quad + \left(\frac{\partial I_{dlin}}{\partial VTH0}\right)_{\sigma=0} \left[R_{VTHO} + F_{VTHO}(\sigma)\right] \\
V_d(\sigma) - V_d(0) &= \left(\frac{\partial V_T}{\partial MULU0}\right)_{\sigma=0} \left[R_{UO}\left(1 - F_{UO}(\sigma)\right) - 1\right] \\
&\quad + \left(\frac{\partial V_T}{\partial VTH0}\right)_{\sigma=0} \left[R_{VTHO} + F_{VTHO}(\sigma)\right]
\end{aligned}
\tag{4.1}
$$

Table 4.4

Parameter	Explanation
I_{dlin}	Drain current in linear region
MULU0	Mobility multiplier
R_{U0}, R_{VTH0}	TSV radius dependent functions of stress components
F_{U0}, F_{VTH0}	Known functions of stress components
VTH0	Long channel threshold voltage
V_d	Drain voltage
σ	Stress value

Stress optimization solves the system of equations shown above, written for all measured transistors (gates), and derives the optimal set of model parameters. Then, for calibration of the model parameters associated with the stress generated by die thinning and stacking, one uses the difference between the characteristics of the devices on the un-thinned (thick) and the thin/stacked dice from the same fab and

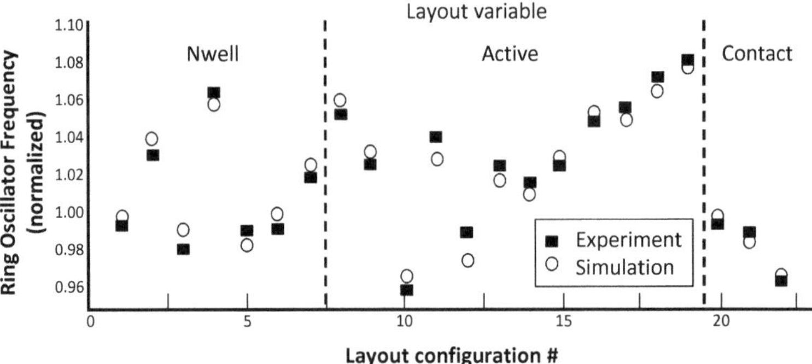

Fig. 4.9 Experimental and predicted ring oscillator frequencies for layout-dependent effects (after [14])

with the same process flow that was used for calibration. The MULU0 and DELVT0 are calculated for every gate in the design:

$$\begin{aligned} \text{MULU0} &= \text{RU}_0\left(1 - \text{F}_{\text{U0}}(\sigma)\right) \\ \text{DELVT0} &= \text{R}_{\text{VTH0}} + \text{F}_{\text{VTH0}}(\sigma) \end{aligned} \tag{4.2}$$

At the simulation, SPICE calculates electrical characteristics of the devices using the annotated netlist.

The predicted electrical characteristics of transistors in the test-chip fit the results of their simulation with the calibrated fab/foundry model. A validation of the model (Fig. 4.9) for ring oscillator frequency shows that the variation of layout-dependent stress effect is captured accurately for a variety of layout configurations [14].

Considering layout-dependent stress effects, one should also take into account wafer pattern inaccuracy due to pattern reproduction. Starting with the 45 nm CMOS technology node, the litho-induced variations [15] are being reduced, but the mechanical stress, while significant, it is not the only cause of variability in transistor characteristics. For best results, one can take into account litho-induced variations at circuit simulations for the zero stress condition. A SPICE-type simulator should then be run on the post LFD ("litho-friendly" design) layout. All as-drawn poly-Si gates should be converted into effective rectangular poly-Si gate shapes, electrically (based on total drain current) equal to the non-rectangular shapes formed by the post-OPC contours [16]. Random variations in device characteristics caused, for example, by dopant fluctuations can also be addressed by averaging of the die-to-die measurements [17]. Process corners should also be included in accordance with the design methodology.

The calibration flow can be improved by fitting the simulated stress components inside the transistor channel with measured intra-channel stress on the test-chip for a sufficient number of transistors at different densities and configurations. One should note that high-resolution strain measurements of test-chip devices are a

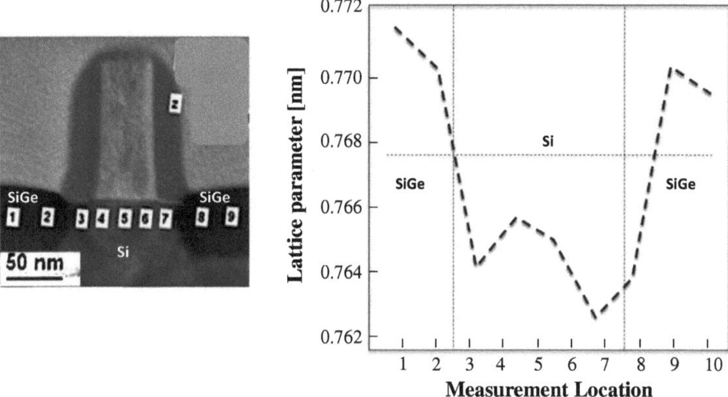

Fig. 4.10 Strain in MOSFET channel based on nano-beam electron diffraction (after [20])

challenging task. Bow measurements at whole wafers lead directly to residual stress, but are not applicable for IC characterization:

- At the micro scale, X-ray diffraction or Raman spectroscopy are suitable to probe the strain at test structures.
- At the nanoscale, transmission electron microscopy (TEM) is the only technique for high-resolution strain measurements, but it might lead to relaxation changing the strain, and hence, the stress state, if not analyzed in detail using e.g., complementary simulations.

TEM-based approaches can be divided into two categories: diffraction-based, like convergent-beam electron diffraction (CBED) or nano-beam electron diffraction (NBED) and image-based methods such as high-resolution electron microscopy (HRTEM) or dark-field electron holography. The most accurate techniques for strain measurements are diffraction-based CBED and dark-field off-axis holography with $\Delta\varepsilon = \pm 0.02\%$ while the accuracy for NBED and HRTEM is $\Delta\varepsilon = \pm 0.1\%$. CBED is adversely affected by specimen bending in devices. HRTEM has a high spatial resolution of 2 nm but is limited in the field of view [18]. Using dark-field inline holography, large field of views can be characterized with a spatial resolution <1 nm [19]. The CBED and MBED are probe-based with a spatial resolution of 5 and 10 nm, respectively, performed at selected points on the specimen (Fig. 4.10) [20]. Imaging techniques allow direct mapping of strain across wide areas of up to $100 \times 1{,}000\ \text{nm}^2$.

4.2.2 Conclusions

Simulation based design verification DfM flow for stress effects in MOSFETs should be capable to analyze any design of 3D IC stacks and determine across-die variations in device electrical characteristics caused by layout and TSV/

package-induced mechanical stress. The limited characterization capabilities of 3D IC stacks and a "good die" requirement make this type of analysis critical for the functional and parametric yield and reliability. Data generation and material characterization approach have to be developed, to generate a database for multi-scale material parameters of wafer-level and package-level structures. Standard DfM methods for characterization of model parameters, either directly (e.g., strain gauges) or indirectly (e.g., electrical characteristics and the associated test structures), will be required.

The simulation stress management flow requires industry acceptance to support a DfM-like solution to enable design entities to model stress implications on their designs.

4.3 Mechanical Reliability of the Chip-Package System

According to the rule of 10, product reliability plays the key role in the hierarchy of DfM implementation. Expanding IC applications have to rely on new, more comprehensive checks of silicon device and packaged product reliability. Therefore, stress simulations are becoming a requirement for single IC's and SIP's at package level.

Advanced packaging techniques caused chip-package interaction (CPI) to become a critical manufacturability issue, especially for flip-chip packaging of Cu/low-k IC products with organic substrate. The thermo-mechanical stress and strain developed inside the package during assembly and subsequent endurance tests cause mechanical reliability issues manifested as solder joint fatigue and underfill delamination due to the mismatch of the coefficients of thermal expansion (CTEs) between the chip materials and the substrate. The deformation of the package translates to the Cu/low-k interconnect local stresses which drive interfacial crack formation and propagation. The reliability risk is aggravated with the implementation of the brittle ultra low-k dielectrics for lower electrical coupling and the switch to hard-melting Pb-free solders for environmental safety.

Unlike the DfM reliability issues related to integrated device scaling along the Moore's shrinkpath, the CPI-induced issues pertain to chip and package level endurance. In many cases, that endurance cannot be electrically tested, at least not at the individual transistor level. High-resolution Moiré interferometry and FEA Analysis are the tools to analyze the thermo-mechanical deformation in flip-chip packages and the effect of properties of the materials surrounding the die such as the underfill, on package warpage.

Package scaling, and reduction of the interconnect dimension are accompanied by the use of more metal levels and the implementation of ultra low-k, porous materials. At the same time, 3D integration (3DI) with through silicon vias (TSV) increases the device density without downscaling of their linear dimensions.

The chip-package interaction originates at the die attach step during assembly and becomes most detrimental to low-k chip reliability because of the high thermal load generated by the solder reflow before underfilling. A multilevel 3D modeling

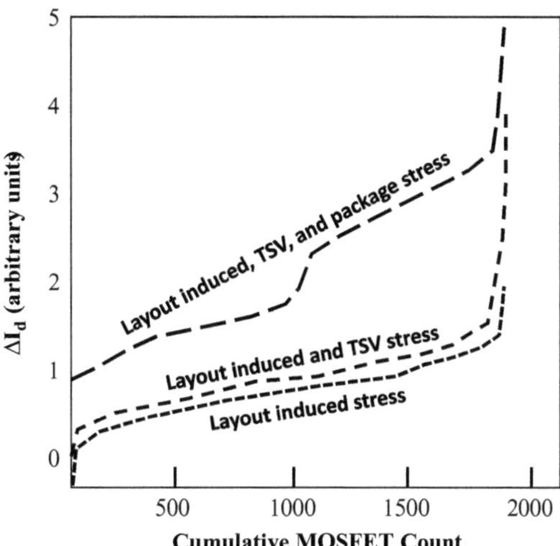

Fig. 4.11 Sorted drain current distributions for TSV – induced stress (after [2])

combined with modified virtual crack closure (MVCC) predicts CPI-induced interfacial delamination in Cu/low-k interconnects. It is first focused on the effects of dielectrics and solder materials and then extended to include the effect of the reduction of interconnect size accompanied with an increased number of metal levels separated by ultralow-k porous and fragile dielectrics.

To increase the device density, especially in flip-chip packages, Chip-Package Integration as 3D integration approach with through silicon vias requires its thermo-mechanical reliability to be optimized for stress reduction. There are numerous risks to package reliability for Cu-low K intercomments reflecting in drawn current distributions (Fig. 4.11), such as:

(a) Underfill delamination caused by thermo-mechanical deformation or moisture
(b) Thermal cycling fatigue (reliability) of Pb-free solder bumps
(c) Modified virtual crack closure (MVCC)
(d) Crack propagation and crack stop design in low-k interconnects
(e) Process-induced residual stress and keep-away zone
(f) Silicon r-cracking, interface debonding between Cu TSV and silicon matrix, silicon z-cracking, mitigated by fracture analysis.

Package and interconnect structural optimization should reduce the risk of dielectric delamination. When combined with stress analysis with FEM, multi-level sub-modeling based on element birth and death technique, one can expect CPI DfM to generate a set of geometrical/material rules for high reliability.

(A) Impact of scaling

Classical semiconductor scaling rules apply to the dimensions of individual transistors and do not consider the mechanical or material properties of the circuit subject to scaling. Other than the general “miniaturization” trend, in the

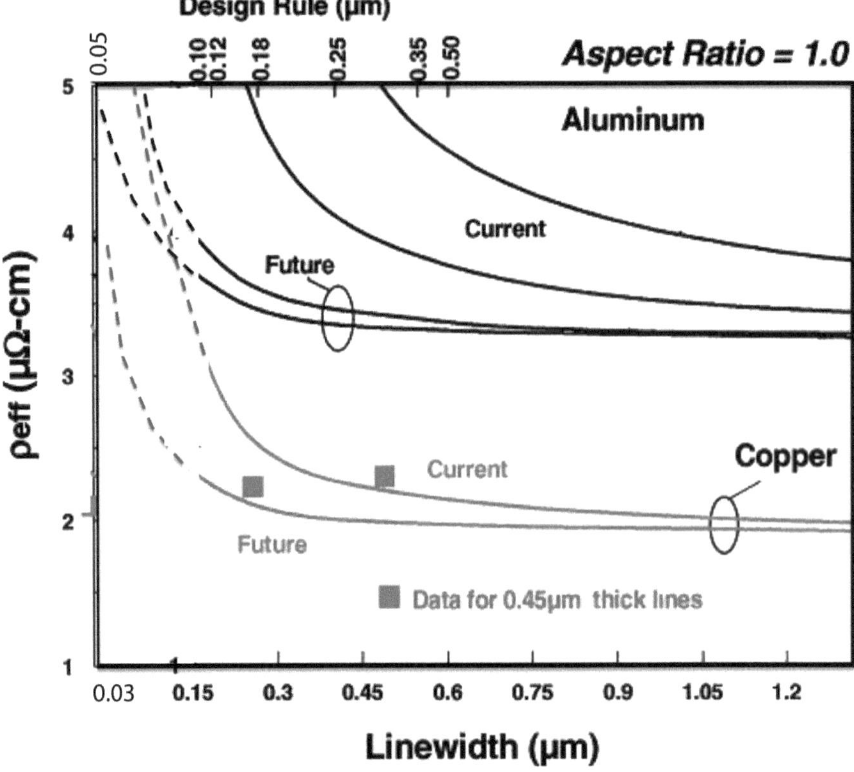

Fig. 4.12 Resistivity change with decreasing metal line width for Cu and Al (after [1])

past there was no strong motivation to similarly reduce the dimensions of IC packages. This is changing now. DfM rules related to package scaling need to address the dynamics of IC interface to the outside world to enable the many new applications. As the device dimensions reduce, resistive-capacitive (RC) delay, cross talk, and power dissipation of the interconnect structure have become performance-limiting factors for IC product design and development requiring new materials and architectures for the interconnect and packaging structures.

Copper (Cu) with a resistivity of ~1.8 μΩ·cm which is ~45 % lower than aluminum-copper (AlCu) alloy (~3.3 μΩ·cm) has been implemented to reduce the RC interconnect delay [8], even if the resistivity of both Al and Cu significantly increases for very narrow line widths (Fig. 4.12). To reduce cross talk, noise, and power dissipation caused by interconnect scaling, one needs to lower the capacitance between the metal lines (interline capacitance $C_{L\text{-}L}$ and line-to-ground capacitance $C_{L\text{-}G}$, Fig. 4.13). Otherwise, the cross – talk would increase with reduction of the feature size due to the to eventual domination of the total capacitance by its line to line component for feature size below 1 μm (Fig. 4.14) [9]. For the same purpose, low-permittivity (low-k) dielectrics, e.g.,

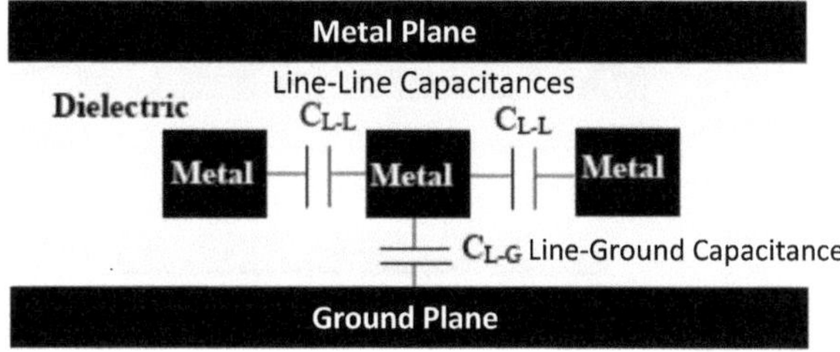

Fig. 4.13 Interline capacitance, CL-L, and line to ground capacitance

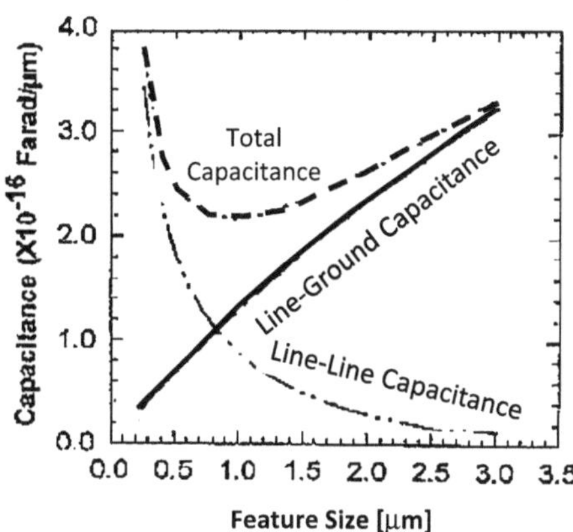

Fig. 4.14 The line to ground, inter-line, and total capacitance versus feature size (after [1])

a carbon-doped oxide (CDO) are required for 45 nm technology [10]. The packaging architecture for ICs below 90 nm technology node is mainly based on the flip-chip solder interconnects (Controlled Collapse Chip Connection, C4) introduced by IBM in 1964. They provide connectivity between the active device side of the silicon die, face-down, and the multilayered wiring substrate through solder bumps. The area-array configuration can support the high input/output (I/O) pad counts with better electrical performance due to the increased device density and shorter interconnection length. The process starts from wafer bumping, where the surface passivation layer on top of a completed wafer is patterned and the under bump metallization (UBM) layers are electroplated into the pattern (Fig. 4.15a). The UBM layers provide good solder wetability, adhesion, and electrical and mechanical connection between the device and the solders. Next, the extra portions of the UBM metal layers are etched away and solder alloy is deposited on top (Fig. 4.15b). The wafer is then heated to reflow the bumps to form spherical solders (Fig. 4.15c) and singulated into individual dice for subsequent packaging. Before bonding, the bumped die is flipped over and aligned with the organic substrate (Fig. 4.15d). All the C4 connections are formed simultaneously by a solder reflow (Fig. 4.15e) and a polymeric underfill is dispensed into the gap between die and substrate to reinforce the solder joints (Fig. 4.15f).

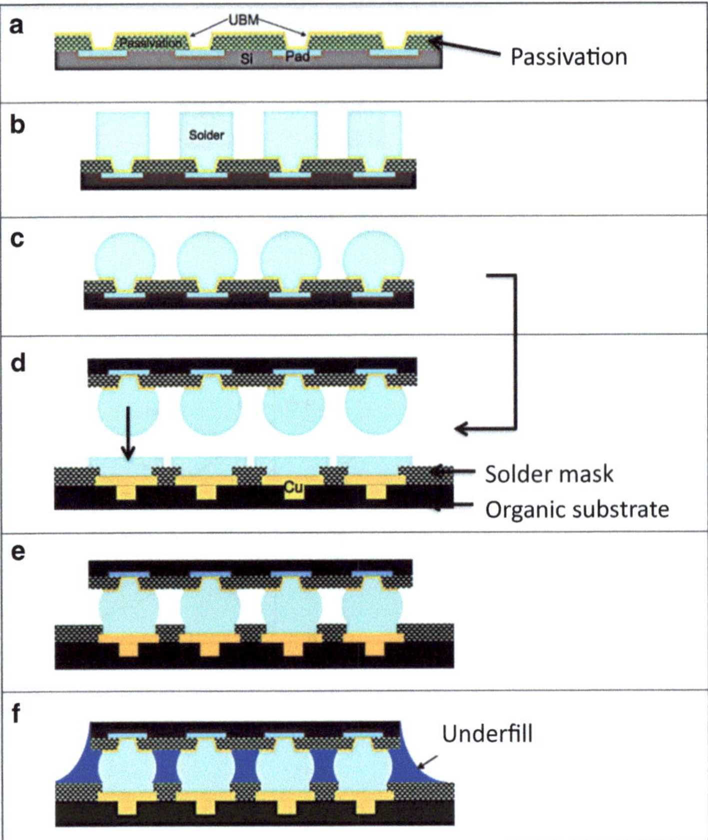

Fig. 4.15 Assembly process for flip chip packages (after [1])

With the implementation of Cu/low-k interconnects, the flip-chip package has evolved to include organic substrates with multilayered high-density wiring and solder bumps with pitch reducing from hundreds to tens of micrometers. While this contributed to the improvement in the electrical performance, it also raised mechanical reliability concerns in solder joints and the Cu/low-k interconnects. The reliability concerns in flip-chip packages due to the mismatch in the coefficient of thermal expansion (CTE) between the Si die and the organic substrate [11, 12] are aggravated by the environmental safety requirement to switch to Pb-free solders prone to thermal cyclic fatigue and electromigration failures [13, 14].

(B) Impact of process
Mechanical properties of the dielectric materials deteriorate with increased porosity, making it difficult to integrate copper with low-k interconnects. Cohesive fracture of the dielectrics [15] and interfacial delamination [16], the former pertaining to the brittleness of low-k materials and the latter resulting

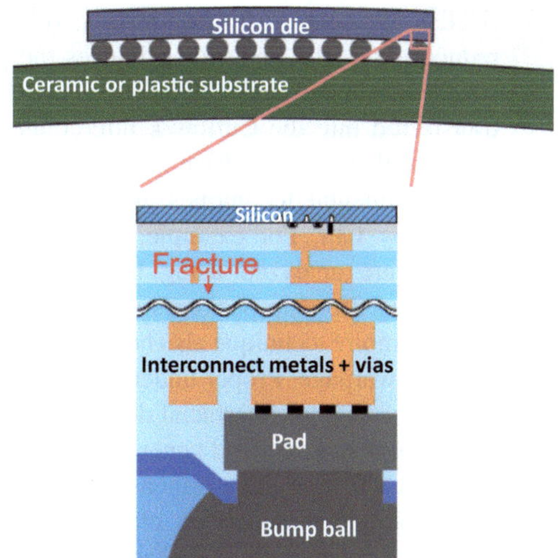

Fig. 4.16 CPI induced fracture in a multilevel interconnect (after [1])

from poor adhesion between the low-k and the surrounding materials, are due to a series of thermal and mechanical processing steps (deposition, annealing, and polishing) during wafer fabrication. The interconnect structure is also subjected to additional thermal stresses induced by the assembly processes, which drive interfacial crack formation and propagation (Fig. 4.16).

Although the origin of the stresses in the in-chip interconnects and packaging structures is similar, i.e., thermal mismatches between the component layers, the reliability impact for the low-k interconnects is different. At the chip level, the interconnect structure during fabrication is subject to a series of thermal steps at each metal layer. During film deposition, patterning, and annealing, the temperature of metal and barrier layers can reach 400 °C. At chemical–mechanical polishing (CMP), the chip is under mechanical stresses and exposed to chemical slurries [16]. When incorporated into the organic flip-chip package, the fabrication of the silicon die containing the interconnect structure has already been completed. The interconnect structure as a whole is subjected to additional stresses induced by the packaging process. Here, the maximum temperature during solder reflow for die attach is about 220 °C or higher for eutectic Pb alloy solders and about 260 °C or higher for Sn-based Pb-free solders. During accelerated or cyclic thermal tests, the temperature varies from −55 °C to 150 °C or from 0 °C to 100 °C. Although package assembly or test temperatures are considerably lower than chip processing temperatures, the combined effect of the flip-chip package introduces very different stresses in the low-k interconnect stack, which are not accounted for in different types of packaging. The thermal stress in the flip-chip package arises from the large mismatch of the CTE between the chip and the substrate; 2.6 ppm/°C for Si and about 17 ppm/°C for an organic substrate [7]. The thermal stress reaches a maximum in the solder bumps at the outermost row of the array, especially at the corners farthest away from the neutral point

(DNP). While by using underfills, the stress at the solder bumps can be reduced [17], the underfill enhances the package warpage resulting in large stresses at the die-underfill interfaces [18]. This warpage force can be directly transferred into the Cu/low-k interconnect structure in the BEOL, inducing large local stresses to drive interfacial crack formation and propagation [19].

Experimental techniques and FEA help investigate reliability problems caused by CPI. For example, packaging-induced thermal deformation and stresses are analyzed using high-resolution moiré interferometry, e.g. by comparing flip-chip packages with and without a heat spreader on top of the die. The moiré technique also helps to study the effect of underfill properties on package warpage. Proper underfill would improve solder fatigue life time and reduce the risks of interfacial delamination.

The high thermal load generated by the solder reflow before underfilling can be simulated by a three-dimensional (3D) multilevel sub-modeling combined with modified virtual crack closure (MVCC) to investigate the CPI-induced interfacial delamination. Packaging-induced crack driving forces in ultra low-k layers are lower for fully dense low-k dielectrics. Therefore, package architecture optimization through DfM has to be aligned with the proper selection of material properties.

4.3.1 Thermo-mechanical Behavior of Microelectronic Packages

Stresses induced by the CTE mismatch between the die and the substrate, and temperature variations in the fabrication process cause the package to bend. Depending on the range of temperatures of operation, the bend may compromise the reliability of solder joints and the stability of Cu/low-k interconnect, especially at the proximity of solder UBM, and raise mechanical reliability concerns in flip-chip packages under thermal loads. High resolution moiré interferometry used to measure the thermal load response of deformation of flip-chip packages is significantly different from the electrical testing which provides inputs for device level DfM. Understanding its fundamentals is important to create DfM rules and guidelines for CPI.

Moiré interferometry is a whole-field optical interference technique with high resolution and high sensitivity for measuring in-plane displacement and strain distributions [19]. The incoming laser beam from the optical fiber is reflected by mirrors and impinging onto the surface of the reference grating, which splits the laser beam into four beams (2U and 2V), reflected onto the surface of the sample grating. A moiré image is formed by the interaction of the virtual grating created by the reference grating with the deformed specimen grating and recorded by the digital camera [20]. A sensitivity of 417 nm per fringe contour can be reached with a specimen grating of 1,200 lines/mm.

For comparison purposes, one may involve two types of packages into sample preparation e.g., 47.5 × 47.5 mm organic substrates and 20 × 20 mm Si dice, Fig. 4.17 (lidless and with lid), on top of the package acting as a heat spreader [19].

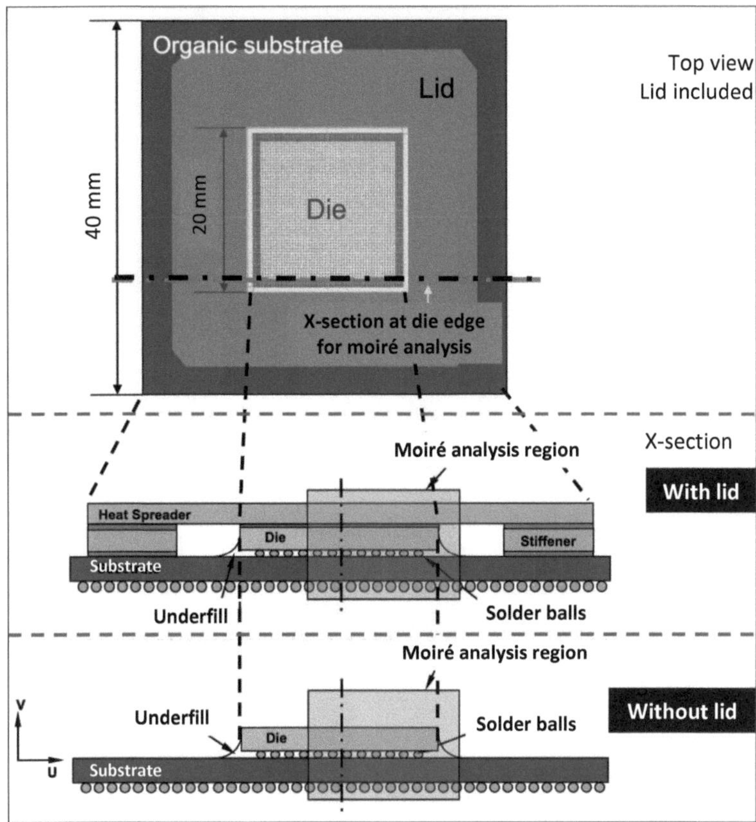

Fig. 4.17 Flip-chip packages with and without lid, top view and cross-sectional view (after [1])

The dashed line in the diagrams indicates the cross-section of the first row of the solder bump array along the polishing direction. The blocks show the areas analyzed during the moiré test.

A low-viscosity, brittle adhesive is used to attach a 1,200 lines/mm grating on the polished surface of the specimen at elevated temperature, such as 82 °C. The deformation at this temperature can be taken as that of reference (zero-stress) state. The moiré test is usually performed at room temperature (22 °C) and providing a thermal load e.g. of −60 °C. Fundamentally, the thermal load for moiré studies is limited by the glass transition temperature (T_g) of the underfill. Under a thermal load of −60 °C the U and V fields show the vertical and horizontal deformation pattern. Package bending can be deduced from the V field by counting the number of fringes, at 417 nm per fringe.

Looking at the deformation near the solder layer, the number of fringes in the die for package with lid, fewer than that of the lidless package indicates a 75 % reduction in die bending (Fig. 4.18). The mismatch in both V and U images (Fig. 4.19) can

Fig. 4.18 X-sectional view of stressed sample (after [1])

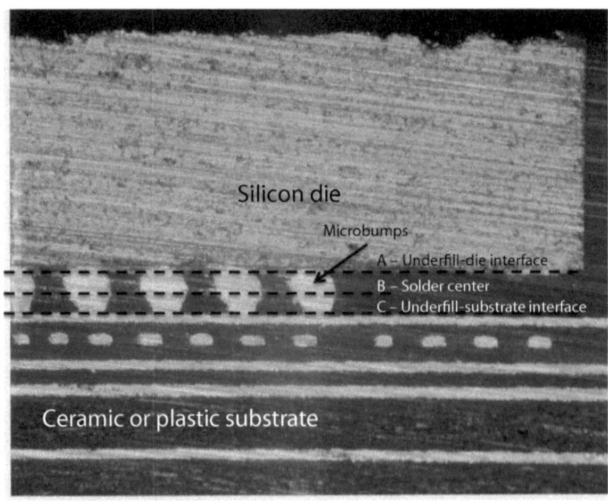

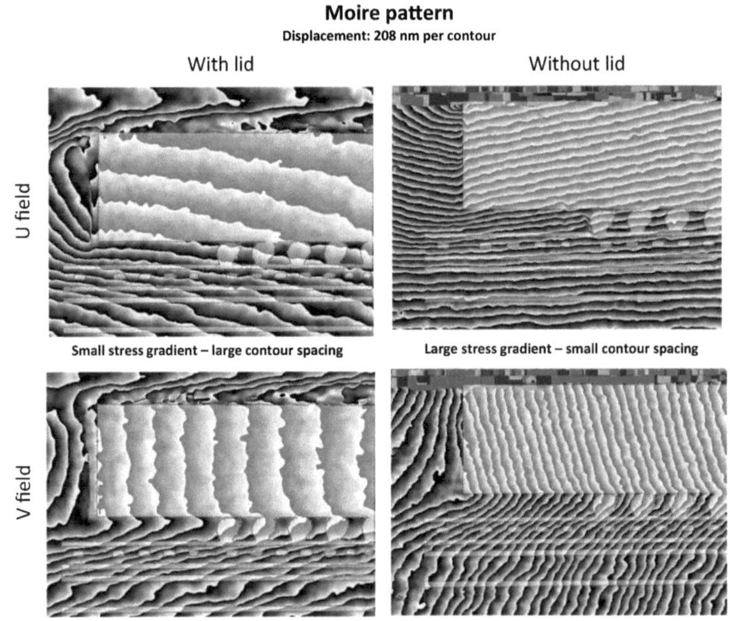

Moire pattern
Displacement: 208 nm per contour

With lid
Without lid
U field
Small stress gradient – large contour spacing
Large stress gradient – small contour spacing
V field

Fig. 4.19 Phase maps for package: U field and V field, with and without lid (after [1])

be attributed to the CTE mismatch across this interface, with the critical area of die corner where the displacement gradient between die and substrate reached maximum and the highest strain was developed [21]. The change of the phase angle can be captured by the phase-shifting technique with four continuous images taken with a phase different by $\pi/2$ and combined to extract the phase angle between fringes. The displacement distribution can then be determined by measuring the change of the phase angle.

Optical images of the cross-section superimposed onto phase maps identified the locations of the deformation at the die edge. A large strain was observed there with

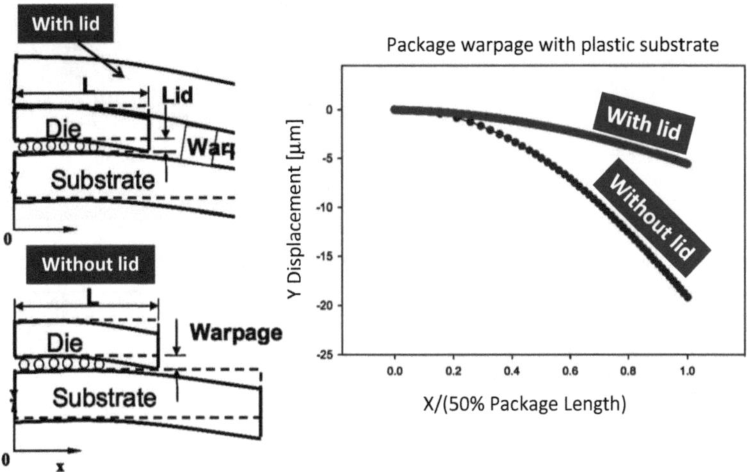

Fig. 4.20 Comparison of warpage in the die for package with and without lid (after [1])

each contour corresponding to a 208 nm displacement. The phase map can be subdivided to obtain displacement contours with resolution of 26 nm, to determine strain along the three lines (Fig. 4.18):

- The silicon-solder interface (Line A),
- The centerline of solder bumps (Line B),
- The solder-substrate interface (Line C).

Shear strains along lines A, B, and C in Fig. 4.18 increase as the die edge is approached and reach a maximum at the die corner making it prone to underfill delamination failures. The shear strain of the package without lid is 1.5 times larger than that of the package with lid, because the bending of the lidless package is much larger (Fig. 4.20). Adding compensating elements to the package structure is a good method for stress reduction.

The CTE of the lid material is usually very close to that of the substrate and it can cancel a good portion of the bending of the substrate. Without the support of the lid, the thermal deformation between the die and the substrate is exerted on the solder and on the underfill buffer layer, which may induce failure in solder joints and interfacial delamination between underfill and the silicon die [22].

4.3.2 *Multi-scale Stress Environment for 3D IC TSV*

Design verification based on simulation should be able to determine across-die variations in device electrical characteristics caused by through-silicon-via (TSV)/package-induced mechanical stress in the layout of 3D IC stacks. The limited test and characterization capabilities of such stacks and a strict "good die" requirement make this type of analysis critical for an acceptable level of functional and parametric yield

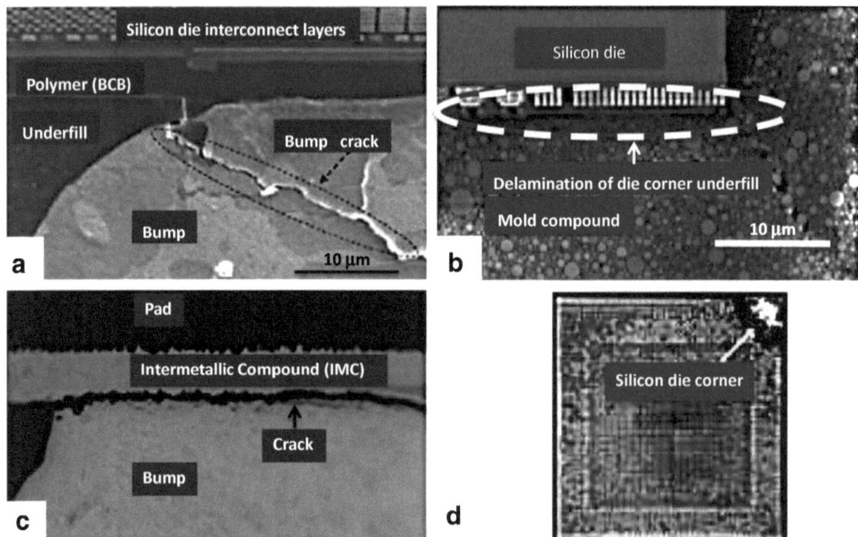

Fig. 4.21 Common failure modes for flip-chip packages (**a**) bump cracking (**b**) failure at IMC layer (**c**) delamination of underfill at die corner (**d**) ultra low-k dielectric delamination (after [1])

and reliability. The DfM methodology for managing mechanical stresses in designs including 3D TSV-based dice, stacks and packages, has to be supported by a set of physics-based compact models for a multi-scale simulation, to assess the stress across the device layers of different dimensions and materials. One needs to establish a database for multi-scale parameters of wafer-level and package-level structures. A model validation and calibration to measured local stress components and to electrical characteristics is becoming critical starting from the 45 nm CMOS technology node, with similar impact as that of the defects caused by poor lithography models.

Low K/Cu interconnect and redistribution layer stacks cause stress-driven reliability problems such as delaminating, cracking and/or fracture of various materials. In the process of stress model development, one should consider a number of process steps employed by the 3D IC technology: through silicon via (TSV) drilling and filling, wafer/die thinning, wafer bumping, high-temperature solder reflow (solder ball solidification), chip stacking, etc., as stress sources that can affect the chip-stack performance. Managing internal mechanical stress is a key task to ensure high performance and high reliability of products manufactured in advanced nodes of CMOS-based semiconductor technology especially for 3D ICs. A full picture of failure modes due to the stress distributed across the device layers is very complicated (Fig. 4.21).

Multi-scale material parameters (Tables 4.5) are critical to calibrate the stress distribution across the device layout and for model validation. Traditional methods such as finite-difference (FDA) and finite-element (FEA) analysis cannot be employed for a simulation of the transistor channel stress distribution across a die, due to the size of a model (hundreds of millions degrees of freedom, DOF) and the multi-scale character (typical dimensions spanned from centimeters package scale to nanometers device scale). The FEA simulators have been used for addressing the traditional chip-package interaction where a silicon chip was modeled as a homogeneous isotropic piece and the problem with calculating variations of transistor-to-transistor intra-channel stress and of electrical characteristics have not been addressed yet. Empirical modeling cannot take into account package-induced variations in transistor characteristics or provide a link to the physics-based package-scale simulation in order to explain a CPI-induced stress loading. A look-up table methodology is not practical due to a large area surrounding the layout of each device (radius of up to 5 μm) that should be accounted for a correct stress prediction. It will require an enormous amount of local layout configurations around a specific gate in order to get a proper representation. Therefore, compact, physics-based DfM models are the only possible solution to simulate transistor-to-transistor stress variation and its conversion into the electrical response of the devices. The flow should include an interface to layout formats (GDS, OASIS, etc.) with entire die layout linked with package-scale models (FEA).

The simulation should enable design verification of any 3D die stack for performance variations, with the key goal of controlling across-die transistor channel stress components impacted by:

1. 3D IC integration: die thinning, bump mounting, packaging, etc., responsible for the redistribution of pre-existing layout-dependent stress and generation of package-induced stress. The output of the FEA depends on package-scale boundary conditions (BC) e.g., displacements on the surfaces of the thinned die change strain distribution at the device layer.
2. Distribution of the CPI-induced strain across the device layer as simulated by conventional FEA. Interconnects, silicon/TSV and BRDL (back-chip redistribution layer) are characterized by the anisotropic mechanical properties linked to GDS II layer information.
3. The transistor-to-transistor stress variation calculated with compact model of a composite character of the device layer, with 3D IC stresses. TCAD/FEA-based stress simulation capabilities should evaluate the physics-based compact models to predict the layout-dependent stress components inside the transistors including all stress sources.
4. Compact models should convert stress values into the corrections of μ_o (low-field mobility) and V_{T0} (zero bias threshold voltage for long channel transistor) for the transistor SPICE netlist.
5. Analogously to the "multi-scale modelling", a set of "multi-scale materials data" is the input for the simulation of stress distribution with process/technology/fab-dependent material properties needed.
6. Calibration/validation of models/flows should be based on test vehicle characterization.

The displacements of IC geometries or volumes, such as MOSFETs and interconnects simulated with package-scale FEA depend on the packaging-induced loads, underfill options, and process temperatures of the thinned die (Fig. 4.18).

Table 4.5 Material parameters required for package-scale simulation (after [2])

Material	Meas. as function of temp. (−50..200 °C)	Underfill (epoxy resin)	Cu pillars (copper)	Molding compound (ceramic-filled epoxy polymer)	Flip-Chip ball (approx. 100 μm)	Micro ball (10s of μm)	BGA bump (>200 μm)	Effective interconnect (low-k/copper composite)	Effective bulk of silicon die (silicon/copper TSV)	Effective package substrate (organic PCB/copper)
CTE (ppm/K)	x	x	x	x	x	x	x	x	x	x
Young's modulus (GPa)	x	x	x	x	x	x	x	x	x	x
Poisson's ratio		x	x	x	x	x	x	x	x	x
Plasticity (MPa)	x		x							
Visco-plasticity (MPa)	x				x	x	x			
Visco-elasticity (MPa)	x	x		x						x
Glass transition temp. (°C)	NA	X								x
Die attach T (°C)	NA		x		x	x	x			

Table 4.6 Examples of material property measurement techniques

Parameter	Function	Measurement method	Resolution (local)	Accuracy	Specimen	Notes
Polymer based materials (organic interposer, PCB, underfill, molding compound)						
E (Young modulus)	T	Tensile test		±20 mN	Milled dog bone specimen (real components)	Customized specimen
		Nanoindentation		±10 nN	Real specimen, cross sections	Local structures affect results accuracy, unable to provide global or macro-scopic results
	t, T	Dynamic Mechanical Analysis (DMA) (–50–200 °C), (double/ single cantilever)		±0,01 mN, 0,1–200 Hz	Milled stripes (real components)	Customized specimen
		Nanoindentation (DMA-Module)		±10 nN	Real specimen, cross-sections	Local structures affect result accuracy Unable to provide global or macroscopic results
K (Thermal conductivity)	t,T	Volumetric dilatometry (PVT)			Milled stripes (real comp.)	Customized specimen
T_g	Glass transition	Dynamic mechanical analysis (DMA)		±0.1 K	Milled stripes (real components)	Customized specimen
CTE (Coefficient of thermal expansion)	T	Thermo mechanical analysis (TMA)			Milled cube (real components)	Customized specimen
		Digital image correlation (DIC)			Milled stripes (real components)	
	Poisson's ratio	Tensile tests and digital image correlation		±20 mN	Milled stripes (real comp.) Milled dog bone (real comp.)	Very accident-sensitive, not validated yet customized specimen

(continued)

Table 4.6 (continued)

Parameter	Function	Measurement method	Resolution (local)	Accuracy	Specimen	Notes
E		Tensile test (20–150 °C)		±50 nm	Cast dog bone (>40 mm)	Customized specimen T > 20 °C
	T	US microscopy (5–50 °C)	8 μm		Solder foils (approx. 250 μm)	Customized specimen Limited T range
		Nanoindentation	20 nm	±0.2 nm	Micro ball (10 s of μm) Cross sections	Limited to small solder joints
Poisson's ratio		US microscopy (5–50 °C)	8 μm		Solder foils (approx. 250 μm)	Customized specimen Limited T range Min. thickness depends on wave length
		Nanoindentation and FEM property extraction	20 nm	±0.2 nm	Microbumps (10 s of μm) Cross-sections	Calculation procedures to be validated
CTE	E, T	Moire technique	10 μm		FC bumps	
		Digital image correlation (20–150 °C)	1 μm		Cast dog bone (>40 mm)	Customized specimen
		Thermal Mechanical Analysis (TMA)		±0.1 K	Cast dog bone or foil	Customized specimen
Viscoplasticity creep	T	Shear test (20–150 °C)		±170 nm	Real solder joint (>200 μm)	
		Tensile test (20–150 °C)		±50 nm	Cast dog bone (>40 mm)	
		Nanoindentation	20 nm	±0.2 nm	Microbumps (10 s of μm) Cross-sections	Limited to small joints
Metals (Cu pillars and Cu wires)						
E(T)		Nanoindentation or AFAM	20 nm	±10 nN ±0.2 nm	Real specimen, cross section or blanket films	–

Poisson's ratio	US microscopy (5–50 °C)	8 μm		Deposited Cu layers (approx. 250 μm)	Customized specimen Min. thickness depends on wave length
	Nanoindentation and FEM property extraction	20 nm	±10 nN ±0.2 nm	Real specimen, cross section or blanket films	Calculation procedure to be validated
Yield strength (T)	Nanoindentation	20 nm	±10 nN ±0.2 nm	Real specimen, cross section	Customized specimen
CTE(T)	Digital image correlation/SEM	tbd	tbd	blanket films	Customized specimen
	X-ray reflectometry	tbd	tbd	Blanked films and bulk Si TSC arrays	Customized samples Test structures

Table 4.7 Mechanical model assumptions. Representation of bulk-silicon die as layers with spatial distributions of elastic (mechanical) properties for TSV implementation. Die properties are represented with global coordinates with a case-specific granularity (compact model)

	Layers	Representation
1	On-chip metal/dielectric interconnect stack (front)	Multilayer interconnect layout
2	Backside redistribution layers (BRDL)	Multilayer BRDL layout

Table 4.8 Stress direction impacting mobility (after [2])

		Stress impacting mobility	
Carriers	Stressed devices	Tensile	Compressive
Electrons	NMOS	Longitudinal, transversal	Out of plane
Holes	PMOS	Transversal, out of plane	Longitudinal

In the standard package simulations, high complexity may require the introduction of additional sub-models, but the chip has always been considered as a homogeneous isotropic piece of silicon. A multi-layer representation of a die should be implemented for advanced processes. The low-k (and ultra low-k) materials have much lower stiffness (Young's modulus) than the silicon, which affects the final stress distribution across a device layer after packaging, therefore these materials should not be treated as part of the silicon die. As a consequence, the way these materials attach to the die needs to be defined in boundary conditions.

For the following die components, different mechanical properties should be used:

- Interconnect stack and BRDL: averaged characteristics of a Cu/low-k "composite" with the fractions of Cu/low-k content,
- The die Si/TSV bulk: averaged characteristics of a Si/metal (TSV fill) "composite" with the fraction of metal (e.g., Cu) content,
- Package substrate: averaged characteristics of the PCB/Cu "composite" with the fraction of Cu content and interposer materials.

In order to evaluate the impact of packaging, the process should be considered a die-attach process, with reliability cycling critical for generating stress in materials (Table 4.7) and device layers (Table 4.8). The employment of several models for package materials (for example elastic vs. elasto-plastic or visco-elastic properties, temperature-dependent versus constant room-temperature values) would significantly impact the accuracy of the predicted strain/stress distributions. Unjustified approximations may lead to grossly incorrect conclusions.

Preferred material characterization methods (Table 4.6) depending on material nature and geometry are applicable to IC stacks in real packages and are temperature-dependent.

Simulation INPUT:

- Packaging geometry
- Stress-free temperature
- Thermal and mechanical load
- Mechanical properties (multi-scale)

Table 4.9 TEM measurements of transistor strain (after [2])

TEM method	TEM mode	Accuracy	Spatial resolution	Field of view
CBED	Diffraction/probe	$2 \cdot 10^{-4}$	5 nm	n.a.
NBED	Diffraction/probe	$1 \cdot 10^{-3}$	10 nm	n.a.
HRTEM	Image	$1 \cdot 10^{-3}$	2 nm	150×150 nm^2
Dark-field off-axis holography	Image	$2 \cdot 10^{-4}$	4 nm	$1{,}500 \times 400$ nm^2
Dark-field in-line holography	Image	$< 1 \cdot 10^{-4}$ (precision)	< 1 nm	$1{,}000 \times 1{,}000$ nm^2

- BC representing packaging-induced displacements on the faces of the thinned die, from the output of the package-scale FEM simulation
- Mechanical properties of the relevant materials: interconnect stack (Cu/low-k) of silicon die (Si/Cu-TSV)
- GDSII for interconnect layers, TSVs and BRDL

Simulation OUTPUT:

- Distribution of the strain components across the device layer
- Displacement components on the faces of the thin die.

Young's modulus, CTE, plasticity (yield stress level and isotropic or kinematic tangent modulus) and visco-elasticity (complex modulus, phase angle, etc.) should be determined as functions of temperature within a temperature range (typically, 30 °C to >200 °C). The thermal load would be considered zero at the die attach temperature, and reach its maximum at room temperature (or below, as needed by the product application).

The following assumptions are typically adopted for simple approach simulations:

- All structures (interconnects, Si die and package substrate) would be modeled as perfectly elastic materials,
- Cu pillars are considered elasto-plastic material,
- Solder balls, BGA bumps and micro bumps are modeled as elastic and viscoplastic materials,
- PCB, underfill and molding compound are modeled as visco-elastic materials.

The accuracy of stress simulation depends on the calibration of the many material coefficients and on their metrology (Table 4.9).

Solder joints release the stress due to visco-plastic creep deformation. At Flip-Chip and BGA-level, an adapted shear microtester needs to be applied, however, the mechanical tests become challenging for solder joints with sizes below 100 μm. Nanoindentation is preferred for high accuracy microbump characterization of real products at sub-um scale. The strain distribution across a device layer caused by packaging is calculated with the FEA tool by implementing the displacement BC. Interconnect, BRDL, and thin silicon layers should be approximated as layers with spatially distributed elastic properties determined by their layouts. Calculations of effective Young's modulus, Poisson ratio, and CTE as the functions of the metal density in all interconnect levels should be based on compact models of mechanical properties of anisotropic composite materials. The position-dependent mechanical characteristics of all layers should be presented in a format readable by the FEA tool.

4.3.3 Effects of Underfill Materials

DfM guidelines for CPI would not succeed without cooperation from packaging manufacturers driving package MfD based on improved selection of package architectures and materials. The introduction of bumps as connections in flip-chip packages between Si die and organic substrate enables high input/output (I/O) pad counts and improves electrical performance. The bumps also serve as mechanical joints but in this role, they are sensitive to package deformation due to the CTE mismatch with the organic substrate and its transfer to the die. This raises mechanical reliability concerns, particularly for the outermost joints, which are subject to maximum deformation [23]. Underfill improves the reliability as it couples the CTE among chip, solder, and substrate, and this coupling reduces thermal stresses in the solder bumps. But bending of the package board together with the large stresses at the die corners can lead to delamination at the passivation layer-to-underfill interface [28]. Such bending, acting on the porous ultra low-k interconnects can also drive their delamination [29]. In addition, the use of fine-pitch solders makes the underfilling and flux cleaning process more difficult, which may lead to void formation and localized underfill delamination [22–24]. Fatigue in bulk solder during thermal cycling, delamination at the intermetallic compound (IMC) layer due to overstress, underfill to die-passivation delamination and ultra low-k delamination at the die corners due to stress concentration further compromise package reliability.

One way to reduce low-k delamination risk is to select mechanically compliant underfills. However, this contradicts the original purpose of underfill as a solder protection layer. The underfill has to meet a number of package DfM requirements to balance designed-in reliability between solder joint reliability and ultra low-k dielectrics integrity:

- Good protection for both low-k materials and solder bumps
- Good adhesion to the passivation layer and solder mask
- Short filling time
- Minimum filler settlement
- Low moisture absorption [25–27].

The underfill selection became more difficult with the implementation of fine-pitch Pb-free solder bumps. Time-dependent behavior of underfill material can be simulated with a two-stage Young's modulus and thermal expansion coefficient model to represent a realistic behavior of polymer materials. Darveaux's strain energy density model [32] would help predict the thermal fatigue behavior of Pb-free solder joints under thermal cycling. The effects of glass transition temperature (T_g), E, and CTE on solder fatigue, Chip/underfill delamination, and low-k/passivation delamination, need to be investigated using virtual underfill properties, for optimal reliability performance at package and chip level.

A qualified underfill should provide low stresses both in the solder joints and in the low-k interconnects. A three-dimensional (3D) multilevel sub-modeling combined with modified virtual crack closure (MVCC) is first focused on the effects of dielectrics and solder materials. Then, it is extended to include effect of the interconnect reduction size accompanied with an increased number of metal levels separated by ultralow-k porous and fragile dielectrics. Due to the symmetry of the package, one-quarter model with symmetric boundary condition is commonly used (Fig. 4.22).

The thermal cycling between 125 °C and −55 °C impacts mechanical behavior of the polymeric underfill depending on the temperature below above T_g(Fig. 4.23). Upon cooling, there is a transition from compressive stress at its bottom surface, to tensile stress at the top surface, a result of the distortion in the solder joint due to CTE mismatch between the silicon die and the polymer substrates (Fig. 4.24). The peeling stress at the top surface transferred into the silicon die via solder UBM and pad, harmful to the passivation low-k interconnects inside chip, can be reduced by underfill material (Fig. 4.25a). While underfill reduces the peeling stress, it generates a shear stress at die corner which is a potential threat for silicon to delaminate, with an opposite T_g impact (Fig. 4.25b).

To verify product reliability, several solder joint fatigue life prediction models have been proposed based on various stress, strain and fracture criteria [32]. In an energy-based method proposed by Darveaux [32], the inelastic strain in solder joints during thermal cycling consists of a time-dependent creep strain and a time-independent plastic strain. To apply Darveaux's model, a relationship representing the elastic–plastic behavior of the solder joint is required. A commercial FEM simulation package, Ansys, has viscoplastic elements, which use Anand's constitutive model [35] consisting of a flow equation and three evolution equations.

The thermal fatigue life time of solder joints was found to decrease with increasing strain energy accumulated ΔW_{ave} [36]. The effect of the thermo-mechanical properties of underfill on solder fatigue life was verified by calculating the change in ΔW_{ave} during thermal cycling test. Because the different combinations of T_g, E, and CTE impacted the modeling results, a sub-modeling technique was introduced to identify the most critical property for improving solder fatigue life while keeping the calculations at manageable level.

The way to study crack driving force at the interfaces between the different materials is by inserting initial cracks into the model followed by the energy release rate calculations by the modified virtual crack closure technique. Energy release rate (ERR) of delamination at the chip/underfill interface showed the effect of T_g with the opposite trend to that of solder fatigue. T_g was the key parameter determining the crack driving force. T_g varying from 139 °C to 30 °C caused about 5.5 times difference in the ERR (Table 4.10). Underfill with high T_g yielded a larger crack driving force due to the larger stress concentration at the die corner compared to that of low T_g underfill. On the other hand, ERR increased by 67 % when the E of underfill increased from 5.4 to 10 GPa, and around 100 %, when CTE increased from 20

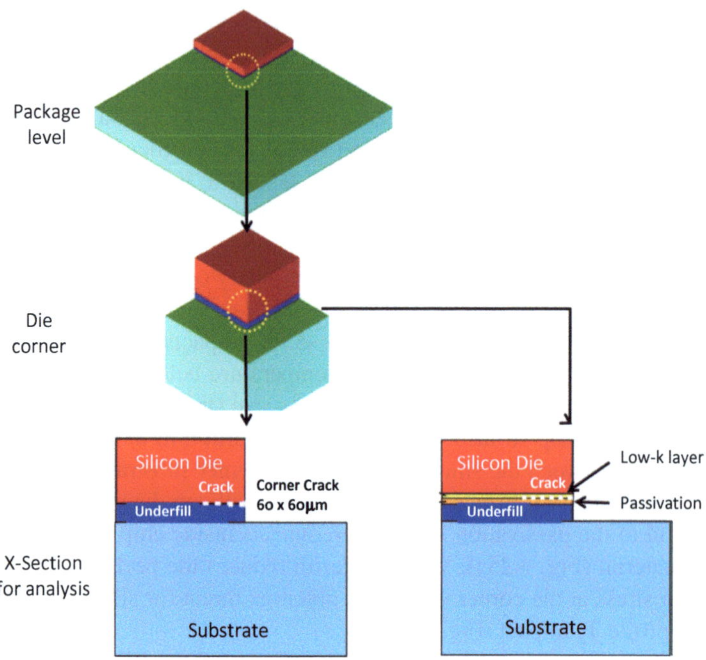

Fig. 4.22 A quarter model of a flip-chip package with cross-sections, for hierarchical sub-modeling to study chip/underfill delamination (after [1])

Table 4.10 Summary of failure parameter response to material properties

Consequential parameter		Increase rate		Primary parameter	Rate of change
Energy Release Rate	Increased	5 times	When	Tg	Increased from 30 °C to 139 °C
Energy Release Rate		100 %		CTE	Increased from 40 to 20 ppm/°C
Energy Release Rate		67 %		E	Increased from 5.4 to 10 GPa
Delamination		5 times		Tg	
Delamination		66 %		E	

to 40 ppm/°C. It was concluded that increasing T_g of underfill provided better protection for solder joints from fatigue under thermal cycling, but increased the hazard of underfill-to-chip delamination. Therefore, E and CTE should be optimized instead of T_g to improve underfill reliability without impacting the solder integrity. Decreasing the E of the underfill can help reduce the ERR for chip/underfill delamination without impacting solder fatigue determined by the creep and plastic deformation in the solder, not very sensitive to E. In both cases, underfill with smaller CTE (~20 ppm/°C) yielded better reliability performance (Table 4.10).

Another common failure for flip-chip packages is "white bump" low-k delamination in the C-Mode Scanning Acoustic Microscopy (C-SAM) test. The fracture

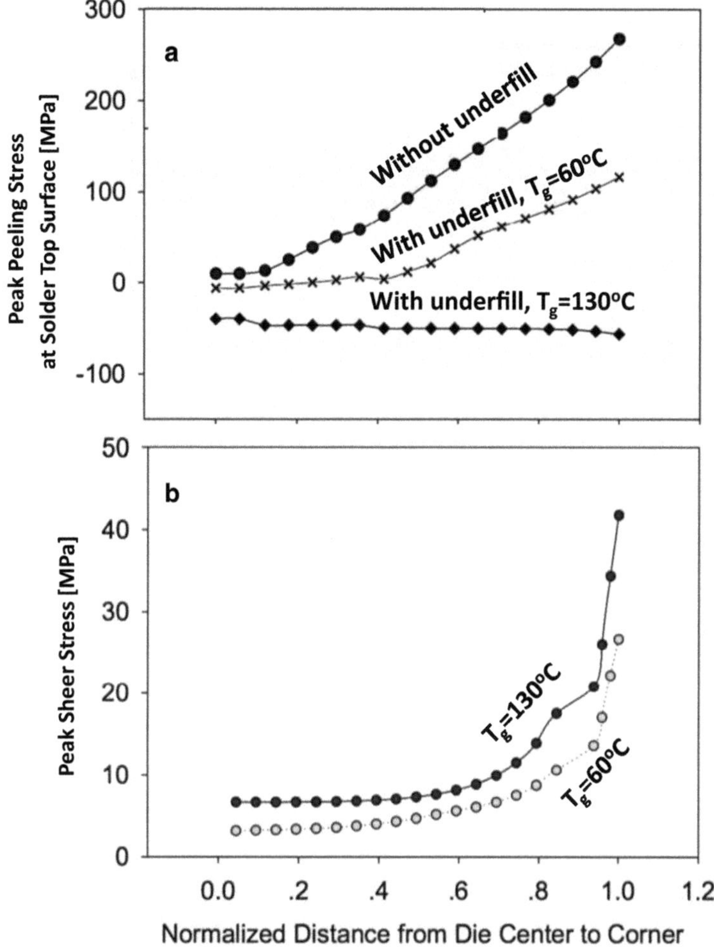

Fig. 4.23 Stress distributions (**a**) normal z-stress in solder, (**b**) maximum peeling stress at solder top surface (after [1])

resistance of porous low-k materials is usually in the range of 2–6 J/m^2, much weaker than the adhesion strength of underfill to solder mask (over 35 J/m^2) due to silane additives to promote adhesion [30]. Moreover, the dicing defects at the periphery of Si die can initialize defects and facilitate crack propagation observed during thermal cycling [37] (Fig. 4.24).

To study the effect of underfill material, a simplified interconnect structure with porous low-k layer was analyzed (Fig. 4.26). A pre-crack was inserted into the structure as an initial defect. Underfill mechanical properties especially T_g were found to have a large effect on the low-k reliability. The ERR of low-k dielectric delamination can increase by five times when T_g of the underfill increases from

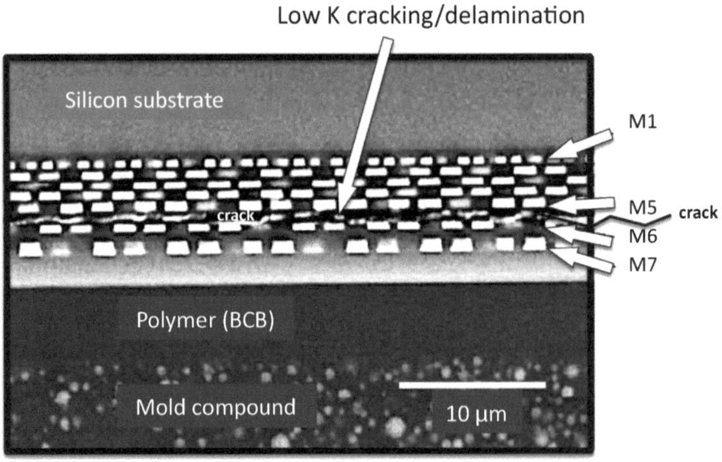

Fig. 4.24 Delamination in low-k layer (after [1])

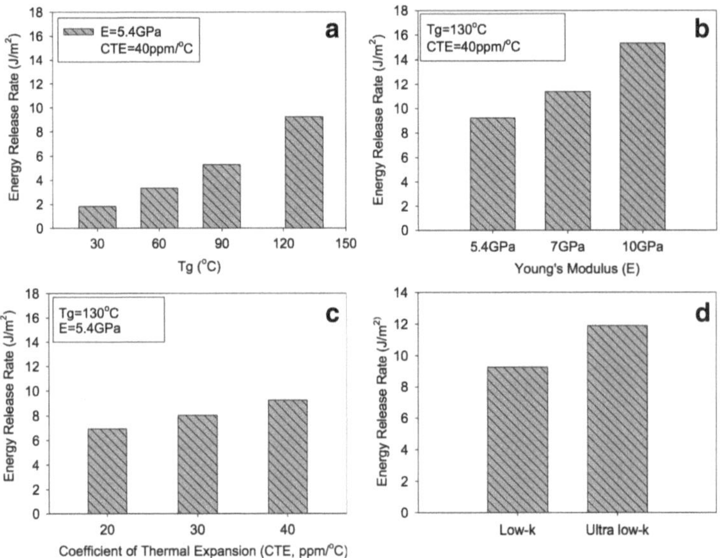

Fig. 4.25 Effect of underfill properties on low-k layer delamination (energy release rate) from passivation layer (**a**) T_g effect (**b**) E effect (**c**) CTE effect (**d**), comparison between low-k materials (after [1])

30 °C to 130 °C. An increase in ERR of 66 % and 33 % was observed by varying E and CTE, respectively (Fig. 4.25). Underfills with lower T_g, smaller E and CTE are required to protect the low-k layer from delamination and provide proper protection for solder bumps.

Material requirements leave a small selection window for a desirable range of mechanical properties of underfills and package DfM has to navigate around them.

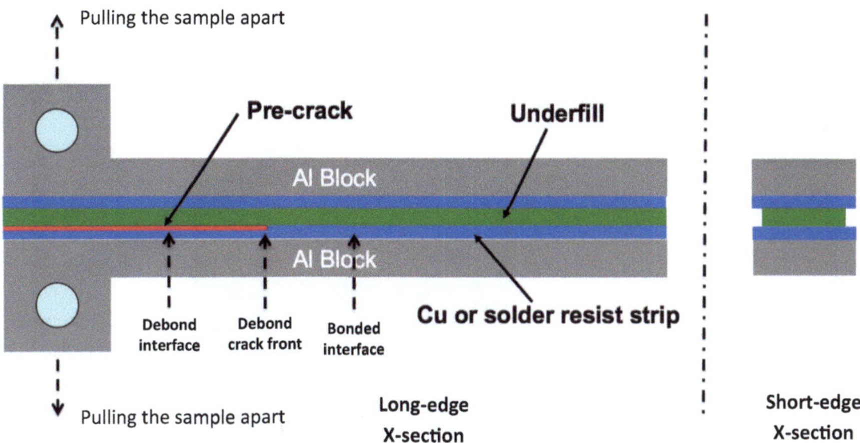

Fig. 4.26 Sample for DCB measurement (after [1])

The dielectric material with a lower k value induces larger driving forces for failure. Also, for the given crack length, the ERR is typically larger than the nominal fracture strength of low-k dielectrics, about 2–6 J/m^2 [30], indicating an unstable crack propagation. For the 32 and 22 nm processes, the choice of underfills is even more difficult due to the use of dielectric materials with lower dielectric constants and Pb-free solder bumps. Here, DfM for packaging represents the optimized choice of materials, because electrical design does not have enough DOF's to improve reliability.

4.3.3.1 Moisture Effect on Underfill Adhesion

The crack growth depends on the amount of strain energy released per unit area of crack growth and on the resistance, i.e. the energy required to break the bonds, create new surfaces, and generate dislocations near the crack tip (fracture toughness of the material). To propose DfM rules for underfill reliability, cracking criteria need to be established first. A fracture criterion is established by comparing the energy release rate with the fracture toughness [32]. If the driving force exceeds the interface adhesion strength, the crack will grow, otherwise, it won't propagate. A double cantilever beam (DCB) method [33] can be used to measure interfacial fracture energy of underfill materials to Cu and solder resist (SR). These interfaces, relatively weak, are the source of failure in packages with Cu pillar structures. Delamination at underfill/solder resist interface may lead to failure in the solder joints due to the crack propagation. The failure at the Cu pillar/underfill interface can create sharp cracks and pose high risk for fracture in low-k ILD when the crack propagates into the die.

For package DfM and design rule definition, failure modes to be avoided could be similar to those emulated in the process of defect metrology. The DCB method

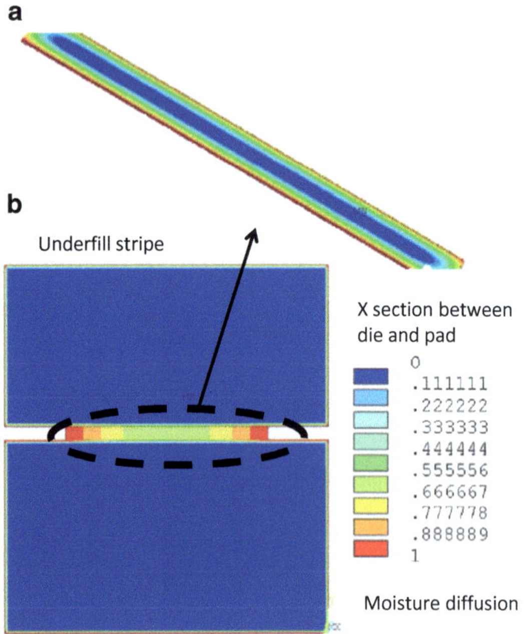

Fig. 4.27 Diffusion after 5 days moisture soak at 85 °C and 85 RH (**a**) overview of underfill (**b**) cross section view (after [1])

measures the adhesion strength between two materials by pulling the sample apart while monitoring the load and the opening displacement. A pre-crack generated by Cu sputter or mold release can be built into the sample in order to initiate failure at the desired interface (Fig. 4.26). For testing in a micro-tensile system at room temperature, DCB samples are first loaded until the pre-crack started to grow, and then unloaded to obtain the sample stiffness repeated until the sample failed. When the crack was not growing, the sample stiffness remained constant and the loading/unloading curve was linear. Once the load reached a critical value, the loading curve started deviating from the unloading curve. The fracture toughness, G_c, is calculated from the critical load for the crack to grow, the sample stiffness and dimensions [30]. XPS surface analysis verifies that the defect was the delamination at the Cu/underfill interface rather than cohesion failure inside underfill.

Reliability rules may also depend on the ambient. Moisture plays an important role in the thermal and mechanical reliability of microelectronic packages, especially for underfill polymer materials which have high moisture diffusivities [34]. In a test, the DCB samples can be soaked in a moisture chamber (85RH) at 85 °C for 5 days and FEA employed to simulate the moisture absorption (with heat conduction equation) [35]. After 5 days of moisture soak at 85 °C/85 RH, for underfill materials fully saturated at the outside and about 60 % saturated in the center (Fig. 4.27), G_c values dropped by ~40 %. Similar trends for the moisture effect were obtained between solder resist film and underfill (Fig. 4.28). A decrease of 34 % was found for samples soaked in the moisture chamber for 5 days due to degradation of adhesion strength of Cu/UF and Cu/SR interfaces (Cu/SR interface was relatively more stable).

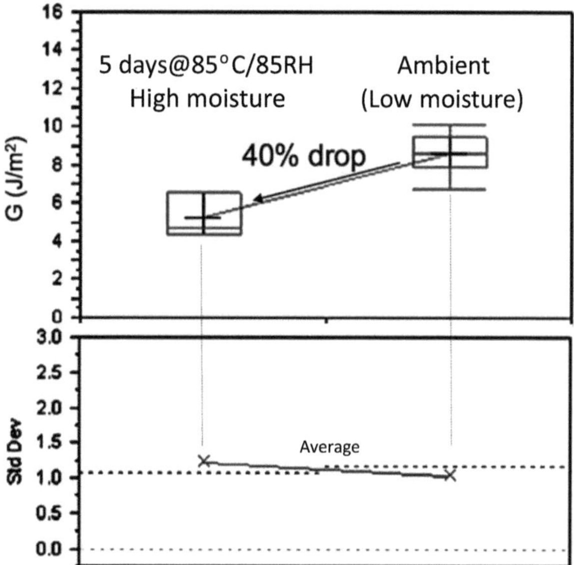

Fig. 4.28 Moisture effect on the adhesion strength of Cu/underfill interface moisture effect on the adhesion (after [1])

The key failure modes to be prevented by package DfM include solder fatigue and layer delamination. Thermal cycling fatigue life of Pb-free solders depends on the T_g and CTE of the underfill, but not on its E. Underfills with high T_g and CTE close to that of solder bumps are preferred for longer fatigue lifetime.

In summary, from the material side of DfM, T_g remains the key parameter for chip/underfill delamination. Delamination of Low-k dielectric/passivation increased by five times when T_g of the underfill increased from 30 °C to 130 °C. A much lower increase, of 66 % and 33 %, was found when E increased from 5.4 to 10 GPa or CTE increased from 20 to 40 ppm/°C, respectively.

Meeting DfM requirements in underfill selection requires material optimization for package design. For example, underfill with high T_g is preferred for better solder fatigue life time, while underfill with low T_g is favored when considering low-k dielectric integrity. The requirements of underfill to provide good protection for both the solder joints and weak low-k dielectrics make underfill selection increasingly difficult.

Moisture content in underfill has significant impact on the interfacial adhesion strength. The adhesion energy was reduced by about 40 % for Cu-underfill and Cu-SR interfaces after soaking the test samples in a moisture chamber at 85 °C-85 RH for 5 days. Therefore, package design has to either prevent moisture penetration or have these values factored into product specifications.

4.3.4 Cu/low-k Interconnect Reliability

Interfacial fracture induced by CPI arise from the mismatch in the CTE's between the chip and the substrate, directly coupled into the Cu/low-k interconnect structure to drive interfacial delamination. Thermal load on the package reaches a maximum

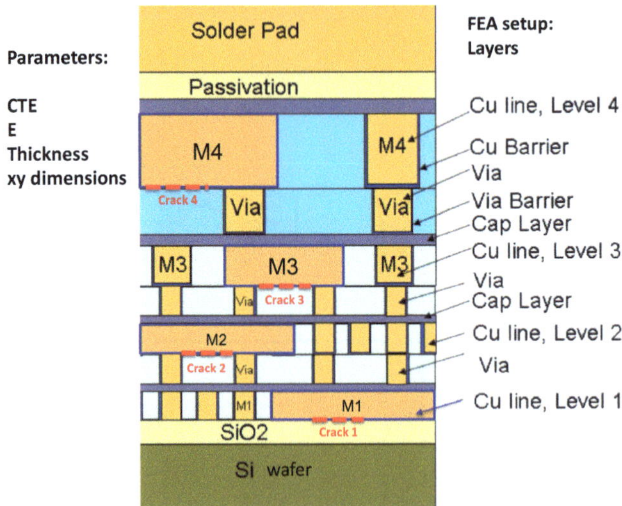

Fig. 4.29 Cu/low-k die structure cross-section with crack locations, for FEA (after [1])

during solder reflow before underfilling in the die attach process. The shear and peeling stresses reach a maximum at the outermost corner bumps, driving interfacial crack formation and propagation in the Cu/low-k interconnect [38–40]. This may dictate MfD rules for special strengthening of these bumps by buffering materials, or DfM rules for locations of these bumps with respect to die corners.

A three-dimensional (3D) multilevel sub-modeling method would help calculate the CPI induced crack driving force for interfacial delamination in Cu/low-k interconnect structures. For stand-alone silicon chips, FEA results show that thermal stresses in Cu interconnect lines depend on the aspect ratio (height to width), and the degree of confinement from the surrounding dielectric materials, barriers, and cap layers (Fig. 4.29) [42]. For a Cu line with an aspect ratio greater than 1, the stress state of the Cu line is triaxial and it behaves almost linear, elastically under thermal cycling, in quantitative agreement with the results from x-ray diffraction [43].

Flip-chip package deformation increases the thermo-mechanical stresses in the interconnect structures. To establish the joint package – chip codesign DfM rules, large dimensional difference between the packaging and interconnect structures requires a multilevel sub-modeling to evaluate the energy release rate, ERR at interfaces in the interconnect structures [45], to bridge the gap in the geometrical differences between the package and the die. A 3D FEA model could be developed based on a five-level sub-modeling (Fig. 4.30), as follows:

Level 1: Package level would cover thermal deformation of the whole flip-chip package. A quarter section of the package is modeled using the symmetry condition. No interconnect structure detail is considered because its thickness is too small compared to the whole package. Simulation results can be calibrated with Moiré interferometry.

Level 2: Critical solder level requires a sub-model focusing on the critical solder bump region with finer meshes and the Ansys cut boundary technique for

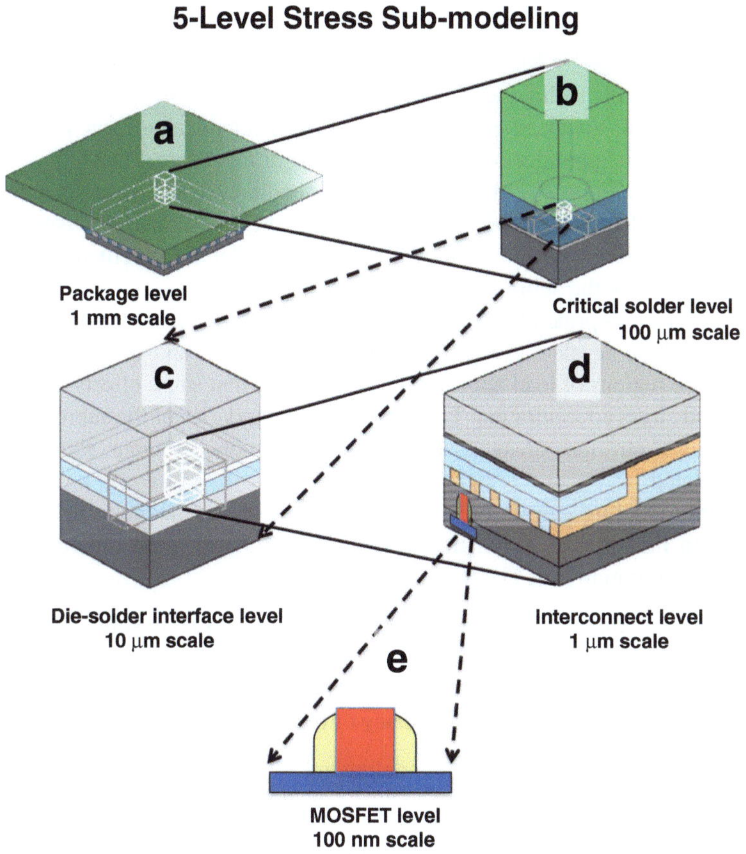

Fig. 4.30 Five-level sub-modeling: (**a**) package level; (**b**) critical solder level; (**c**) die-solder interface level; (**d**) detailed interconnect level, (**e**) transistor level (after [1])

sub-modeling [44]. A uniform ILD layer at the die surface would be considered with no detailed interconnect structure.

Level 3: Die-solder interface level requires a sub-model based on Level 2 using the cut boundary technique. Level 3 model focuses on the peeling stress at die-solder interface region with portion of the die with uniform ILD layer and a portion of the solder bump, but with no detailed interconnect structure included.

Level 4: Detailed interconnect level uses a sub-model zoomed in from Level 3, focusing on the die-solder interface, an interconnect structure with four metal levels and the effects of multilevel stacks. A crack with a fixed length is introduced along the center axis at several interfaces of interest.

Level 5: Device level focusing on individual MOSFETs.

The J-integral method as a standard option of FEA would calculate both the energy release rate and the mode mixity for 2D and 3D interfacial cracks, using fine meshes near the crack tip to achieve convergence [45].

4.3.5 Sub-modeling of Stress Propagation with Ansys

(A) Principle of the technique

Ansys is an advanced tool for simulating mechanical effects in materials. Setting up its input decks is becoming a part of mechanical DfM.

Submodeling uses cut boundaries (or specified boundary displacements) of the submodel, which represent a cut through the coarse model. Displacements calculated on the cut boundary of the coarse model are specified as boundary conditions for the submodel.

St. Venant's principle states that if an actual distribution of forces is replaced by a statically equivalent system, the distribution of stress and strain is altered only near the regions of load application. This implies that stress effects are localized around stress concentration. Therefore, if the boundaries of the submodel are far enough from the stress concentration, reasonably accurate results can be calculated.

Submodeling can be used effectively in a stress, magnetic field analysis, etc., in the systems requiring stress modeling across wide dimension spans, e.g., from mm to nm. The technique:

- Reduces the need for complicated transition regions in solid finite element models
- Enables experiments with different designs for the region of interest
- Helps demonstrating the adequacy of mesh refinements.

Restrictions for the use of submodeling are:

- Valid only for solid elements and shell elements
- Assumes that the cut boundaries are far enough from the stress concentration region.

The process for using submodeling is as follows:

- Create and analyze the coarse model
- Create the submodel
- Perform cut boundary interpolation
- Analyze the submodel
- Ensure adequate distance between the cut boundaries and the stress concentration.

The following detailed actions are related to these steps:

1. The coarse model should be created for the entire structure. Only solid and shell elements support the submodeling technique.

   The coarse model need not include local details such as fillet radii, but the FE mesh must be fine enough to produce a reasonably accurate degree of freedom (DOF) solution.
2. The submodel is completely independent of the coarse model (the database needs to be cleared) and assigned a different jobname for the submodel so

that the coarse-model files are not overwritten). Using the same element type (solid or shell) that was used in the coarse model, specify the same element real constants (such as shell thickness) and material properties. The location of the submodel (with respect to the global origin) must be the same as the corresponding portion of the coarse model.

3. Cut-boundary Interpolation consists of the following steps:

- Identify and write the cut-boundary nodes of the submodel
- Restore the full set of nodes
- Point to the coarse results file
- Initiate cut-boundary interpolation
- In the process of submodel analysis, duplicate any other loads and boundary conditions on the submodel that existed on the coarse model.

(B) Element birth and death technique

If a material is added to or removed from a system, certain elements in the model may become "existent" or "nonexistent" [44]. One can employ element birth and death options to deactivate or reactivate selected elements. The birth and death feature is useful for analyzing excavation (as in mining and tunneling), staged construction (as in shored bridge erection), sequential assembly (as in fabrication of layered computer chips), and other applications in which one can identify activated or deactivated elements by their known locations.

The Ansys program does not actually remove "killed" elements, but it deactivates them by multiplying their stiffness (or conductivity, or other analogous quantity) by a severe reduction factor. Similarly, when elements are "born", they are not actually added to the model, but reactivated. Thermal strains are computed for newly-activated elements based on the current load step temperature and the reference temperature. Thus, newborn elements with thermal loads may not be stress-free as intended.

One can apply element birth and death behavior to most static and nonlinear transient analyses using the same basic procedures:

1. Build the Coarse Model
2. Apply Loads and Obtain the Solution

- Define the First Load Step
- Define Subsequent Load Steps
- In the remaining load steps, one should deactivate and reactivate elements as desired, apply and delete constraints and nodal loads.
- Sample Input for Subsequent Load Steps

3. Review the Results

Follow standard procedures when postprocessing an analysis containing deactivated or reactivated elements.

An output of such die-scale simulation should contain at least three strain components distributed across the device layer or the entire stress tensor.

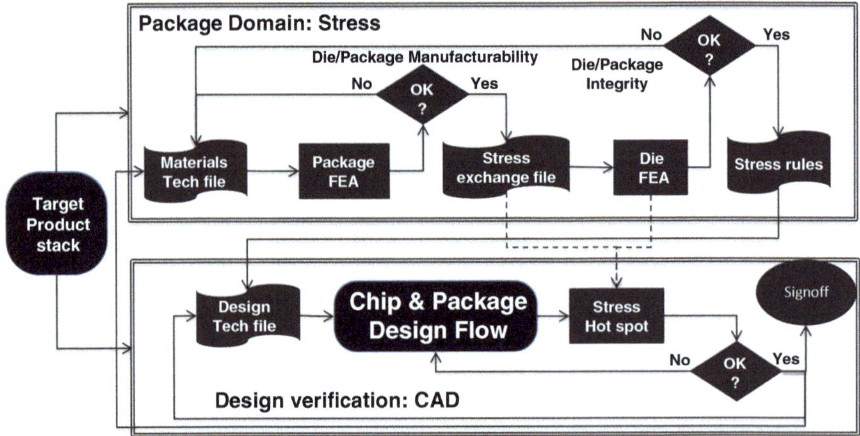

Fig. 4.31 3D stress analysis and optimization: (1) estimate target TSS, (2) Package analysis, (3) Stack analysis, (4) Design tech file, (5) Chip design, (6) stress hot spot analysis, (7) final iteration (after [3])

Both BEOL and BRDL for FEM simulation should be represented by spatial distributions of their mechanical properties also based on the multilayer interconnect/BRDL layout (GDS II). For the package scale simulation, elastic properties of the BEOL interconnect stack, BRDL layers, and the die bulk are implemented with volume averaging. For TSV related structures, FEM simulation requires the elastic–plastic properties to be represented by a spatial distribution, depending on the TSV layout.

CTE values for Cu TSV or Cu pillars can be determined either for individual structures using the digital image correlation technique in a scanning electron microscope (after separation of the structures using e.g., the focused ion beam, FIB technique) or, for arrays of test structures, using X-ray reflectometry. However, since CTE is a material property, which may only to a small amount depend on the material dimensions, diffraction measurements may be required to determine its value.

4.3.6 Stress Simulation Environment

To become a part of standard IC design process, simulation tools and techniques integrated into a DfM flow, starting from initial exploration, ending on final verification and stack sign off (Fig. 4.31), including:

Step 1: Target 3D Through Si Stack: die, interconnect and package materials. It is assumed that Si chips and the package substrate are not (yet) designed, but the target die sizes and global interconnect schemes and constraints are defined so that basic dimensions and material properties are known.

Step 2: Package level analysis with incumbent FEA tools such as Ansys or Abaqus for the global stress modeling. It is assumed that package materials are fairly soft and plastic, with relatively low T_g points and characteristics non-linear with temperature. Si dice are monolithic bricks and the mesh is set according to the usual packaging considerations driven by bump dimensions, consistent with die sizes or the basic material properties for TSS implementation. When convergence is reached and the proposed implementation is considered manufacturable, a 'Stress Exchange Format' (SEF) file is extracted. It transfers boundary conditions from package level to feature level analysis and decouples the choice of the simulation environments for different domains, thereby allowing the use of optimum tools for each step. SEF is a matrix of stress fields, expressed in terms of displacement, on every face of every die in the package, with the granularity of a given matrix determined by the feature size.

Step 3: A TSS stack feature analysis, using the boundary conditions imported from the package level analysis (SEF) and the target TSS material characteristics. A specialized FEA tool (e.g., SNPS FAMMOS) relates mechanical stress to electrical device characteristics (charge carrier mobility), compatible with the materials set in Si processing, and capable to interpret Si layout parameters (an FEA based tool may model limited layout configurations at a selected granularity based on the analysis of a GDS for the whole chip). The 'designability' of a proposed stack is should be verified:

- First, assessed by modeling of the local stress distribution,
- Then, explored in more detail about the physical and electrical interactions for specific layout configurations under package boundary conditions.

Basic design rules and target stack design statistics fed into layout would be calibrated to performance criteria (e.g., mobility shift $< \sim \times$ %). If the layout constraints, such as die sizes or placement and floor planning restrictions, are inconsistent with the desired stack target performance, then a different configuration is required. When convergence is reached and the proposed implementation is considered designable, a set of stack-specific stress design rules is extracted. This type of analysis can be applied to derive the compact models for the whole chip analysis.

Stack Specific Stress Design Rules are an incremental set of layout and placement to chip and stack design: Keep Out Area (KOA) around a TSV, placement constraints for FC bumps or μ-bumps, alignment constraints from die-to-die, etc., based on criteria for maximum allowed deviation in electrical performance, as well as physical integrity constraints (e.g., CPI) for a given set of packaging and assembly boundary conditions.

Step 4: Upgrading the design technology (tech) file with the stack-specific stress design rules to converge on chip, package substrate and stack designs. While rules by definition tend to be 'one-size-fits-all' solutions which include some margins, they would be specific to a given 3D TSS stack and the excess margin would be limited. The advantages of relying purely on a verification based methodology would be retained. The simulation and optimization would be done 'off line' to develop a set of Stack Specific Stress Rules, while the 'one-line' implementation of stress design rules would be accomplished via a verification approach. Eventually, an integrated and interactive design-for-stress methodology would emerge.

Step 5: Standard Si TC design flow, including the stack specific stress design rules and constraints to produce minimal perturbation to the standard design practices. The mechanical stresses would be handled through a set of layout rules. Later, the 3D TSS stack design will be implemented in a series of quasi-independent 2D chip designs, where the constraints from one layer will be imported into the next level. This transfer would evolve from quasi-manual to automated supported by upgrades in EDA tools. One should note that this approach is adequate for heterogeneous stack design with the partitioning along functional lines, e.g., memory-on-logic or analog-on-logic integration. The design of a fully optimized logic-on-logic type of stacks will require a significant change to the design flows and EDA tools.

Step 6: Si chip analysis for "Stress Hot Spots". Stresses from several sources result in complex interactions that produce net cumulative effect on a design feature. For example, a combination of stresses driven by 2D layout, the TSV stress, the stress from μ-bumps and FC bumps, the stresses from die-to-die interactions in the TSS stack, etc., would differ depending on location on a given die. Some features, such as analog blocks, may be more sensitive to the stress configuration than other features, such as standard cells. However, all current design methodologies that evaluate device electrical performance and timing across process corners assume temperature or stress to be uniform across the die. A "stress hot spot" would be an interaction that accounts for both the cumulative stress gradients at a given location on a die, and the specific sensitivity of a circuit at that site, to those gradients. Thus, a 'stress hot spot' checker that evaluates a complete GDS would have to rely on compact models, and the tools which cannot be FEA class but conceptually similar to the DfM tools, to analyze a layout for 'printability hot spots', by fragmenting it into a series of features. Their stress-response characteristics are described as behavioral models, the effects are accumulated, and the design is reconstituted with adjusted performance characteristics. Note that the specialized FEA tool in step 3, which is to derive the stress design rules, can analyze specific layout configurations and produce the compact behavioral models.

Step 7: Sign Off. If hot spots are detected, performance effects are analyzed and the layout is altered. Once hot spots are removed, the design of the Si die and the stack is signed off. The flow is to be iterative, and full final sign off may require going back to step 1, with suitable corrections in the design of the stack or the package substrate.

The TTS DfM flow requires model infrastructure to support, because simulations will be only as good as the models and the models are as good as the data extracted. A difficult task ahead is to align hierarchies between design and sub-modeling volumes.

4.3.7 Stress Simulation in Packages

Significant effort is expected to optimize package simulation and calibrate it to the experimental data, due to the variety of parameters. To improve the numerical accuracy of the simulation without requiring fine meshes, the singular element method [46], the extended (XFEM) [47], and an enriched finite element method [48, 49] could be implemented, with limited acceptance to problems with simple geometry and material combinations. Stress intensity can be calculated by comparing the crack surface displacement with the analytical crack-tip solution [50], from which the energy release rate and mode mixity were determined based on fine meshes near the crack tip, not readily applicable for 3D problems.

A modified virtual crack closure (MVCC) technique [51] uses standard FEA and coarse meshes, by calculating the components of the energy release rate separately corresponding to the three basic fracture modes I, II and III. With the local stress/strain and displacement distributions obtained by the finite element modeling, both the energy release rate and the mode mixity for the rate can be expressed as:

$$G = G_I + G_{II} + G_{III} \tag{4.3}$$

with silicon toughness components G_I, G_{II}, and G_{III} relate to the x, y, and z directions, i.e. opening, shearing, and tearing.

The critical features which impact both stress propagation and the accuracy of the simulation, include:

- Large volumes of materials ending with features of small areas. Example: metal pad attached to a single via.
- Sharp corners (in 3D) and reentrant angles. Example: Breadloafing at the bottom of metal stack would give rise to crack origination site.
- Adjoining materials with largely different CTE's, especially when the contact area is also large.
- Stacks of thick and rigid layers next to stacks of thinner and elastic layers can give rise to torque and shearing as well as normal stress components.

The criterion for interfacial delamination can be established by comparing the total energy release rate to the experimentally measured mode-dependent interface toughness. Due to the oscillatory singularity at the interfacial crack tip, the calculated energy release rate and mode mixity might depend on the element size at that location. The element size should be small enough to assure a converged solution by the FEM, but also large enough to avoid oscillating results for the energy release rate. The dividing up of the energy release rate components is therefore dependent on the element size and on the phase angles.

The total energy release rate was found to be less sensitive to the element size [52, 53]. Several approaches have been suggested to extract consistent phase angles of the element size [54]. The FEA for the package level were found to be in good agreement with Moiré interferometry i.e. z-displacement (package warpage) distribution along the die centerline.

In a 4-level, 3D wiring structure to analyze the effect of porous low-k, the pitch and line dimensions in the first two metal layers (M1 and M2) are often doubled in the third layer (M3), and doubled again in the fourth layer (M4). To simulate the dimensional hierarchy in real interconnect structures, cracks were introduced at

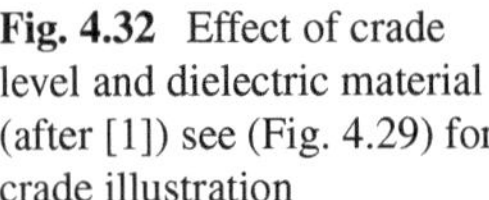

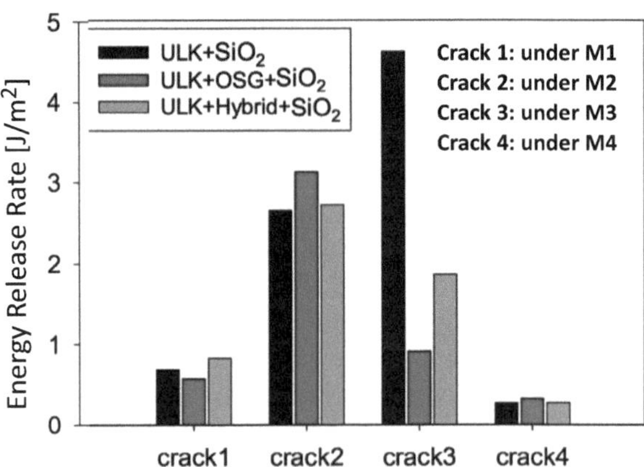

Fig. 4.32 Effect of crade level and dielectric material (after [1]) see (Fig. 4.29) for crade illustration

several interfaces with a width of 0.1 μm and a length of 2 μm extending in the multiple wiring directions. The calculated ERRs (Fig. 4.32) were the highest for ultra low-k dielectric especially closest to the solder pad. The ERR was the lowest for OSG, which had the highest E.

4.3.7.1 Impact of Solder Materials

The processing step causing the highest thermal load in flip-chip package assembly is the die attach step before underfilling. The solder reflow occurs at a temperature higher than the solder melting point. Afterwards, the package structure is cooled down to room temperature. Without the underfill serving as a stress buffer, the thermal mismatch between the die and the substrate can generate a large stress at the solder/die interface near the die corner, promoting interfacial delamination. As the semiconductor packaging shifts from Pb-based solders to Pb-free solders, the effects of solder material properties on CPI reliability became a concern, for the solders with different reflow cycles: 220 °C to 25 °C for eutectic solder and 260 °C to 25 °C for lead-free solder, respectively. As expected, the eutectic solder package had a lower crack driving force for interfacial delamination, due to its low reflow temperature and more compliant solder properties, compared to the lead-free counterpart.

The CPI-induced delamination driving force was compared for several porous MSQ materials (A to G, Table 4.1). Some of them were artificial, only for the purpose of separating the effect of E and CTE, which can't be achieved by comparing real MSQ materials. Comparing porous MSQ-A (k~ 2.3) with dense MSQ-C (k~3.0) with similar CTE, their ERR values were quite different due to the variation in E. While Fig. 4.33a, shows a good correlation between ERR and E, the likelihood of low-k delamination under CPI increased rapidly as the E of low-k dielectrics was reduced. Interestingly, the porous MSQ-D to MSQ-F with very different CTE had ERR values about the same as shown in Fig. 4.43b. In contrast, the ERR increased considerably with decreasing E. The fracture toughness of ultra low-k dielectrics is

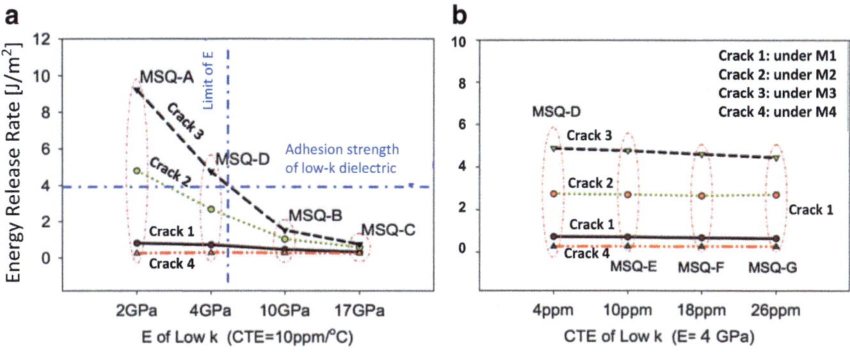

Fig. 4.33 Effect of low-k material properties on ERR (**a**) E (**b**) CTE (after [1])

usually below 4 J/m^2 for modeI delamination [1]. In order to maintain low-k integrity under CPI, the crack driving force has to be below 4 J/m^2, which corresponds to E values greater than 5 GPa.

According to the Moiré experiment and FEA results, the Cu/low-k interconnect directly on top of the corner bump had the highest interfacial peeling and shear stresses and it was most prone to fracture during packaging or subsequent stressing tests.

For simplification, the multi-layer Cu/low-k interconnect can be simulated as a uniform layer with average material properties deduced based on the Cu density in the interconnect structure, to calculate average stress as an indicator of the stability of low-k dielectrics for structural optimization at package level.

The thermal stresses induced by packaging are transferred into low-k interconnects via the Cu UBM solder pad and passivation, which act as stress buffers. DfM rules should optimize the geometric factors like UBM diameter, UBM Al pad and passivation thicknesses (Fig. 4.34). For example, a ~5 % decrease in the stress of the low-k material was observed by reducing the UBM thickness by ~50 %. A 17 % reduction in the stress can be achieved by shrinking the UBM diameter by 20 % (Fig. 4.34a). Such decease in the low-k stress was obtained for a ~80 % thickness increase in the pad, while the stress remained constant as the diameter of Al pad was increased from 90 to 120 μm. Increasing passivation thickness from 0.8 to 2.3 μm (Fig. 4.34c) demonstrated a 10 % decrease in the stress. The final reliability DfM rules need to be worked out by aligning with process capability rules.

In summary, structural optimization in the Cu UBM, Al pad, and passivation layer is quite effective in maintaining the mechanical stability of Cu/low-k interconnects during assembly and also very important when ultra low-k material with weak mechanical properties and Pb-free solder bumps are incorporated into the package.

4.3.7.2 Other DfM Rules for CPI

DfM response to mitigate the impact of CPI on the reliability of low-k interconnect during assembly should include pad shift design in which the thermal mismatch

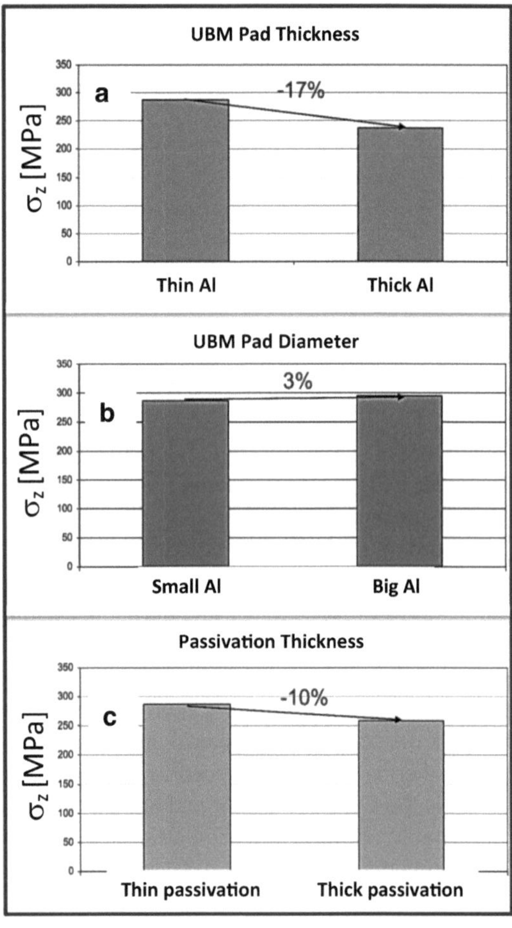

Fig. 4.34 Effect of UBM and geometry passivation layer thickness on vertical stress in low-k layer (**a**) UBM pad thickness (**b**) UBM pad diameter (**c**) Passivation thickness (after [1])

between the die and the organic substrate is compensated by adjusting metal pads on the substrate and redistribution layer (RDL) [54]. Placing the weak low-k dielectrics directly on top of solder bump, especially on corner bumps with large stress concentration during reflow, is avoided by adding redistribution lines to electrically connect the solder and low-k interconnects without direct contact right at the peripheral pads. DfM rule definitions need to identify the problem features by either layout context (preferred) or id layers and provide logical relationships among the geometries involved.

Energy release rates for horizontal cracks at each metal level for interfacial delamination in the four-level interconnect model (Fig. 4.29) indicate that, for low-k materials between all layers except metal 4, the interfacial crack at level 3 had the largest ERR being closest to the solder bump. For a fully dense low-k OSG at level 3 (mechanically stronger than ULK), the ERR of crack 3 was reduced and the effect of elastic mismatch shifted the largest ERR to crack 2. For a hybrid structure, OSG/ULK, the ERRs for crack 2 dropped by a small amount while ERR for crack 3

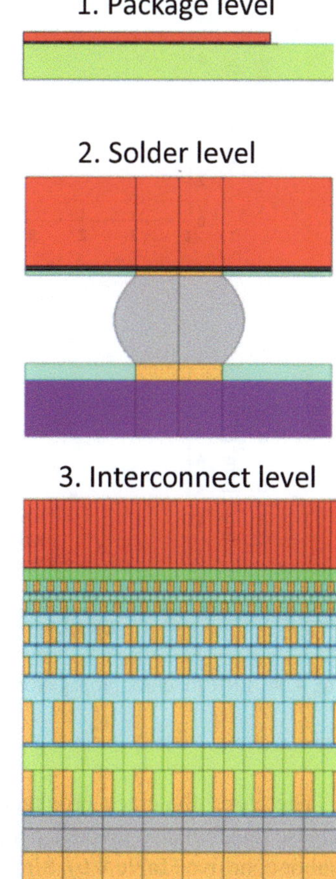

Fig. 4.35 Three-level interconnect model for crack propagation sub-modeling (after [1])

increased slightly, indicating better reliability than a full low k structure. In summary, the multilevel stacking structure can be optimized to minimize the CPI effect on ULK reliability.

The crack does not always propagate along one interface. As a crack propagates in a multilevel interconnect structure, both the energy release rate and the mode mixity at the crack tip vary. Depending on the local material combination and geometry, an interfacial crack can kink out of the interface, causing cohesive fracture of the low-k materials. Similarly, a cohesive crack may deflect into a weak interface. The crack propagation path depends on the loading conditions as well as material properties (including interfaces) and geometrical features in the interconnect structure. A general rule of crack propagation for anisotropic materials and composites, may be stated as follows [55]:

A crack propagates along a path that maximizes G/T i.e., the ratio between the energy release rate and the fracture toughness.

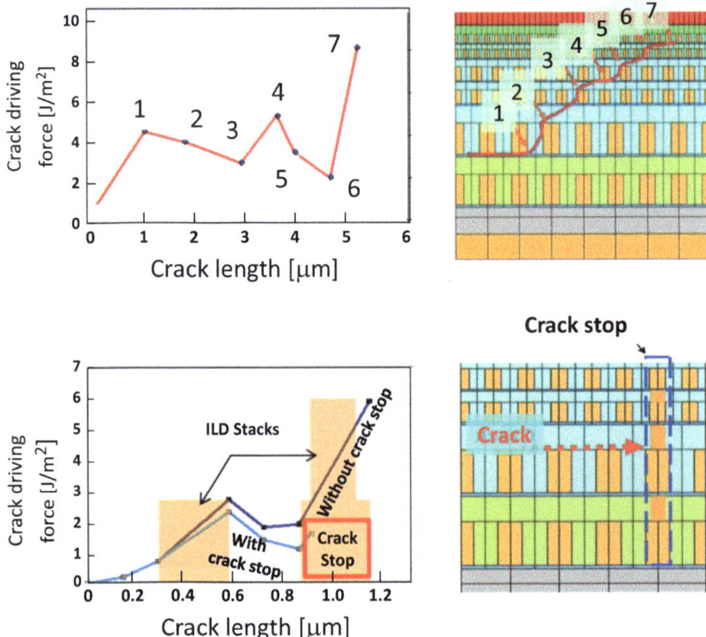

Fig. 4.36 ERR for crack propagation showing the effect of crackstop structure (after [1])

Therefore, the crack propagation not only seeks a path with the largest energy release rate, but also favors a path with the lowest fracture toughness, either interfacial or cohesive (Table 4.6). Simulations to define DfM methodology should predict which of those paths are close to devices, eventually causing die cracking (Fig. 4.36).

The following model levels:

- Level 1 (package) for investigating the overall thermal deformation for the flip-chip package, verified with results from Moiré interferometry,
- Level 2 (a sub-model with finer meshes for critical solder bump region at the outermost chip corner),
- Level 3 (detailed features for the interconnect structure),

are compared against the crack tip opening displacement (CTOD) method and the maximum hoop stress criterion to study the crack propagation behavior [55, 56]. In a 2D multi-level FEA model, the crack was assumed to initiate at the global-level interface, where a higher energy release rate was shown to exist than at local-level interfaces.

The CTOD method was employed to calculate the mode mixity at each crack tip by stress intensity factor K as the crack extended (Fig. 4.36). The maximum hoop stress criteria were then used to predict the crack propagation direction. The crack propagation path from the upper levels to the lower levels obtained by simulations may not be exactly as shown since it can be affected by process defects and material inhomogeneity. But as the crack propagated and the total crack length increased, the energy release rate continued to increase, indicating an unstable crack growth.

A major challenge in packaging Si die is to prevent cracks propagating from the die edge to the active area of a chip. One design/DfM response to suppress crack propagation is to implement crack stop structures into the interconnect. Dummy Cu crack stops at via levels can be added into Cu/low-k interconnect as local reinforcements. The simulation showed that the crack driving force was suppressed by the crack stop structure (Fig. 4.36) even as it increased with the crack length. Therefore, the most effective way to use the crack stop is to define rules about how to embed it close enough to the location where cracks initiate, such as the die edge or the dense material contacts the ultra low-k material.

A Modified Edge Liftoff Test (m-ELT) was used to determine the effect of fracture toughness of crack-stop structures built into the Cu/low-k interconnects [57, 58] (Fig. 4.37). To prepare the specimen, a thick layer of epoxy was deposited on top of Si wafer and then cured for one hour at 177 °C. After curing, the sample was diced into 1 × 1 cm coupons placed in a chamber with liquid nitrogen to cool down until the epoxy started to peel off from the wafer. The temperature at which debonding occurred was recorded and the corresponding residual stress extrapolated using a calibrated residual stress versus temperature curve. Placing crack stop structures resulted in fracture resistance from 1 to 4 J/m^2 [56].

The crack driving force for low-k dielectric fracture increases with crack length quickly at the beginning stage and then saturates (Fig. 4.38). If the fracture resistance of crack stop is larger than the crack driving force, the crack growth will be halted. Therefore, the location of the crack stop structure is crucial. The most effective way is to embed the crack stop as close as possible to the crack initiation points.

In summary, the chip-package interaction critical for package level DfM, can be investigated using 3D finite element analysis (FEA) based on a multilevel submodeling approach, to deduce the packaging induced crack driving force for interfaces in Cu/low-k structures. The die attach, a critical step, and the energy release rate, depends on the material properties of solder and low-k dielectrics. The implementation of lead-free solder and ultra low-k material compromises the mechanical stability of the Cu interconnect, due to the increased driving force for fracture. Structural optimization such as changing the geometry and structural layout at both package and interconnect level were found to be effective in retaining the mechanical reliability of Cu/ultra Low k interconnects under CPI. Simulation results demonstrated that the crack would propagate from the global-level interconnect towards Si substrate under CPI, which agreed well with the experimental observations. As DfM recommendation, decrease in the energy release rate can be achieved by adding dummy Cu structures into the low-k interconnect. Meanwhile, m-ELT results indicated that the fracture resistance of the structure was increased by implementing crackstops. A DfM set of rules leading to proper layout and design of the crackstop structure is critical in order to effectively suppress crack propagation and improve the mechanical reliability of Cu/ultra low-k interconnect.

4.3.8 Thermo-mechanical Reliability of 3-D Integration with TSVs

Three-dimensional (3D) integrated circuits with through silicon vias (TSVs) have emerged as a promising approach to improve the device density in macro-scale, independently of the continuous downscaling of the on-die interconnect structure. 3D structures enable shorter interconnection paths for better electrical performance and heterogeneous integration of different subsystems such as logic and digital circuits.

The basic TSV manufacturing process consists of:

- Deep reactive ion etching or laser drilling of the silicon substrate,
- Deposition of electrical isolation and diffusion barrier layers,
- Deposition of a Cu seed layer,
- Cu metallization.

The via configuration and process sequence (via first or via last) vary with the TSV materials and geometries. Polysilicon, tungsten, and copper are commonly used for via filling; with copper having lowest resistivity. For Cu TSV, Chemical vapor deposition (CVD) is preferred for vias with diameter less than 5 μm while electroplating is favored for vias with larger diameter [59]. Process optimizations for the fabrication of 3D interconnects include deep silicon etching, optimized seed layer and electroplating profile.

Stress evolution in 3D interconnects during fabrication is related to the CTE mismatch between the Cu TSV and Si with CTE of Cu being 6–7 times higher. Cu-filled TSVs have higher residual stress compared with tungsten-filled TSVs [58]. Additionally, the thin die (25–100 μm) and the high aspect ratio (>5:1) of TSV lead to a complex stress which may be sufficient to drive crack and interconnect failure [61, 62], but also impact the performance of stress-sensitive devices. In-plane stresses of only 100 MPa can degrade the device mobility by several percent [37]. The stress-related problem has necessitated the creation of DfM guidelines for a proper device keep-away zone around the TSVs, to deal with process-induced residual stress, based on fracture analysis of 3D interconnects with Cu TSVs.

The properties of copper required for FEA modeling of TSV structures can be obtained using a dynamic mechanical analysis (DMA) and bending beam technique [63]. Failure modes include silicon z-cracking, silicon r-cracking, and interface debonding between the TSV and silicon matrix. The energy release rate and critical crack stress are compared to the fracture toughness of silicon and adhesion of TSV/silicon interface to propose design guidelines of 3D interconnects.

The thermal stresses of a 2×2 TSV array (Fig. 4.39) were investigated with FEA. Process-induced stresses in the Si matrix were calculated by employing the element birth and death technique on a simplified fabrication process (Fig. 4.40), starting with TEOS oxide layer deposition at 400 °C with thickness varying from 0.15 to 0.5 μm, followed by the deposition of a thin barrier layer and a seed layer at 400 °C, varying with the TSV diameter. After the barrier deposition, copper was electroplated at room temperature, heated to 200 °C and cooled back down to room temperature [59]. This led to thermal stress development due to the CTE mismatch in constituent materials.

Fig. 4.37 Cross-section of an m-ELT sample (after [1])

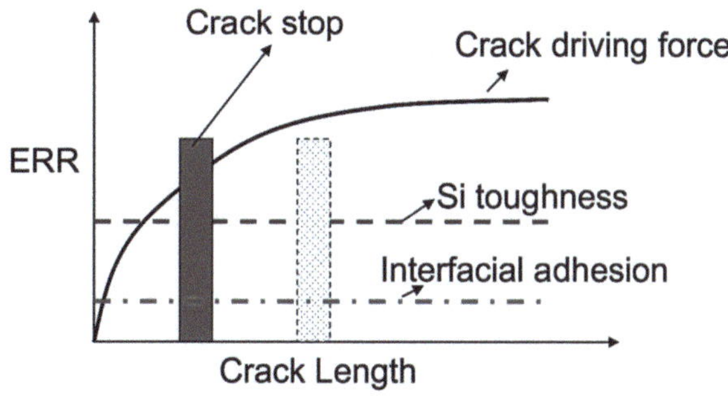

Fig. 4.38 Effect of crack stop on suppressing crack growth (after [1])

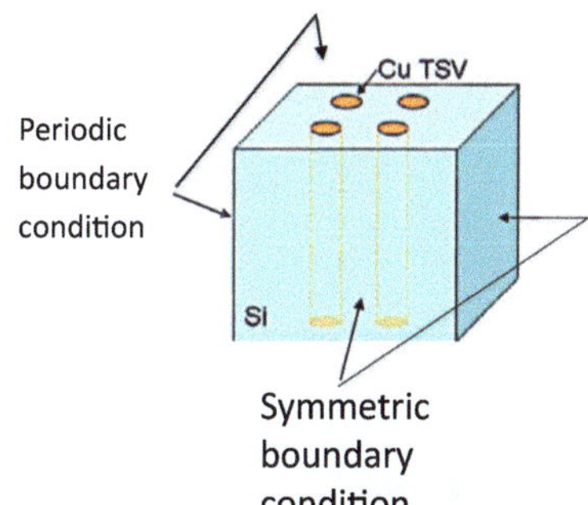

Fig. 4.39 FEA model of TSV interconnect (after [1])

Large stresses in the TSV structure at the thermal treatment after Cu electroplating were sufficient to induce silicon r-cracking and interface debonding between the Cu TSV and the silicon matrix. After cooling down to room temperature, the silicon would still contain residual radial and hoop final stress (Fig. 4.41) very close to the Cu TSVs making it important design the keep-away zone for devices.

The radial stress distributions at the top surface (Fig. 4.42) over the two adjacent TSVs for TSV diameters of 5, 10, and 15 μm (fixed pitch of 20 μm) show a significant increase as the TSV diameter increased (Fig. 4.43). The area affected by the residual stress in silicon was enlarged as well.

For two TSV arrays with the same diameter, stress was enhanced in structures with smaller pitch-to-diameter ratio (Fig. 4.42a, b). The maximum normal stress of silicon increased from 275 to 320 MPa, impacting the keep-away zone and the hazard of silicon cracking. The stress can be reduced when the pitch-to-diameter ratio is greater than two, which can be proposed to be a minimum DfM reliability rule to avoid significant stress enhancement among adjacent TSVs. After fabrication, all

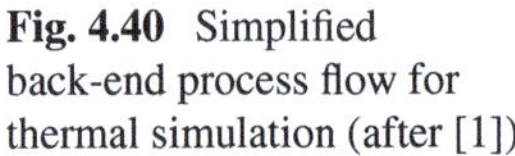

Fig. 4.40 Simplified back-end process flow for thermal simulation (after [1])

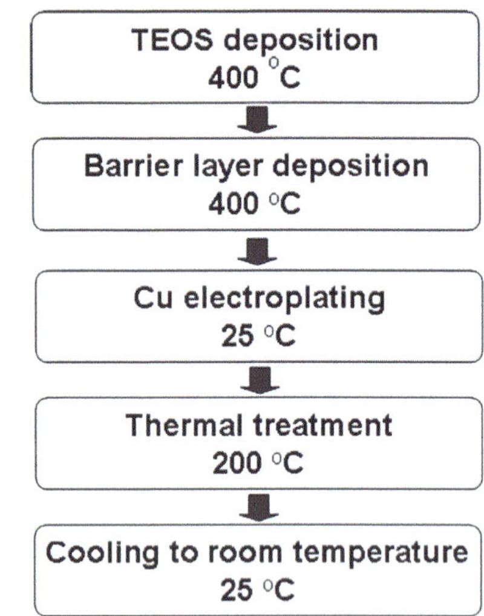

Fig. 4.41 Stress components (after [1])

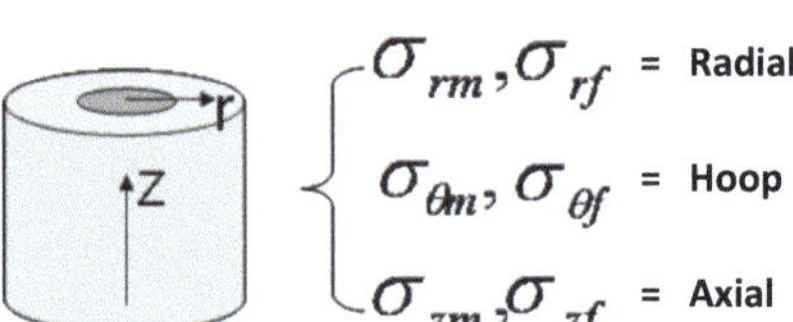

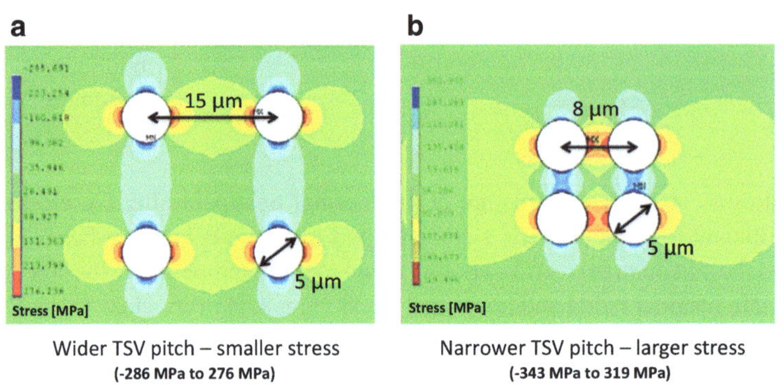

Fig. 4.42 Residual radial stress distribution in TSV interconnects at top surface (after [1])

stress components were reduced as the TSV diameter became smaller except for the axial stress. Another way to reduce the residual stress by as 50 % is to decrease the Cu filling ratio, by replacing solid Cu TSV with hollow structures.

Increasing the Cu volume ratio reduces the thermal residual radial stress and the keep-away zone. However, a silicon substrate with a dense metal configuration is

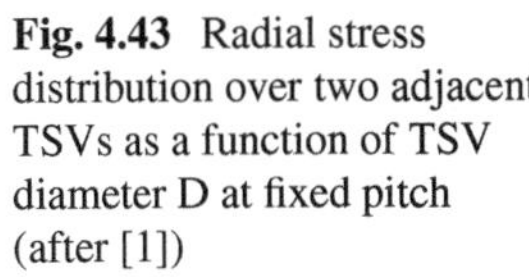

Fig. 4.43 Radial stress distribution over two adjacent TSVs as a function of TSV diameter D at fixed pitch (after [1])

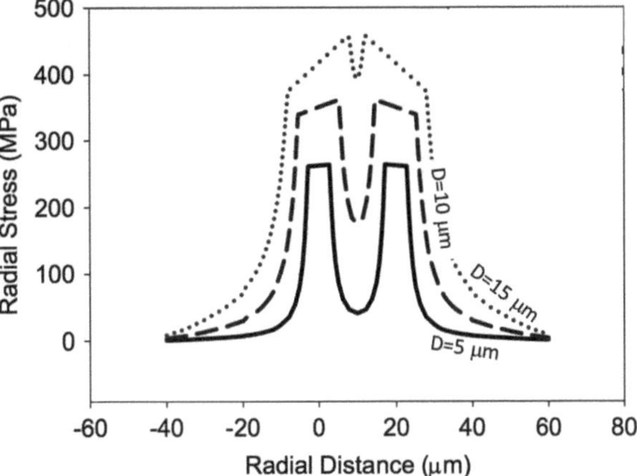

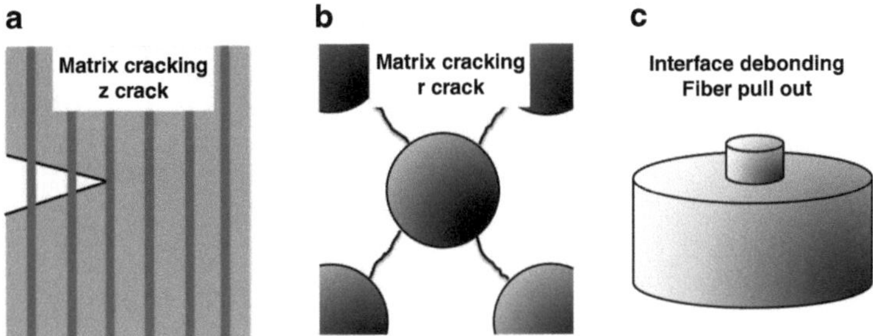

Fig. 4.44 Failure modes for 3D interconnects with Cu TSVs, (**a**) z-crack, (**b**) r-crack, (**c**) debonding and pull out (after [1])

more prone to cracking due to the increase of hoop stress in silicon. The axial stress in silicon increases with metal density. An increase in the thickness of buffer layer between Cu TSV and silicon can reduce all TSV stress components. Placing SiO_2 or soft polymer layers between Cu TSV and silicon can buffer the CTE mismatch. The process-induced residual stress causes cohesive cracking in silicon and delamination at TSV interfaces, due to the various thermal and mechanical loads during wafer thinning, handling, bonding processes, and subsequent thermal cycling. The Cu TSV/Si matrix system is analogous to the classical fiber-reinforced composite with the three failure modes proposed above: silicon matrix r-cracking, silicon matrix z-cracking, and interface debonding between TSV and silicon (Fig. 4.44 [63, 64]).

(A) Silicon r-cracking

For silicon r-cracking, the crack propagates along the radial direction. The stress intensity factor for crack propagation induced by thermal stress can be calculated by superposition for a single TSV in infinite silicon matrix (Fig. 4.45). For a system containing multiple TSVs (Fig. 4.46), each TSV close

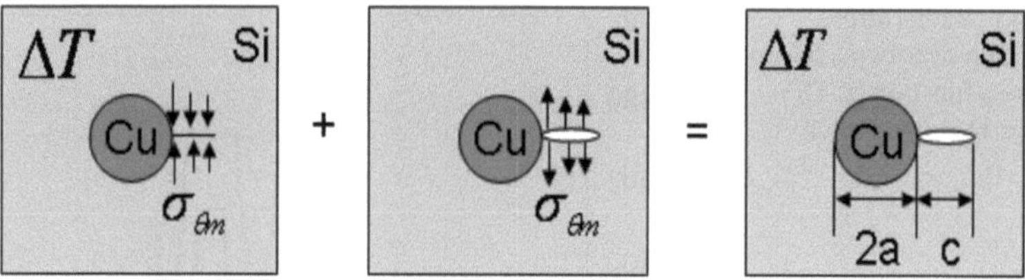

Fig. 4.45 Stress intensity factor for silicon r-cracking (after [1])

Fig. 4.46 Stress intensity factor calculation for r-crack in a TSV array (after [1])

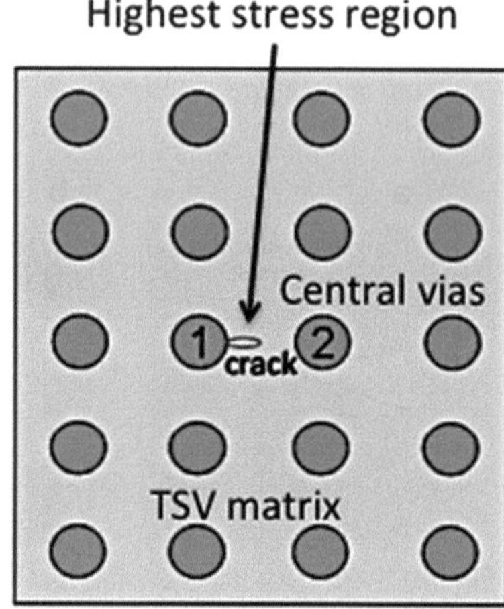

to TSV #1 contributes to the stress intensity factor for the crack. (TSV #1 and #2 were included in the calculation while the contributions from other adjacent TSVs were neglected because they were further away from the pre-crack and had little impact on the K_I [64]).

The total stress intensity factor K_I from both TSV #1 and #2 is a function of Cu via diameter, Cu volume ratio, crack size, and thermal load. The energy release rate of the crack can be a function of G. The trend of G needs to be compared with the silicon toughness G_{IC} to determine whether the crack will propagate or stagnate. If silicon toughness G_{IC} along the different silicon directions smaller is than G, the initial crack will start to grow. For small vias with a radius of less than 5 μm, the energy release rate for r-crack propagation was less than or comparable to the fracture resistance of silicon. However, the crack driving force increased with via diameter quickly, especially for high Cu volume ratio (up to 26 J/m^2 for a radius of 50 μm), much larger than the silicon fracture resistance.

Comparison of the results among r-cracks showed that the TSV with a larger diameter generated a larger crack driving force, which also increased with the thermal load. G increased as the Cu volume ratio increased for all four cases, therefore, smaller via diameter and smaller Cu volume ratio are preferred to improve the resistance of 3D interconnect to silicon r-cracking.

(B) Interface debonding
Another important failure mode is interface debonding between TSV and silicon. Cu TSV tends to thermally shrink or expand more than silicon and a large thermal stress can develop at the interface, raising concerns of interfacial debonding and the TSV popping-out.

As the crack propagates downwards from debonding interface, the stress state near the crack front changes. The steady-state strain energy release rate for crack propagation equals the interfacial adhesion strength.

The energy release rate in the absence of friction for interface debonding [66] increased rapidly with TSV radius, up to 20 J/m^2 at 50 μm. Comparing the adhesion strength of Cu to SiO_2 interface in the range of 0.7–10 J/m^2 with the G values, it can be concluded that interface debonding and TSV pull out can be a serious reliability problem for 3D interconnects with large Cu TSVs.

(C) Silicon z-cracking
For silicon matrix z-cracking, we investigated two cases (Fig. 4.44):

- Case I: no debonding between Cu TSV and silicon
- Case II: with debonding and sliding between Cu TSV and silicon

Similar to the TSV/silicon interface debonding, the z-cracking growth in silicon will change the potential energy in the system. For Case I, without debonding and sliding, the potential energy change is balanced only by the silicon cracking energy. The critical stress for silicon z-cracking depends on G_m, the fracture resistance of silicon, the TSV radius, and Cu volume ratio. The critical stress can reach 1 GPa for a perfect bonding case, while higher stress values would indicate that the system is more resistant to z-crack damage.

For Case II (debonding and sliding between TSV and silicon), the potential energy change in the system is balanced not only by the silicon crack energy release but also by the debonding energy release and frictional energy dissipation. The critical stress for z-cracking for $\mu = 0.1$ (Fig. 4.44a) was reduced to below 250 MPa compared to Case 1 (debonding and sliding, Fig. 4.44c).

Debonding and sliding between Cu TSV and silicon exacerbated silicon z-cracking by reducing the critical stress for a crack growth. The critical stress decreased as TSV radius increased, indicating silicon z-cracking problem is more critical for 3D interconnects with larger vias. In contrast, increasing the Cu volume ratio increased the critical stress for z-cracking, i.e., making silicon more resistant (Fig. 4.47).

In general, the thermal residual stress induced by TSV fabrication is not large enough to cause silicon z-cracking. However, loads applied to the z-direction in the structure during the wafer handing, bonding, and subsequent packaging process may raise the likelihood of z-cracking. Increasing Cu TSV diameter, a key variable in controlling the mechanical stability of 3D TSV structure, will induce larger crack driving force for silicon r-cracking and interfacial debonding.

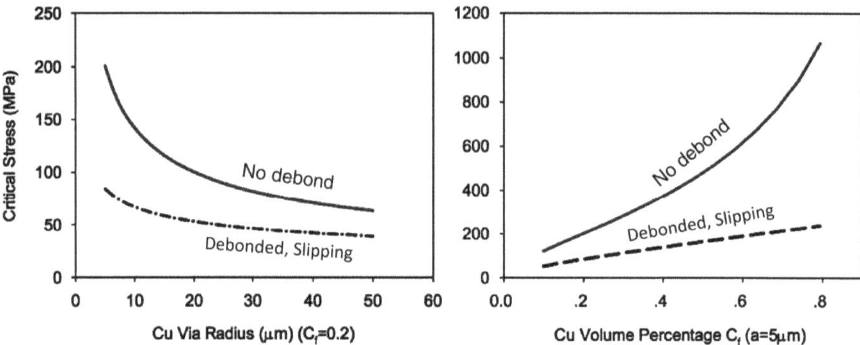

Fig. 4.47 Comparison of critical cracking stress for increasing via radius and Cu volume percentage (after [1])

Increasing Cu metal density (Cu volume percentage) can improve the fracture resistance to z-cracking of silicon and reduce the energy release rate for Cu/TSV interfacial delamination. Strong adhesion at the interface between Cu TSV and silicon can inhibit the debonding and prevent TSV extrusion, but can also substantially increase the critical stress for silicon z-cracking, making 3D interconnects more robust during wafer handling and packaging.

4.3.9 Summary

Stress analysis aimed at defining DfM rules for packaging is based on the critical observations from the FEA study verified by high-resolution moiré interferometry. The key observations are:

1. Large thermal strain was observed in the solder/underfill layer, especially at the die corner. The strain concentration could lead to delamination between underfill and solder resist or silicon passivation. Selection of proper underfill materials becomes critical, especially when Pb-free solder and ultra low-k dielectrics were introduced into the flip-chip packages.
2. Packaging-induced crack driving forces for porous low-k layer delamination, when compared with those for dense low-k dielectrics (smaller interconnect dimension) accompanied by more metal levels, show that the implementation of ultralow-k porous materials affects CPI-induced crack propagation in the low-k interconnect and that crack-stop structures improve chip reliability.
3. FEA and bending beam experiments employed to investigate process-induced residual stresses show that via diameter, Cu fill ratio, and Cu density were the most important parameters in controlling the thermal stress distribution inside and around the TSVs. They impact three failure modes silicon z-cracking, r-cracking, and interface debonding between Cu TSV and silicon. Their energy release rates and critical crack stresses were comparable to the fracture toughness of silicon and adhesion of TSV/silicon interface.

4. The introduction of Pb-free solder would impact the reliability of Cu/low-k interconnects due to the higher stresses generated in the package compared to high Pb and eutectic solders. A bump structure called Cu pillar enhances the electromigration performance of solder joints, mechanically stiffer than Pb-free solders and thus more stresses can be transferred into the Cu/low-k interconnects during assembly.

A set of DfM for stress and fracturing rules need to be integrated into the packaging flow to correlate the parameters of interconnect architecture, materials, process, and stress conditions, to avoid costly reliability failures.

References

1. Zhang, X.: Chip-package interaction and its impact on the reliability of flip-chip packages. Ph.D. Thesis, UT Austin (2009)
2. Sukharev, V., Zschech, E.: Stress management for 3D IC's using through-silicon vias. AIP Conf. Proc. **1378**, 21–49 (2011)
3. Radojcic, R., Novak, M., Nakamoto, M.: TechTuning: stress management for 3D through-silicon via stacking technologies. AIP Conf. Proc. **1378**, 5–20 (2011)
4. Thompson, M.S.E., Armstrong, M., Auth, C., Cea, S., Chau, R., Glass, G., Hoffman, T., Klaus, J., Zhiyong, M., Mcintyre, B., Murthy, A., Obradovic, B., Shifren, L., Sivakumar, S., Tyagi, S., Ghani, T., Mistry, K., Bohr, M., El-Mansy, Y.: IEEE Electron. Devices. Lett. **25**, 191–193 (2004)
5. Flachowsky, S., Wei, A., Illgen, R., Hermann, T., Hocnischel, J., Horstmann, M., Klix, W., Stenzel, R.: IEEE Trans. Electron. Devices **57**, 1343–1354 (2010)
6. Wang, G., Ho, P.S., Groothuis, S.: Microelectron. Reliab. **45**, 1079–1093 (2005)
7. Sheu, Y.M., Yang, S.J., Wang, C.C., et al.: Modeling mechanical stress effect on dopant diffusion in scaled MOSFETs. IEEE Trans. Electron. Devices **52**, 30–38 (2005)
8. Sukharev, A. Kteyan, N. Khachatryan, et al. 3D IC TSV-based technology: stress assessment for chip performance. In: AIP Conference on Proceedings of 11th International Workshop on Stress-Induced Phenomena in Metallization, 1300, 1143 (2010)
9. COMSOL Multiphysics: Multiphysics modeling and simulation software. http://www.comsol.com
10. Ryu, S., Lu, K., Zhang, X., Im, J., Ho, P.S., Huang, R.: Impact of near-surface thermal stresses on interfacial reliability of through-silicon-vias for 3-D interconnects. IEEE Trans. Device Mater. Reliab. **11**, 35–43 (2011)
11. Love, A.E.N.: The stresses produced in a semiinfinite solid by pressure on part of the boundary. Phil. Trans. Roy. Soc. **228**, 377–420 (1929)
12. Smith, C.S.: Piezoresistance effect in germanium and silicon. Phys. Rev. **94**, 42–49 (1954)
13. Aziz, M.J.: Thermodynamics of diffusion under pressure and stress: relation to point defect mechanisms. Appl. Phys. Lett. **70**, 2810–2812 (1997)
14. Joshi, V., Sukharev, V., Torres, A., Agarwal, K., Sylvester, D., Sylvester, D., Blaauw, D.: Closed-form modeling of layout-dependent mechanical stress. In: Proceedings of DAC, pp. 673–678 (2010)
15. Pang, I.T., Nikolic, B.: Measurement and analysis of variability in 45 nm strained-Si CMOS technology. In: Proceedings of IEEE CICC, pp. 129–132 (2008)
16. Peter, K., Marz, R., Torres, A., Attia, M.: White paper. The roadmap to LFD value: quantifying a return on investment in Calibre LFD (2011)
17. Orshansky, M., Nassif, S.R., Bonning, D.: Design for Manufacturability and Statistical Design. Springer, New York (2008)
18. Hytch, M., Houdellier, F., Snoeck, E., Claverie, A.: Strain metrology of devices by dark-field electron holography: a new technique for mapping 2D strain distributions. In: IEEE IEDM, pp. 1–4 (2009)

19. Koch, C.T., Ozdöl, V.B., van Aken, P.A.: Appl. Phys. Lett **96**, 091901-1–091901-3 (2010)
20. Engelmann, H.J., Geisler, H., Huchner, R., Potapov, P., Utess, D., Zschech, E.: Challenges to TEM in high-performance microprocessor manufacturing. In: Proceedings of 4th EMC, vol. 2, pp. 13–14 (2008)
21. van Vroonhoven, J.C.W.: Effects of adhesion and delamination on stress singularities in plastic packaged integrated circuits. Trans. ASME J. Electron. Pack. **15**, 28–33 (1993)
22. Ho, P.S., et al.: Reliability issues for flip chip packages. Microelectron. Reliab **44**, 719–737 (2004)
23. Chen, K., et al.: Effect of underfill materials on the reliability of low-k flipchip packaging. Microelectron. Reliab. **46**(1), 155–163 (2006)
24. Shang, J.K., zeng, Q.L., Zhang, L., Zhu, Q.S.: Mechanical fatigue of Snrich Pb-free solder alloys. J. Mater. Sci. Mater. Electron. **18**, 211–227 (2007)
25. Darveaux, R., Banerji, K.: Fatigue analysis of flip chip assemblies using thermal stress simulations and a Coffin-Manson relation. In: Proceedings of 41st IEEE ECTC, pp. 797–805 (1991)
26. Liu, X.H., Shaw, T.M., Lane, M.W., Liniger, E.G., Herbst, B.W., Questad, D.L.: Chip-package interaction modeling of ultra low-k/copper back end of line. In: International Interconnect Technology Conference (2007)
27. Kang, Bansal T., Li, Y.: Reliability of high-end flip-chip package with large 45 nm ultra low-k die. In: Electronic Components and Technology Conference (2008)
28. Ong, X., et al.: Underfill selection methodology for fine Pitch Cu/low-k FCBGA packages. Microeletron. Reliab. **49**, 150–162 (2009)
29. Anand, L.: Constitutive equations for the rate-dependent deformation of metals at elevated temperatures. ASME J. Eng. Mater. Technol. **104**, 12–17 (1982)
30. Zhang, X., Im, S., Huang, R., Ho, P.S.: Chip-package interaction and reliability impact on Cu/ Low-k Interconnects. In: Electrical, Optical and Thermal Interconnections for 3D Integrated Systems (2008)
31. Yao, Q. et al.: Adhesion enhancement of underfill materials by silane additives. In: Proceedings of the International Symposium on Advanced Packaging Materials: Processes, Properties and Interfaces, p. 165 (1999)
32. Suo, Z.: Reliability of interconnect structures. In: Gerberich, W., Yang, W. (eds.) Interfacial and Nanoscale Failure of Comprehensive Structural Integrity (Milne, I., Ritchie, R.O., Karihaloo, B. Editors-in-Chief), vol. 8, pp. 265–324 (2003)
33. Suo, Z., Hutchinson, J.W.: Sandwich specimens for measuring interface crack toughness. Mater. Sci. Eng. **A107**, 135–143 (1989)
34. Lu, K., et al.: Moisture transport and its effects on fracture strength and dielectric constant of underfill materials. In: ECTC (2007)
35. Wong, E.H., Rajoo, R.: Moisture absorption and diffusion characterization of packaging materials advanced treatment. Microelectron. Reliab. **V43**, 2087–2096 (2003)
36. Mercado, L., Goldberg, C., Kuo, S.-M.: A simulation method for predicting packaging mechanical reliability with low k dielectrics. In: International Interconnect Technology Conference, pp. 119–121 (2002)
37. Mercado, L., Kuo, S.-M., Goldberg, C., Kuo, S.-M., Lee, T.-Y.: Analysis of flip-chip packaging challenges on copper low-k interconnects. In: Proceedings of 53rd Electronic Components and Technology Conference, vol. 166, pp. 1784–1790 (2003)
38. Wang, G.T., Merrill, C., Zhao, J.H., Groothuis, S., Ho, P.: Packaging effects on reliability of Cu/low k interconnects. IEEE Trans. Device Mater. Reliab. **3**, 119–128 (2003)
39. Zhao, J.H., Wilkerson, B., Uehling, T.: Stress-induced phenomena in metallization. In: Ho, P.S., Baker, S.P., Nakamura, T., Volkert, C.A. (eds) AIP Conference Proceedings of the 7th International Workshop, vol. 714, pp 52–61 (2004)
40. Wang, G.T.: Ph.D. thesis, The University of Texas, Austin (2004)
41. Rhee, S.H., Du, Y., Ho, P.S.: Thermal stress characteristics of Cu/oxide and Cu/low-k submicron interconnect structures. J. Appl. Phys. **93**(7), 3926–3933 (2003)
42. ANSYS Advanced Guide Manual, Chapter 9, in ANSYS Version 9.0 Documentation, ANSYS, Inc. (2006)

43. Shih, C.F., Asaro, R.J.: Elastic–plastic analysis of cracks on biomaterial interfaces: part I-small scale yielding. J. Appl. Mech. **55**, 299–316 (1988)
44. Hughes, T.J.R., Stern, M.: Techniques for developing special finite element shape functions with particular references to singularities. Int. J. Numer. Methods. Eng **15**, 733–751 (1980)
45. Sukumar, N., Huang, Z., Prevost, J.-H., Suo, Z.: Partition of unity enrichment for bimaterial interface cracks. Int. J. Numer. Methods. Eng. **59**, 1075–1102 (2004)
46. Ayhan, A.O., Nied, H.F.: Finite element analysis of interface cracking in semiconductor packages. IEEE Trans. Compon. Packag. Technol. **22**, 503–511 (1999)
47. Ayhan, A.O., Kaya, A.C., Nied, H.F.: Analysis of three-dimensional interface cracks using enriched finite elements. Int. J. Fract. **142**, 255–276 (2006)
48. Liu, X.H., Lane, M.W., Shaw, T.M., Simonyi, E.: Delamination in patterned films. Int. J. Solids. Struct. **44**(6), 1706–1718 (2007)
49. Bucholz, F.G., Sistla, R., Krishnamurthy, T.: D and 3D applications of the improved and generalized modified crack closure integral method. In: Atluri, S.N., Yagawa, G. (eds.) Computational Mechanics'88. Springer, New York (1988)
50. Krueger, R.: The virtual crack closure technique: history, approach and applications. NASA/ CR-2002 211628 (2002)
51. Sun, C.T., Jih, C.J.: On strain energy release rates for interfacial cracks in biomaterial media. Eng. Frac. Mech. **28**, 13–20 (1987)
52. Raju, I.S., Crews, J.H., Aminpour, M.A.: Convergence of strain energy release rate components for edge delaminated composite materials. Eng. Frac. Mech. **30**, 383–396 (1988)
53. Chai, T.C., et al.: Impact of Packaging Design on Reliability of Large Die Cu/low- (BD) Interconnect, ECTC, Orlando (May 2008)
54. Hutchinson, J.W., Suo, Z.: Mixed-mode cracking in layered materials. Adv. Appl. Mech. **29**, 63–191 (2002)
55. Suo, Z.: Reliability of interconnect structures, Interfacial and nanoscale failure. In: Gerberich, W., Yang, W. (eds) Comprehensive Structural Integrity (Milne, I., Ritchie, R.O., Karihaloo, B. Editors-in-Chief), vol. 8, pp. 265–324
56. Im, J., Shaffer, E., Stokich, T., Strandjord, A., Hetzner, J., Curphy, J., et al.: On the mechanical reliability of photo-BCB-based thin film dielectric polymer for electronic packaging applications. J. Electro. Packag. **122**(1), 28–33 (2000)
57. Chiang, M., Wu, W., He, J., Amis, E.J.: Combinatorial approach to the edge delamination test for thin film reliability—concept and simulation. Thin. Solid. Films. **437**(1–2), 197–203 (2003)
58. Garrou, P., et al.: Handbook of 3D Integration. Wiley-VCH, 20 Oct 2008
59. Ramm, P., et al.: Through silicon via technology – processes and reliability for wafer-level 3D system integration. In: ECTC (2008)
60. Thompson, S., et al.: Uniaxial-process-induced strained-Si: extending the CMOS roadmap. IEEE Trans. Electron. Devices **53**(5), 1010–1020 (2006)
61. Savastiouk, S.: Through silicon vias (TSV): Physical design and reliability. In: Semetech 3D ICs Workshop, San Diego (September 2008)
62. Zhao, J., et al.: Measurement of elastic modulus, Poisson ratio, and coefficient of thermal expansion of on-wafer submicron films. J. Appl. Phys. **85**(9), 6421 (1999)
63. Lu, T.C., et al.: Matrix cracking in intermetallic composites caused by thermal expansion mismatch. Acta. Metall. Mater. **39**(8), 1883–1890 (1991)
64. Eldrige, J., et al.: Fiber push-out testing apparatus for elevated temperatures. J. Mater. Res **9**(4), 1035–1042 (1994)
65. Hutchinson, J.W., et al.: Models of fiber debonding and pullout in brittle composites with friction. Mech. Mater. **9**, 139–163 (1990)
66. Bagchi, A., Evans, A.G.: Measurements of the debond energy for thin metallization lines on dielectrics. Thin. Solid. Films. **286**, 203–212 (1996)
67. Budiansky, B., Hutchinson, J.W., Evans, A.G.: Matrix fracture in fiber-reinforced ceramics. J. Mech. Phys. Solids. **34**(2), 167–189 (1986)
68. Eneman, P., Verheyen, A., De Keersgieter, M., Juczak, K.D., De Meyer, K.: Scalability of stress induced by contact-etch-stop layers: a simulation study. IEEE Trans. Electron. Devices **54**(6), 1446–1453 (2007)

Chapter 5
Closure and Future Work

We have discussed three key trends in the IC Design for Manufacturability approaches of the early 2010's, when key IC makers are moving along the Moore's shrinkpath, passing the 28 nm technology node, on the way down to 22 nm and then, 15 nm.

Trend #1 – keep using the many generations-old, "classic" DfM with rule-based and model-based approaches to mask pattern corrections. Many of these approaches are still valid and important for the layouts with high degree of randomization.

Trend #2 – simplify DfM by requiring the layout to adhere to strict templates, in other words, remove the randomization. This approach is particularly true for device layers of active and poly, but may not work too well for the connecting layers that still need to have much more freedom.

Trend #3 – expand into new dimensions, such as the silicon volume, in other words, make sure the device would not be subject to 3D surprise defects. Add time as 4th dimension to prevent reliability failures. Engage a new realm of stress simulations based on the many material properties, neglected so far, but critical for the new device applications.

That latter aspect is of particular importance for the future devices that are being built now in such way that they would become our ears and eyes. They should see what a human eye can see, rain or shine. For that, they need to reliably operate in a wide range of temperatures, humidities, and pressures, especially when involved in the works of another piece of heavy machinery (automobiles).

But we are getting close to the point we would expect the IC's to get us information from the future. It is not necessarily the "future in time", but the locations where the humans cannot control the machine in real-time and have instead to rely on its intelligence.

Today it is the domain of pure research. An automated vehicle immersed in liquid methane on one of Jupiter's moons cannot be remotely controlled from Earth due to the signal travel taking many hours. So it has to be designed to deal with the unknown.

A. Balasinski, *Design for Manufacturability: From 1D to 4D for 90–22 nm Technology Nodes*, DOI 10.1007/978-1-4614-1761-3_5,

Appendix

Table A.1 Glossary of abbreviations

Term	Explanation
1D, 2D, 3D, 4D	1, 2, 3, 4 – dimensional
AI	Artificial intelligence
ALUT	Adaptive lock-up tables
ASIC	Application – specific integrated circuit
BoM	Bill of materials
BRDL	Backside redistribution layer
BSIM	Berkeley short channel IGFET model: MOSFET model
CAD	Computer – aided design aka EDA (electronic design automation)
CBC	Correct-by-construction
CD	Critical dimension
CTE	Coefficient of thermal expansion
CLY	Circuit limited yield
(C)MOS	(Complementary) metal-oxide-semiconductor
CMP	Chemical–mechanical polishing
CMPY	CMP yield
COO	Cost of ownership
CPI	Chip-package integration
CPU	Central processing unit
D2D	Die-to-die
DOF	Degrees of freedom or depth of focus
DfM	Design for manufacturability – the subject matter of this book
DfR	Design for reliability
DfT	Design for test
DfY	Design for yield
DPT	Double patterning techniques
DSL	Dual stress liner
DRC	Design rule check
DUF	Design unified format
ECO	Engineering change order
EDA	Electronic design automation (see CAD)

(continued)

A. Balasinski, *Design for Manufacturability: From 1D to 4D for 90-22 nm Technology Nodes*, 275
DOI 10.1007/978-1-4614-1761-3,

Table A.1 (continued)

Term	Explanation
EFR	Early failure rate
EPE	Edge placement error
ERR	Energy release rate
ESiGe	Embedded Silicon Germanium
e-SiGe	Embedded stressor for SiGe
ESL	Electronic System Level
EUV(L)	Extreme ultraviolet (lithography)
FDSOI	Fully depleted SOI
FET	Field effect transistor
Fill pattern	Wafer pattern not performing design function, required only for technology reasons. A.k.a. waffles (waffling), dummy fill
FMEA	Failure mode and effect analysis
FPGA	Field programmable gate arrays
G_c	Fracture toughness
GIDL	Gate-induced drain leakage
GPU	Graphic processing unit
HCI	Hot carrier injection
HDL	Hardware description language
Hotspot	Contextual defect of layout, allowed by nominal design rules but limiting reliability, functionality, or yield
HRTEM	High resolution TEM
ILD	Inter-layer (or inter-metal) dielectric
ILT	Inverse lithography technology
In Sb	Indium antimoxide
IO, I/O	Input - output
IP lib	Intellectual property (design) library
ISO	Isolated (e.g., MOSFET layout)
ITRS	International technology roadmap for semiconductors
LBU	Layout base unit
LEE	Line end extension
Left	Effective length of MOSFET channel
LEs	Logic elements
LfD	Litho-friendly design
LEF	Library exchange format
LfL	Litho-friendly layout
L_{gate}, L	MOSFET gate length
LUP	Litho unfriendly pattern
LVL	Layer vs. layer: layout comparison algorithm
Mask	Lithographic stencil used to transfer design pattern to wafer. A.k.a. reticle (due to the fine grid of lines/spaces), photomask
mb, MB	Model-based
MC	Monte Carlo (simulations)
MCMM	Multi-corner, multi-mode
MDP	Mask data preparation
MfD	Manufacturability for design: a methodology to meet design goals by investing in fab process or equipment

(continued)

Table A.1 (continued)

Term	Explanation
MR	Mask ratio
MRC	Mask rule check
MVCCC	Modified virtual crack closure
MTTF	Mean time to fail
MUL	Multi-Finger MOSFET
NAND	Not-and cell
NBTI	Negative bias temperature instability
NBED	Nano-beam electron diffusion
N_{ch}	Channel doping density
NCP	Number of critical paths
NDR	Non-default rules
NGL	Next generation lithography
NIL	Nano-imprint lithography
NIST	National Institute of Standards and Technology
NRE	Non-recurring expenses
OAT	Off axis illumination
OCV	On-Chip Variation
OIP	Open Innovation Platform
OPC	Optical proximity correction
PAC	Probably approximately correct
PFA	Physical Failure Analysis
PHR	Process Hotspot Repair
PLY	Process limited yield
P&R	Place and route
PSM	Phase shifting masks
PWOPC	Process window OPC
P2C	Printability – predicting classifier
QA	Quality analysis
RAM	Random access memory
RDF	Random Dopant Fluctuation
RDR	Restricted design rules (driving restricted layout)
ROI	Return on investment
rb, RB	Rule-based
RCP	Reliability critical path
RET	Resolution enhancement techniques
RoI	Return on investment. Money earned vs. money invested
RH	Relative humidity
RR	Recommended (design) rule (non-mandatory)
SERDES	Serializer-deserialiser
SI	Signal integrity
SILC	Stress-induced leakage current
SiP	System in package
SMO	Source-mode optimization
SoC, SOC	System on chip
SOI	Silicon-on-Insulator

(continued)

Table A.1 (continued)

Term	Explanation
SRAF	Sub-resolution assist features
SRAM	Static random access memory
STI	Shallow trench isolation
SVRF	Standard verification rule format
t_{ox}, T_{ox}	Gate oxide thickness
TAT	Total turnaround time
TDDB	Time dependent dielectric breakdown
TSV	Through silicon via
TSS	Through silicon stack
ULK	Ultra – low k (dielectrivity)
V_{dd}	Drain power supply
V_T, V_{th}	MOSFET threshold voltage
W	MOSFET gate width
WID	Within – die
WPE	Well Proximity Effect

If you have any concerns about our products,
you can contact us on
ProductSafety@springernature.com

In case Publisher is established outside the EU,
the EU authorized representative is:
Springer Nature Customer Service Center GmbH
Europaplatz 3, 69115 Heidelberg, Germany

Printed by Libri Plureos GmbH
in Hamburg, Germany